ATZ/MTZ-Fachbuch

MAHLE GmbH
Editor

Pistons and engine testing

2nd Edition

Springer Vieweg

Editor
© MAHLE GmbH
Stuttgart, Germany

The book is based on the 2nd edition of the German book "Kolben und motorische Erprobung" edited by MAHLE GmbH.

ATZ/MTZ-Fachbuch
ISBN 978-3-658-21507-1   ISBN 978-3-658-09941-1 (eBook)
DOI 10.1007/978-3-658-09941-1
Springer Heidelberg Dordrecht London New York

Springer Vieweg
© Springer Fachmedien Wiesbaden 2011, 2016
Softcover reprint of the hardcover 1st edition 2016

Printed on acid-free paper

This Springer Vieweg imprint is published by Springer Nature
The registered company is Springer Fachmedien Wiesbaden GmbH
(www.springer.com)

Preface

Dear readers,

This is the second revised edition of the second volume of the MAHLE Knowledge Base, a multivolume set of technical books.

The volume "Pistons and engine testing" expands on and adds greater depth to the first volume, "Cylinder components". In this book, MAHLE experts share their broad-based, extensive technical knowledge of pistons, including layout, design, and testing. The large number of pictures, charts, and tables provide a good visual overview of the entire subject, making your day-to-day work in this field easier.

Never before have such high and sometimes contradictory demands been placed on modern engines, and therefore on their pistons, by international legislation as well as customers. Therefore, the following pages are filled with detailed information on everything to do with pistons: their function, requirements, types, and design guidelines, as well as simulation of operational strength using finite element analysis, and piston materials, cooling, and component testing. Engine testing is, however, still the most important element in component development, as well as for validating new simulation programs and systematically compiling design specifications. In the meticulous scientific depth you have come to expect, you will find out more in the detailed section "Engine testing".

This second volume in the series of technical books is again primarily intended for engineers and scientists in the fields of engine development, design, and maintenance. However, professors and students in the subjects of mechanical engineering, engine technology, thermodynamics, and vehicle construction, as well as any readers with an interest in modern gasoline and diesel engines, will also find valuable information on the following pages.

We wish you much enjoyment and many new insights from this reading.

Stuttgart, October 2015

Wolf-Henning Scheider
Chairman of the Management Board and CEO

Heinz K. Junker
Chairman of the Supervisory Board

Acknowledgments

We wish to thank all the authors who contributed to this volume.

Dipl.-Wirt.-Ing. Jochen Adelmann
Dipl.-Ing. Ingolf Binder
Dipl.-Ing. Karlheinz Bing
Dr.-Ing. Thomas Deuß
Dipl.-Ing. Holger Ehnis
Dr.-Ing. Rolf-Gerhard Fiedler
Dipl.-Ing. Rudolf Freier
MSc Armin Frommer
Dipl.-Ing. Matthias Geisselbrecht
Dr.-Ing. Wolfgang Ißler
Dipl.-Ing. Peter Kemnitz
Dr.-Ing. Reiner Künzel
Dipl.-Ing. Ditrich Lenzen
Dr. Kurt Maier
Dipl.-Ing. Olaf Maier
Dr.-Ing. Uwe Mohr
Dipl.-Ing. Helmut Müller
Dr. Reinhard Rose
Dipl.-Ing. Wilfried Sander
Dipl.-Ing. Volker Schneider
Dr.-Ing. Wolfgang Schwab
Dipl.-Ing. Andreas Seeger-van Nie
Dipl.-Ing. Bernhard Steck
Peter Thiele
Dr.-Ing. Martin Werkmann

Table of contents

1 Piston function, requirements, and types

1.1 Function of the piston

1.1.1 The piston as an element of power transmission

In the cylinder of an engine, the energy bound up in the fuel is converted into heat and pressure during the expansion stroke. The heat and pressure values increase greatly within a short period of time. The piston, as the moving part of the combustion chamber, has the task of converting part of this released energy into mechanical work.

The basic structure of the piston is a hollow cylinder, closed on one side, with the segments piston crown with ring belt, piston pin boss, and skirt; **Figure 1.1**. The piston crown transfers the compression forces resulting from the combustion of the fuel-air mixture via the piston pin boss, the piston pin, and the connecting rod, to the crankshaft.

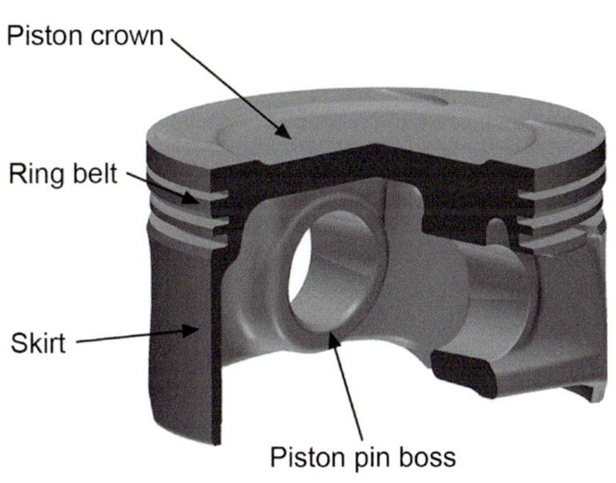

Figure 1.1: Gasoline engine pistons for passenger cars

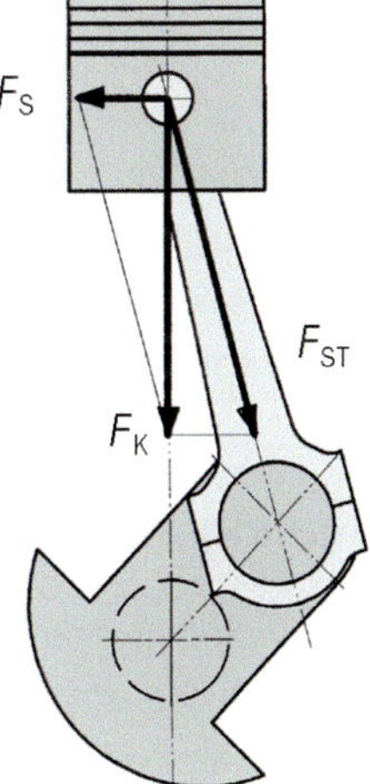

Figure 1.2: Forces on the piston

The gas pressure against the piston crown and the oscillating inertial forces, referred to in the following as the inertia force, of the piston and the connecting rod constitute the piston force FK; **Figure 1.2**. As a result of the redirection of the piston force in the direction of the connecting rod (rod force FST), an additional component arises–following the force parallelogram–, namely the lateral force FS, also known as the normal force. This force presses the piston

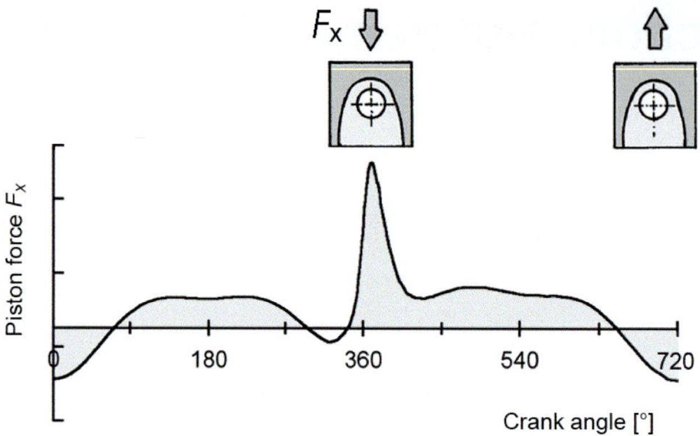

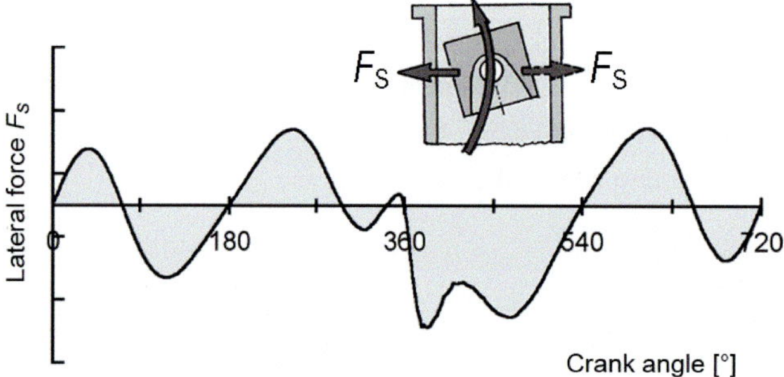

Figure 1.3:
Force curves

skirt against the cylinder bore. During an operating cycle, the lateral force changes direction several times, which presses the piston from one side of the cylinder bore to the other. **Figure 1.3** shows the piston force and lateral force curves as a function of the crank angle.

1.1.2 Sealing and heat dissipation

As a moving and power-transmitting component, the piston, together with the piston rings, must reliably seal the combustion chamber against gas passage and the penetration of lubricating oil in all load cases. A prerequisite is that the materials, geometries, and surfaces are carefully matched. Long piston service life requires good wear behavior, which in turn requires that the running partners are sufficiently supplied with lubricating oil. The oil-retaining capacity of the components due to their surface structure, particularly that of the cylinder, plays a decisive role. This is particularly challenging near the dead center points of the piston, because the hydrodynamic lubricating film is less significant here and mixed friction predominates. Average sliding speeds are typically 10 to 12 m/s.

In four-stroke engines, the piston crown also supports the mixture formation. For this purpose, it has a partially jagged shape, with exposed surfaces (such as the bowl rim) that absorb heat and reduce the load carrying capacity of the component. In two-stroke engines with outlet ports, the piston also acts as a sliding valve and is thermally highly loaded as a result of the combustion gases flowing out at high speed.

In order for the piston to be able to withstand the briefly occurring, extreme combustion temperatures, it must dissipate the heat sufficiently; Chapter 5.6. The heat in the cylinders is primarily dissipated by the piston rings, but also by the piston skirt. The inner contour transfers heat to the air in the housing and to the oil. Additional oil can be applied from underneath to the piston for improved cooling; Chapter 5.

1.1.3 Variety of tasks

The most important tasks that the piston must fulfill are
- transmission of power from and to the working gas;
- variable bounding of the working chamber (cylinder);
- sealing off the working chamber;
- linear guiding of the conrod (trunk piston engines);
- heat dissipation;
- support of gas exchange by air intake and exhaust (four-stroke engines);
- support of mixture formation (by means of suitable shape of the piston surface on the combustion chamber side);
- controlling gas exchange (in two-stroke engines);
- guiding the sealing elements (piston rings);
- guiding the conrod in longitudinal direction of the crankshaft (for top-guided conrods).

As the specific engine output increases, so do the requirements on the piston.

1.2 Requirements on the piston

Fulfilling different tasks such as
- adaptability to operating conditions;
- seizure resistance and simultaneous running smoothness;
- low weight with sufficient shape stability;
- low oil consumption;
- low pollutant emissions values; and
- lowest possible friction losses

result in partly contradictory requirements, in terms of both design and material. These criteria must be carefully coordinated for each engine type. The optimal solution can therefore be quite different for each individual case.

Table 1.1: Operating conditions and solution approaches for piston design and materials

Operating conditions	Requirements on the piston	Solution Design	Solution Material
Mechanical load a) Piston crown/combustion bowl Max. gas pressure, two-stroke gasoline engine: 3.5–8.0 MPa Max. gas pressure, four-stroke gasoline engine: Naturally aspirated engine: 6.0–9.0 MPa Turbo: 9.0–13.0 MPa Max. gas pressure, diesel engine: Naturally aspirated engine: 8.0–10.0 MPa Turbo: 14.0–24.0 MPa b) Piston skirt Lateral force: approx. 10% of the max. gas force	High static and dynamic strength, even at high temperatures	Sufficient wall thickness, shape-stable design, consistent "force flow" and "heat flow"	Various AlSi casting alloys, warm-aged (T5) or age-hardened (T6), cast (partly with fiber reinforcement), or forged; forged steel
c) Piston pin boss Permissible contact pressure temperature-dependent	High contact pressure in the pin bores; low plastic deformation	Pin bore bushing	Specialty brass or bronze bushings
Temperatures: gas temperatures in the combustion chamber over 2,300°C, exhaust gas up to 1,050°C Piston crown/bowl rim, 200–400°C, for ferrous materials 350–500°C Piston pin boss: 150–260°C Piston skirt: 120–180°C	Strength must also be retained at high temperature. Characteristics: strength at elevated temperatures, fatigue resistance, high thermal conductivity, scale resistance (steel)	Sufficient heat-flow cross sections, cooling galleries	As above
Acceleration of piston and connecting rod at high rpm: at times much higher than 25,000 m/s^2	Low mass, resulting in small inertia forces and inertia torques	Lightweight construction with ultrahigh material utilization	AlSi alloy, forged
Sliding friction in the ring grooves, on the skirt, in the pin bearings; at times poor lubrication conditions	Low frictional resistance, high wear resistance (affects service life), low seizing tendency	Sufficiently large sliding surfaces, even pressure distribution; hydrodynamic piston profiles in the skirt area; groove reinforcement	AlSi alloys, tin-plated, graphited, coated skirt, groove protection with cast-in ring carrier or hard anodizing
Contact alteration from one side of the cylinder to the other (primarily in the area of the top dead center)	Low noise level, moderate "piston tipping" in cold or warm engine, low impulse loading	Low running clearance, elastic skirt design with optimized piston profile, piston pin offset	Low thermal expansion Eutectic or hypereutectic AlSi alloys

The operating conditions of the piston and the resulting design and material requirements are summarized in **Table 1.1**.

1.2.1 Gas force

The piston is subjected to an equilibrium of gas, inertia, and supporting forces. The supporting forces are the resultant of the conrod and lateral forces. The maximum gas force in the expansion stroke has critical significance for the mechanical loads. The maximum gas forces that occur depending on the combustion process (gasoline, diesel, two-stroke, four-stroke) and charge intake (naturally aspirated/turbocharger) are shown in **Table 1.1**. At an engine speed of 6,000 rpm and a gas force resulting from a peak cylinder pressure of 90 bar in the expansion stroke, each piston (D = 90 mm) of a gasoline engine is subjected to a load of about 6 t, 50 times a second!

In addition to the maximum gas force, the rate of pressure increase also affects the stress on the piston. The values for diesel engines are about 6 to 12 bar/1 CAD (crank angle degree), but can be significantly higher in case of combustion faults. The rates of pressure increase in gasoline engines are in the range of 3 to 6 bar/1 CAD. Especially if unsuitable fuels are used (octane number too low), combustion faults can occur under high load, known as "knocking." Pressure increase rates of up to 30 bar/1 CAD are possible. Depending on the knocking intensity and period of operation, it can lead to significant damage to the piston and thus failure of the engine. As a prevention method, modern gasoline engines are equipped with knock control systems.

1.2.2 Temperatures

The temperature of the piston and cylinder is an important parameter for operational safety and service life. The peak temperatures of the working gas, even if present only for a short time, can reach levels of over 2,300°C. The exhaust gas temperatures range from 600 to 850°C for diesel engines, and 800 to 1,050°C for gasoline engines.

The temperature of the fresh intake charge (air or mixture) can exceed 200°C for turbocharged engines. Charge air cooling reduces this temperature level to 40–60°C, which in turn lowers the component temperatures and improves filling of the combustion chamber.

As a result of their thermal inertia, the piston and the other parts in the combustion chamber do not exactly follow the cyclic temperature fluctuations of the working gas. The amplitude of these temperature fluctuations at the bowl rim of an aluminum diesel engine piston, for example, is only about 50 K, and this drops off quickly toward the interior. The piston crown, which is exposed to the hot combustion gases, absorbs different amounts of heat, depending on the operating point (rpm, torque). For non-oil-cooled pistons, the heat is primarily conducted to the cylinder wall by the compression rings, and to a much lesser degree, by the piston skirt. For cooled pistons, in contrast, the engine oil carries off a large portion of the accumulating heat; Chapter 7.2.

Heat flows that lead to characteristic temperature fields result from the material cross sections that are determined by the design. Typical temperature distributions for gasoline and diesel engine pistons are shown in **Figures 1.4** and **1.5**.

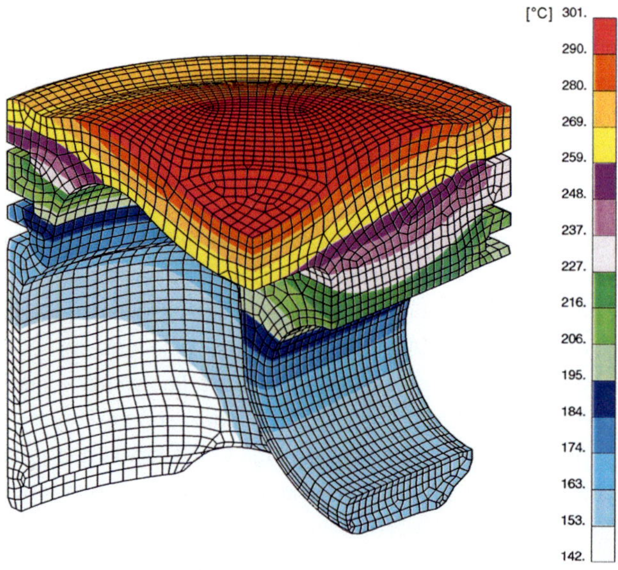

Figure 1.4:
Temperature distribution in a gasoline engine piston

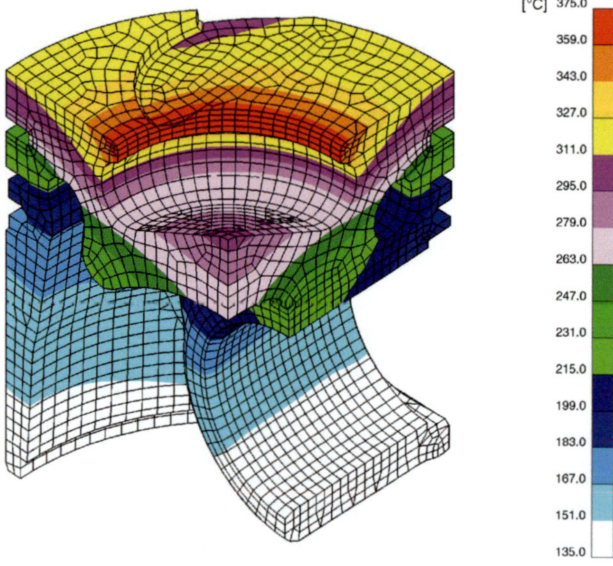

Figure 1.5:
Temperature distribution in a diesel engine piston with cooling gallery

The temperature levels and distributions in the piston essentially depend on the following parameters:

- Operating process (gasoline/diesel)
- Operating principle (four-stroke/two-stroke)
- Combustion process (direct/indirect injection)
- Temperature of the incoming fresh charge (depending on turbocharging and charge air cooling)
- Operating point of the engine (speed, torque)
- Engine cooling (water/air)
- Design of the piston and cylinder head (location and number of gas channels and valves, type of piston, piston material)
- Piston cooling (yes/no)
- Intensity of cooling (spray jet cooling, cooling gallery, cooling gallery location, etc.)

The strength properties of the piston materials, particularly of light alloys, are very dependent on the temperature. The level and distribution of temperatures in the piston essentially determine the mechanical stress that it can withstand. High thermal loads cause a drastic reduction in the fatigue resistance of the piston material. The critical locations for diesel engines with direct injection are the boss zenith and the bowl rim; for gasoline engines, the transition area from the boss connection to the piston crown and to the skirt.

The temperatures in the top ring groove are also significant in terms of oil carbonization. If certain limit values are exceeded, the piston rings tend to "lock up" (carbonization) as a result of residue buildup in the piston ring groove, which leads to an impairment of their functionality. In addition to the maximum temperatures, the piston temperatures are significantly dependent on the engine operating conditions (such as speed, brake mean effective pressure, ignition angle, injection quantity). **Table 7.2** shows typical values for passenger car gasoline and diesel engines in the area of the top ring groove.

1.2.3 Piston mass

The piston, equipped with piston rings, piston pin, and circlips, together with the oscillating connecting rod portion, constitute the oscillating mass. Depending on the engine type, free inertia forces and/or inertia torques are thus generated that can no longer be compensated at times, or that require extreme efforts to do so. This characteristic gives rise to the desire to keep the oscillating masses as low as possible, particularly in high-speed engines. The piston and the piston pin account for the greatest proportion of the oscillating masses. Any weight reduction undertaking must therefore start with these components.

About 80% of the piston mass is located in the area from the center of the piston pin to the upper edge of the crown. The remaining 20% is in the area from the center of the piston pin to the end of the skirt. The determination of the compression height KH is therefore of great significance, because it predetermines about 80% of the piston mass.

For pistons of gasoline engines with direct injection, the piston crown can be used to support mixture formation and can thus be shaped accordingly. These pistons are taller and heavier, which means the center of gravity is shifted upward.

Piston mass mK can best be compared (without piston rings and piston pin) when related to the comparative volume D^3, as shown in **Figure 1.6**. The compression height, however, must always be taken into consideration. Mass figures ("X factors") for proven piston designs mK/D^3 are shown in **Table 1.2**.

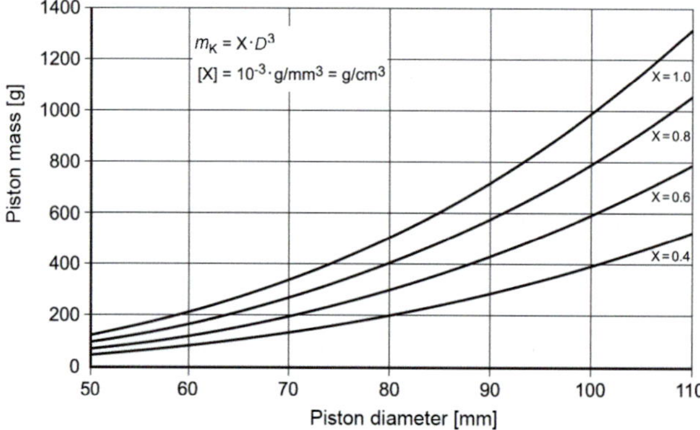

Figure 1.6:
Piston mass mK (without piston rings and piston pin) for passenger car engines, as a function of the piston diameter

Table 1.2: Mass figures for passenger car pistons <100 mm in diameter, made of aluminum base alloys

Operating principle	Mass figure mK/D^3 [g/cm^3]
Two-stroke gasoline engine with port fuel injection	0.50–0.70
Four-stroke gasoline engine with port fuel injection	0.40–0.60
Four-stroke gasoline engine with direct injection	0.45–0.65
Four-stroke diesel engine	0.90–1.10

1.2.4 Friction power loss and wear

In times of increasing environmental awareness, in connection with climate change and the CO_2 emissions legislation put in place to tackle it, the subject of reducing friction power loss in the engine is increasingly in focus. As a moving element, the piston can make a significant contribution here, as described below.

Installation clearance:
One major parameter that influences piston friction is the Installation clearance. An appropriate installation clearance can significantly reduce the typical thermal warping between piston

and cylinder in warm operating conditions. Excess clearance is counterproductive, however, because it can lead to greater secondary motion and wedging of the piston in the cylinder (the "drawer effect"). Under these circumstances, the hydrodynamic lubricating film can be penetrated and mixed friction can arise. Too large an installation clearance can also have a negative acoustic effect. The lower thermal expansion coefficients of steel pistons in comparison with aluminum pistons can result in frictional advantages.

Piston profile (cf. Chapter 2.2):

For low friction power loss values, the local tapering of the piston profile toward the piston axis (mantle curve decrease values) and the ovality are designed in such a way that the contact pressures in the desired support area are uniform. As a rule, this can limit the disturbance of the hydrodynamic lubrication behavior by the change in the direction of movement of the piston to the points where the piston reverses direction (top and bottom dead center). This is where mixed friction conditions arise in the support area. **Figure 1.7** shows, as an example, the lubricant gap between the piston and cylinder for this case.

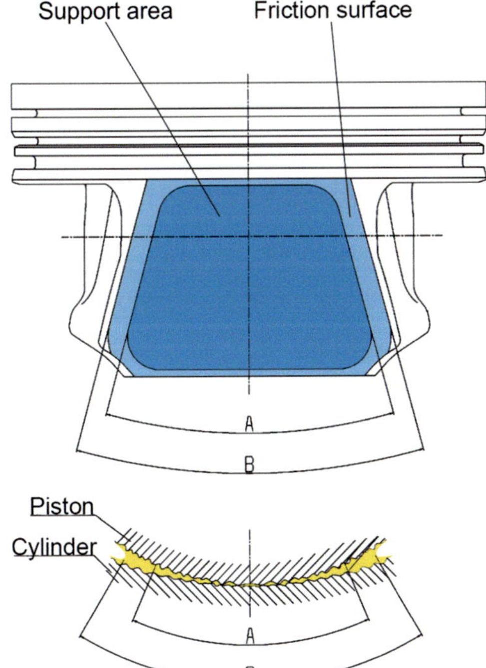

Figure 1.7:
Lubricant gap between
piston and cylinder

Surface roughness:

Besides the skirt profile, the surface of the skirt running surface also has a great influence on the sliding behavior of the piston. Excessive skirt roughness increases friction power losses. In addition to friction forces on the piston skirt, the lubrication of the skirt also plays a decisive role in the proper functioning of the piston in the cylinder. Certain minimum surface roughness values of the piston skirt and the honed cylinder surface

- enhance running-in characteristics;
- prevent abrasive wear;
- contribute to the formation of a hydrodynamic lubricating film between the piston skirt and the cylinder wall; and
- prevent the piston from seizing, i.e., local fusing between the piston and the cylinder due to a lack of clearance or lubricating oil.

If the skirt roughness is too great, friction power losses are increased, while if the skirt roughness is too low, then the piston may not run in as well. A good compromise between these requirements yields piston skirt surface roughness levels in the range $R_a = 1.5–5$ µm.

Protective coatings for the running surface:
Protective coatings for the running surface, such as MAHLE GRAFAL® or EvoGlide, have a positive effect on friction losses in the boundary lubrication conditions, increase wear resistance, and improve resistance to seizing.

1.2.5 Blow-by

One of the primary functions of the piston and the piston rings is to seal off the pressurized combustion chamber from the crankcase. During the motion sequence, combustion gases can escape between piston, cylinder, and piston rings and enter the crankcase (blow-by). In addition to the resulting energy loss, escaping leakage gas also poses a risk to the piston and piston ring lubrication due to contamination and displacement of the lubricating film, and due to oil carbonization as a result of excessive temperatures at the locations in contact with the combustion gases. Increased blow-by values also require greater crankcase ventilation.

Sealing against gas penetration is mainly accomplished by the first piston ring, which is a compression ring. For naturally aspirated engines, the quantity of blow-by is a maximum of 1%; for turbocharged engines, it is a maximum of 1.5% of the theoretical air intake volume.

1.3   Piston types

The various operating principles of combustion engines give rise to a wide variety of engine types. Each engine type requires its own piston variant, characterized by its construction, shape, dimensions, and material.

The most significant piston types in engine design are described in the following. There are also new development paths, such as pistons for very low-profile engines, or pistons made of composite materials with local reinforcement elements.

1.3.1 Pistons for four-stroke gasoline engines

Modern gasoline engines employ lightweight designs with symmetrical or asymmetrical skirt profiles and potentially different wall thicknesses for the thrust and antithrust sides. These piston types are characterized by low weight and particular flexibility in the central and lower skirt areas.

1.3.1.1 Controlled-expansion pistons

Controlled-expansion pistons are pistons with struts that control thermal expansion. They are installed in gray cast iron crankcases. The main target of controlled-expansion piston designs, and many inventions in this field, was and still is to reduce the relatively large differences in thermal expansion between the gray cast iron crankcase and the aluminum piston. The known solutions range from Invar strip pistons to autothermic or Autothermatik pistons.

Because of various adverse properties–notch effects resulting from cast-in struts, increased piston mass, and higher cost–controlled-expansion pistons have faded more and more into the background. For completeness' sake, however, older piston types are addressed briefly.

Autothermic pistons

Autothermic pistons, **Figure 1.8**, are slotted at the transition from the piston crown to the skirt, at the height of the oil ring groove. They are characterized by their particularly quiet running behavior. The unalloyed steel struts cast in between the skirt and the piston pin boss, together with the light alloy that surrounds them, form control elements. They reduce the thermal expansion of the skirt in the direction that is critical for the guiding of the piston in the cylinder. On account of their relatively low load carrying capacity (slits), however, autothermic pistons are no longer up to date.

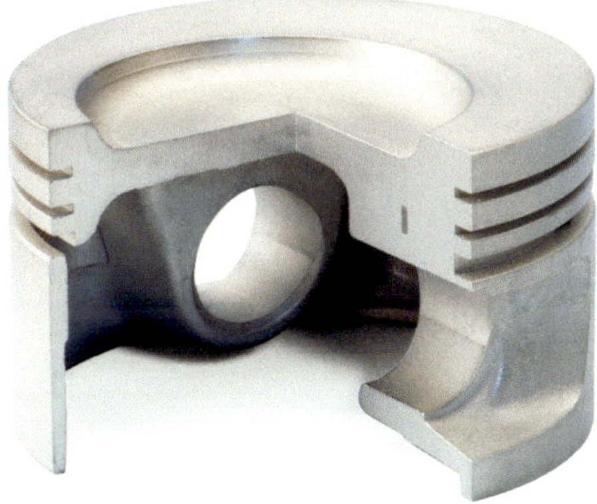

Figure 1.8:
Autothermic piston

Figure 1.9:
Autothermatik pistons

Autothermatik pistons

Autothermatik pistons, **Figure 1.9**, operate according to the same control principle as auto-thermic pistons. In the case of Autothermatik pistons, however, the transition from the crown to the skirt is not slotted. The transition cross sections are dimensioned such that they barely constrain the heat flow from the piston crown to the skirt and, on the other hand, the effectiveness of the steel struts remains unaffected by the connection of the skirt to the rigid crown. This piston design thus combines the high strength of the nonslotted piston with the advantages of the strut design.

1.3.1.2 Box-type pistons

Compared with controlled-expansion pistons, this piston type, **Figure 1.10**, is characterized by its reduced mass, optimized support, and box-like, often slightly oval skirt design. The box-type piston is compatible with both aluminum and gray cast iron crankcases. With a flexible skirt design, the difference in thermal expansion between the gray cast iron crank-

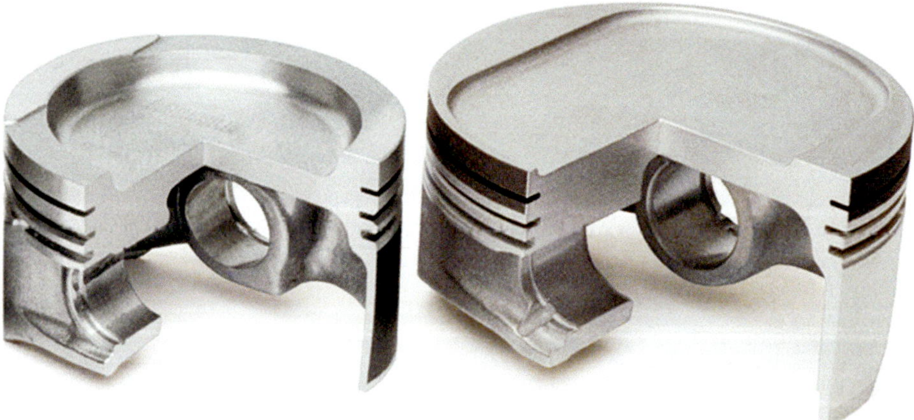

Figure 1.10: Asymmetrical duct pistons

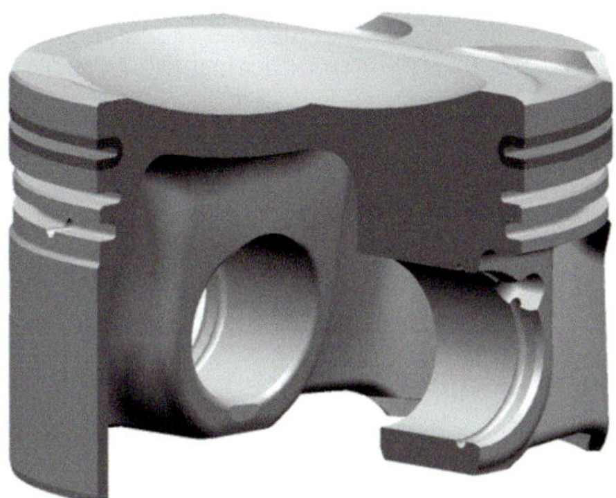

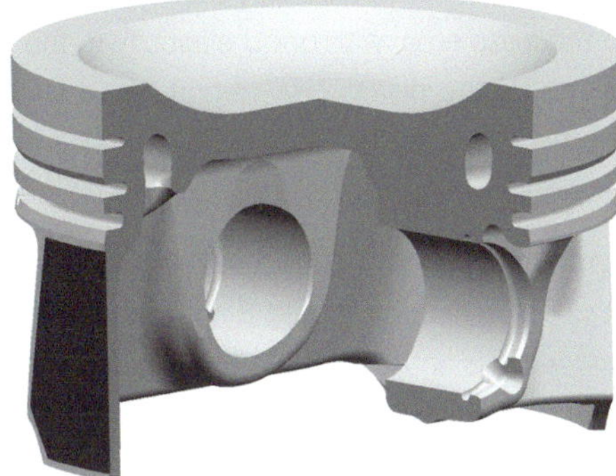

Figure 1.11: Ring carrier piston for gasoline engines

Figure 1.12: Piston with cooling gallery for gasoline engines

case and the aluminum piston can be compensated very well in the elastic range. If the box width is different on the thrust and antithrust sides, the piston is referred to as an asymmetrical duct piston. Box-type pistons are cast or forged.

In addition to the classical box-type piston with vertical box walls, new shapes have recently been established, with box walls that are tapered toward the top. One example is the EVOTEC® piston; Chapter 1.3.1.3.

A ring carrier can protect the compression ring groove in turbocharged engines with high peak cylinder pressures against wear and deformation; **Figure 1.11**.

Pistons for engines with very high specific power output (>100 kW/L) may have a cooling gallery; **Figure 1.12**.

1.3.1.3 EVOTEC® pistons

The greatest current potential for reducing the piston mass in four-stroke gasoline engines is the EVOTEC® design, which is primarily used in conjunction with trapezoidal supports; **Figure 1.13**.

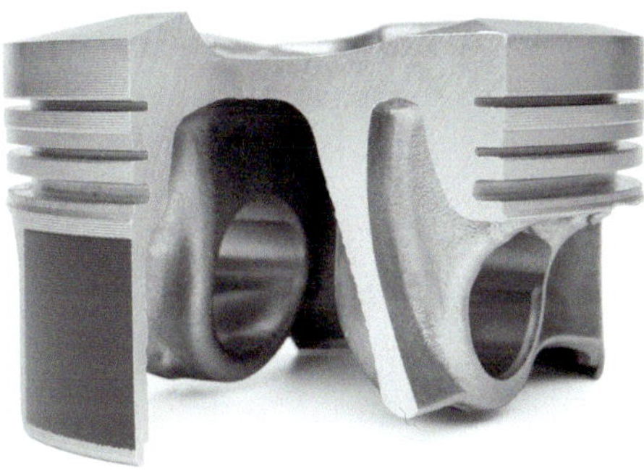

Figure 1.13: EVOTEC® piston

Box walls set at a steep angle allow particularly deep cast undercuts behind the ring grooves in the boss area, with good elasticity in the lower skirt area. The connection of the box walls far inside the piston crown–combined with supporting ribs in the piston window between the ring area and the box wall–provides excellent structural stiffness with very small cross sections.

Another significant feature of this piston concept is the asymmetric design of the box walls. In order to accommodate the higher lateral force load on the thrust side, the spacing of the box walls is smaller on the thrust side. The shorter lever arm between the box wall and the contact area between the piston and the cylinder reduces the bending moment load. This allows smaller cross sections, even with extremely high lateral forces, which are preferred for turbocharged gasoline engines with direct injection. In order to provide the required elasticity and good guiding properties, the antithrust side, which experiences significantly lower loads, features greater spacing between the box walls. Careful design of the piston profile (see Chapter 2.2) ensures uniformly low contact pressures on the thrust side, even under ultrahigh loads.

1.3.1.4 EVOTEC® 2 pistons

The next stage of development of the EVOTEC® piston is the EVOTEC® 2 piston; **Figure 1.14**. The EVOTEC® 2 piston is characterized by the
- new, load-optimized shape of the skirt-to-wall joint to increase strength; and
- reduced floor thickness to decrease weight without compromising strength as a result of temperature reduction.

The sum total of all measures taken in developing the MAHLE EVOTEC® piston into the EVOTEC® 2 is a weight reduction of about 5% at a neutral cost, as well as local increases in structural strength to meet the increasing requirements of future generations of engines.

Figure 1.14: EVOTEC® 2 piston

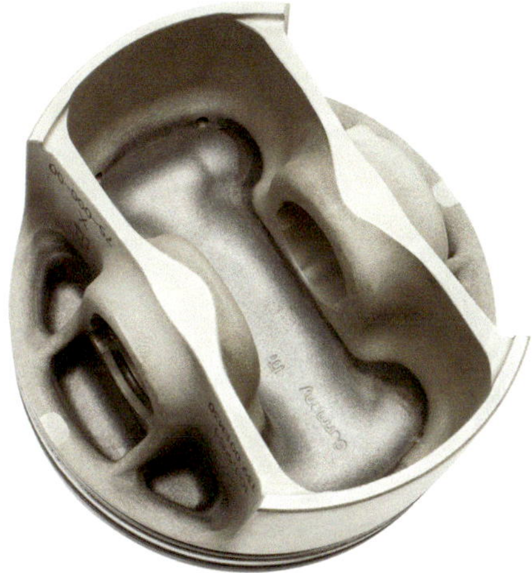

Figure 1.15: Forged aluminum piston

Figure 1.16: Forged piston for Formula 1

1.3.1.5 Forged aluminum pistons

In engines with very high power densities–such as highly loaded turbocharged gasoline engines–cast pistons reach their limits. MAHLE forged pistons are a particularly good fit for this area of application; **Figure 1.15**. Their strength advantage in the temperature range of up to about 250°C improves the load carrying capacity for lateral forces, and increases the load carrying capacity of the pin bore and the fracture toughness. Forged pistons are therefore especially well suited for high-speed concepts and turbocharged engines. Because of the high ductility of the forged material, they also react more tolerantly to peak pressures that can arise if an engine is operated very close to the knock limit. This allows lower ring land widths, among other things, and therefore lower compression heights. Since the manufacturing process is very stable, the forged pistons can be designed to the limit in order to minimize component weight.

One disadvantage, compared with cast counterparts, is the higher product cost of the forged piston. Limited design flexibility is another. Undercuts or inserts such as ring carriers or salt cores, in particular, cannot be formed.

Motorsport pistons are all special designs; **Figure 1.16**. The compression height KH is very low, and the overall piston is extremely weight-optimized. Only forged pistons are employed in this field. Weight optimization and piston cooling are critical design criteria. In Formula 1, specific power outputs of greater than 200 kW/L and speeds of more than 19,000 rpm are common. The service life of the pistons matches the extreme conditions.

1.3.2 Pistons for two-stroke engines

For pistons of two-stroke engines, **Figure 1.17**, the thermal load is particularly high as a result of the more frequent heat incidence–one expansion stroke for every revolution of the

Figure 1.17:
Piston and cylinder for a
two-stroke engine

crankshaft. It also needs to close off and expose the intake, exhaust, and scavenging ports in the cylinder during its up-and-down motion, thus controlling the gas exchange. This leads to high thermal and mechanical loads.

Two-stroke pistons are equipped with one or two piston rings, and their external design varies from open window-type pistons to full-skirt piston models. This depends on the shape of the scavenging ports (long channels or short loop-shaped scavenging passage). The pistons are typically made of the hypereutectic AlSi alloy MAHLE138.

1.3.3 Pistons for diesel engines

1.3.3.1 Ring carrier pistons

Ring carrier pistons, **Figure 1.18**, have been in use since 1931. The first, and at times even the second piston ring are guided in a ring carrier that is securely joined to the piston material by metallic bonding.

The ring carrier is made of an austenitic cast iron with a similar coefficient of thermal expansion to that of the piston material. The material is particularly resistant to frictional and impact wear. The top ring groove, which is the most vulnerable, and the piston ring inserted in it are thereby effectively protected against excessive wear. This is particularly advantageous at high operating temperatures and pressures, which are particularly prevalent in diesel engines.

1.3.3.2 Cooling gallery pistons

In order to cool the area around the combustion chamber most effectively, and thereby address the increased temperatures that result from higher outputs, there are various types of cooling galleries and cooling cavities. The cooling oil is generally fed through fixed ports in the crankcase. Chapter 5 gives an overview of possible cooling variants.

Figure 1.18: Ring carrier piston for a diesel engine

Figure 1.19: Salt-core cooling gallery piston with ring carrier for a passenger car diesel engine

Figure 1.19 shows a cooling gallery piston with ring carrier for a passenger car diesel engine. The annular cavities are formed by casting around salt cores, which are then dissolved and washed away with water.

1.3.3.3 Pistons with cooled ring carrier

The piston with cooled ring carrier, **Figure 1.20**, is another cooled piston variant. The cooled ring carrier significantly improves the cooling of the top ring groove and the thermally highly loaded combustion bowl rim. The intensive cooling of this ring groove makes it possible to replace the usual full keystone ring with a rectangular ring.

Figure 1.20:
Piston for passenger cars with cooled ring carrier

1.3.3.4 Pistons with bushings in the pin bore

One of the most highly stressed areas of the piston is the piston pin bearing. Temperatures of up to 240°C can occur in this area, a range at which the strength of the aluminum alloy starts to drop off considerably.

For extremely stressed diesel engine pistons, measures such as shaped pin bores, pin bore relief, or oval pin bores are no longer sufficient to increase the load carrying capacity of the pin bore.

For this reason, MAHLE has developed a reinforcement of the pin bore, using shrink-fit bushings made of a material with higher strength (e.g., CuZn31Si1).

1.3.3.5 FERROTHERM® pistons

In FERROTHERM® pistons, **Figure 1.21**, the guiding and sealing functions are implemented separately. The two parts, piston crown and piston skirt, are movably connected to each other through the piston pin. The piston crown, made of forged steel, transfers the gas pressure to the crankshaft via the piston pin and connecting rod.

The light aluminum skirt only bears the lateral forces that arise because of the angle of the connecting rod, and can therefore support the piston crown with an appropriate design. In addition to this "shaker cooling" via the skirt, closed cooling cavities can also be incorporated in the piston crown. The outer cooling cavity of the steel piston crown is closed by split cover plates.

Thanks to its design, the FERROTHERM® piston exhibits good wear rates in addition to high strength and temperature resistance. Its consistently low oil consumption, small dead space, and relatively high surface temperature provide good conditions for maintaining low exhaust emissions limits. FERROTHERM® pistons are used in highly loaded commercial vehicle engines.

Figure 1.21:
FERROTHERM® piston

Figure 1.22: MONOTHERM® piston for a commercial vehicle engine

Figure 1.23: Optimized MONOTHERM® piston for a commercial vehicle engine

1.3.3.6 MONOTHERM® pistons

The MONOTHERM® piston, **Figure 1.22**, emerged from the development of the FERRO-THERM® piston. This piston type is a single-piece forged steel piston that is greatly weight-optimized. With a small compression height (to less than 50% of the cylinder diameter) and machining above the pin boss spacing (internal), the piston mass, including the piston pin, almost corresponds to the mass of a comparable aluminum piston with piston pin. In order to improve the piston cooling, the outer cooling cavity is closed off by two cover plate halves. The MONOTHERM® piston is used in passenger car and commercial vehicle engines with peak cylinder pressures of up to 20 MPa.

1.3.3.7 Optimized MONOTHERM® pistons

In optimized MONOTHERM® pistons, **Figure 1.23**, the piston skirt is connected to the pin boss on the side, just as a conventional MONOTHERM® piston is. The upper edge of the piston skirt is additionally connected to the inner contour of the piston.

The advantages of the optimized MONOTHERM® piston are

- stiffening of the structure, resulting in reduced deformation and greater load carrying capacity;
- reducing the secondary piston motion, resulting in both a
- reduced cavitation propensity and
- improved guide properties, particularly for the piston rings;
- smoothing of contact pressure on the skirt;
- additional surface and additional cross section for heat dissipation;
- advantages in forging and machining.

The optimized MONOTHERM® piston is used for peak cylinder pressures of up to 25 MPa.

1.3.3.8 MonoWeld® pistons

The portfolio of steel pistons was complemented with the new friction-welded MonoWeld® piston; **Figure 1.24**. The structure is stiffer than the optimized MONOTHERM® piston, making for an increased thermal and mechanical load carrying capacity. In contrast to the MonoX-comp® piston, the MonoWeld® piston has no cooling cavity in its interior.

The MonoWeld® piston is suitable for peak cylinder pressures of up to 25 MPa.

Figure 1.24:
MonoWeld® piston for a
commercial vehicle engine

1.3.3.9 Electron beam-welded pistons

The basis is a forged aluminum-base alloy piston for highly loaded engines. Forging provides significantly higher and more consistent strength values than casting. On the other

Figure 1.25:
Electron beam-welded piston with cooling
gallery for a high-performance diesel engine

hand, the casting process allows the use of ring carriers and the incorporation of salt cores to form cooling galleries.

Electron beam welding makes it possible to combine forged piston bases with cast ring bands, in order to unite the advantages of both production processes. **Figure 1.25** shows an example of an electron beam-welded piston with ring carrier and cooling gallery.

1.3.4 Composite pistons for large-bore engines

1.3.4.1 Areas of application and design types

The composite piston enables the incorporation of cooling cavities and the combination of the properties of various materials in one piston. The performance range of four-stroke engines with composite pistons extends from 500 to 30,000 kW, with up to 20 cylinders. Areas of application include gensets, main ship drives and auxiliary ship drives, and heavy construction and railroad vehicles.

There are many variants of composite pistons. Common to them all is a design consisting of two main constituents: the piston crown with the ring belt (upper part of piston) and the piston skirt with the piston pin boss (piston skirt). The two parts are screwed together with appropriate threaded fasteners.

The connection between the upper part and skirt of the piston can be designed as a central screw joint–with only one screw, **Figure 1.30** and **Figure 1.26 left**–or as a multiple screw joint. The multiple screw joint is essentially divided into models with four single screws, of which each has its own thrust piece, **Figure 1.26 center**, and twin screw joints, i.e., two screws for each thrust piece, **Figure 1.26 right**.

The selection of the screw joint type is based on the geometric ratios of the piston. For example, the component size, combustion bowl geometry, cooling principle, and design of the upper part (bore cooling or shaker cooling) have a great influence.

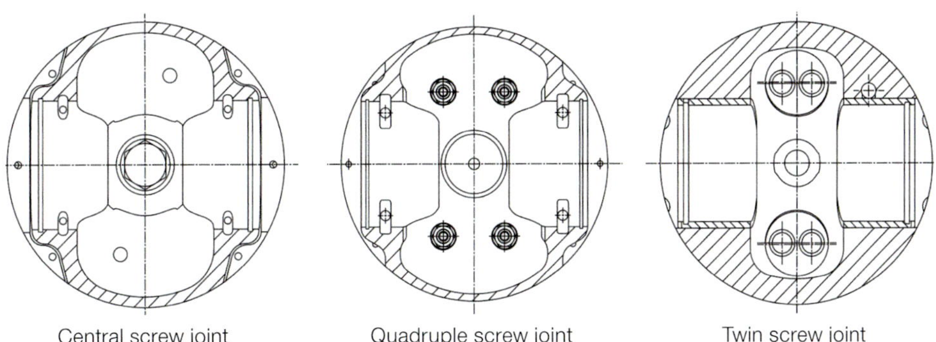

Figure 1.26: Types of screw joints–bottom views

1.3.4.2 Piston upper part

The piston upper part is made of forged steel and can be designed with shaker cooling or bore cooling; **Figure 1.27.** The bore-cooled variant is characterized by increased stiffness, with the efficient heat transfer remaining unchanged. The increased stiffness makes a design with only one contact surface possible.

If heavy fuel oil is used, the piston ring grooves are inductively hardened or chrome-plated to reduce wear.

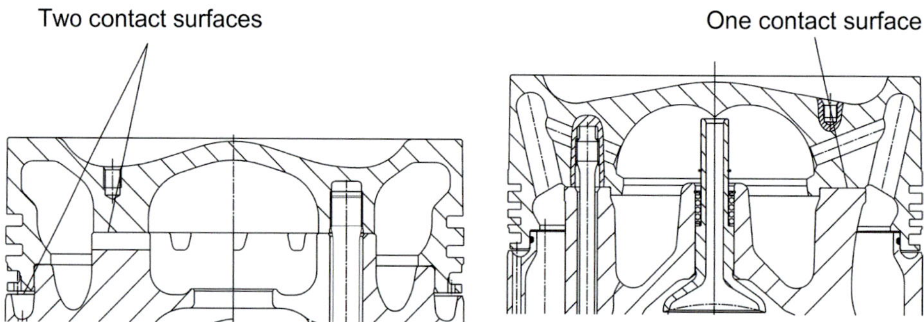

Figure 1.27: Piston upper part with shaker cooling (left) and bore cooling (right)

1.3.4.3 Piston skirt made of forged aluminum-base alloy

Forged aluminum skirts, **Figure 1.28**, are suitable for low and medium peak cylinder pressures, exhibit a low mass, and are easy to machine.

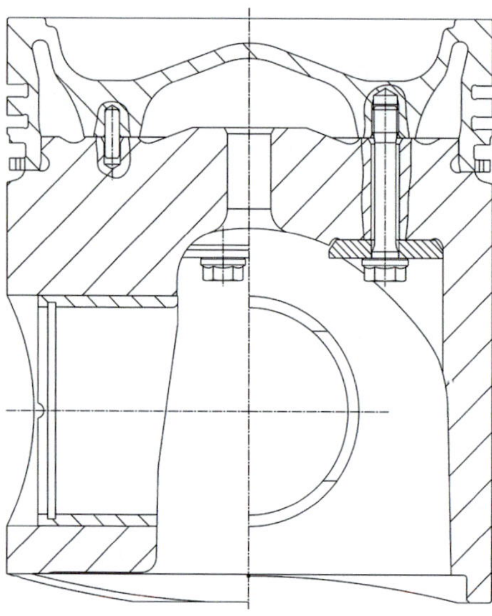

Figure 1.28:
Composite piston with steel crown, aluminum skirt, pin bore bushings, thrust piece, and anti-fatigue bolt

1.3.4.4 Piston skirt made of nodular cast iron

Key features of a composite piston with nodular cast iron piston skirt, **Figure 1.29**, are low cold piston clearance and the resulting low secondary piston motion as well as high seizure resistance. The casting process, in contrast to forged steel, allows for undercuts and therefore a lighter design. With an appropriate design, it is suitable for peak cylinder pressures greater than 20 MPa. Compared with pistons with aluminum skirts, however, the mass is increased as a result of the higher material density.

1.3.4.5 Piston skirt made of forged steel

Piston skirts made of forged steel, **Figure 1.30**, provide ultrahigh component strength and, related to the process, a material with extremely few defects. They are suitable for ultrahigh stresses greater than 24 MPa. Similar to pistons with nodular cast iron piston skirts, they provide the advantage of low cold piston clearance, resulting in low secondary piston motion.

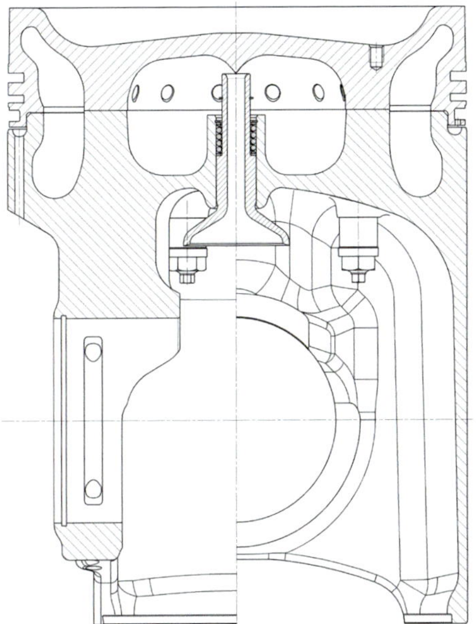
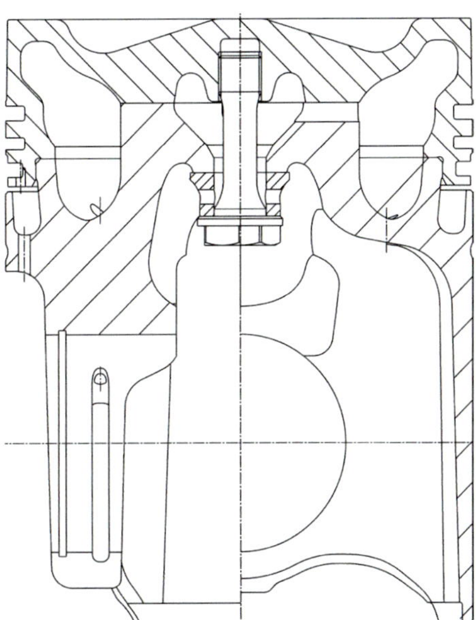

Figure 1.29: Composite piston with steel crown and nodular cast iron piston skirt, thrust piece, and antifatigue bolts

Figure 1.30: Composite piston with steel crown and forged steel piston skirt, thrust piece, and antifatigue bolt

2 Piston design guidelines

In view of the operational requirements of typical internal combustion engines (two-stroke, four-stroke, gasoline, and diesel engines) aluminum-silicon alloys are generally the most appropriate piston materials. Large-bore pistons and commercial vehicle pistons, or their crowns or upper parts, however, are often made of steel.

2.1 Terminology and major dimensions

Functional divisions of the piston are the piston crown, the ring belt with top land, the piston pin boss, and the piston skirt; **Figure 2.1**. Additional functional elements, cooling galleries, and ring carriers indicate the piston type. The piston rings, piston pin, and–depending on the design–the pin retaining system are all part of the piston assembly.

In order to keep the masses as low as possible, a careful design of the piston and good piston cooling are necessary. Important dimensions and typical values are shown in **Figure 2.2** and **Table 2.1**.

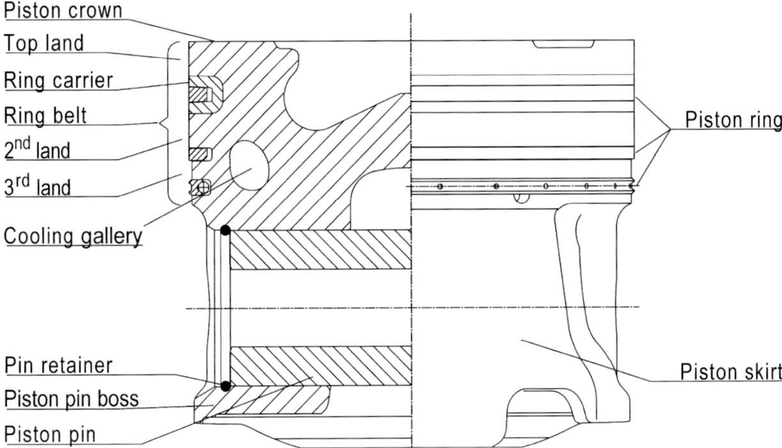

Figure 2.1: Important piston terminology

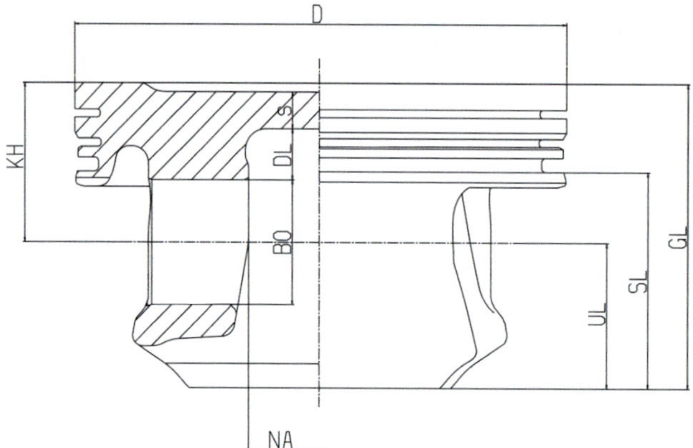

Figure 2.2:
Important piston dimensions
BO: Pin bore Ø (piston pin Ø)
KH: Compression height
NA: Pin boss spacing
D: Piston Ø
s: Crown thickness
DL: Elongation length
SL: Skirt length
GL: Total height
UL: Lower height

Table 2.1: Major dimensions of light-alloy pistons

	Gasoline engines		Diesel engines*
	Two-stroke	Four-stroke (passenger cars)	Four-stroke (passenger cars)
Diameter D [mm]	30–70	65–105	65–95
Total height GL/D	0.8–1.0	0.6–0.7	0.8–0.95
Compression height KH/D	0.4–0.55	0.30–0.45	0.5–0.6
Pin diameter BO/D	0.20–0.25	0.20–0.26	0.3–0.4
Top land height [mm]	2.5–3.5	2–8	6–12
1st ring land height St/D*	0.045–0.06	0.040–0.055	0.055–0.1
Groove height for 1st piston ring [mm]	1.2 and 1.5	1.0–1.75	1.75–3.5
Skirt length SL/D	0.55–0.7	0.4–0.5	0.5–0.65
Pin boss spacing NA/D	0.25–0.35	0.20–0.35	0.25–0.35
Crown thickness s/D or s/DMu, max**	0.055–0.07	0.06–0.10	0.14–0.23

* Values for diesel engines apply to ring carrier pistons, ** Diesel

2.1.1 Crown shapes and crown thickness

The piston crown forms part of the combustion chamber. Pistons for gasoline engines can be flat, raised, or sunken. For diesel engine pistons, the combustion chamber bowl is usually located in the piston crown. The geometry of the piston crown is also affected by the number and location of the valves; **Figure 2.3**. The maximum gas pressure and the quantity of heat to be dissipated determine the thickness of the piston crown (crown thickness). The piston crown, or the bowl rim for diesel engine pistons, is the part of the piston that is exposed to the greatest thermal stress.

The values listed in **Table 2.1** for the crown thickness s apply in general to pistons with flat, convex, or concave crowns.

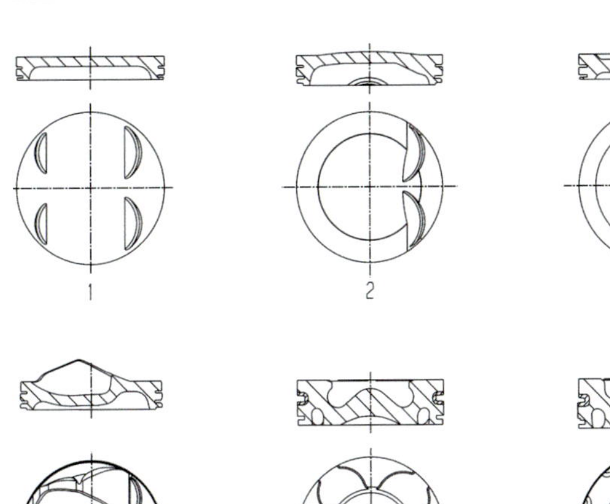

Figure 2.3:
Examples of piston crown shapes of various pistons for gasoline and diesel engines (1 to 3 for four-stroke gasoline engines with port fuel injection, 4 for four-stroke gasoline engines with direct injection, 5 to 6 for four-stroke diesel engines with direct injection)

2.1.2 Compression height

The compression height is the distance between the center of the piston pin and the upper edge of the top land. The goal is to have as low a compression height as possible, in order to keep the piston mass and the height of the engine as low as possible. The number and height of piston rings, the required ring lands, the piston pin diameter, and the top land width, however, result in a minimum compression height; a lower height is not possible. For diesel engine pistons, in addition to the combustion chamber depth, the conrod bore radius and the required minimum crown thickness below the bowl generally determine the compression height.

Reducing the compression height also has disadvantages. For high power output and gas pressures, higher temperatures in the pin bore and higher stresses on the piston crown are the result of the low compression height. Cracks in the pin bore or the piston crown may then occur. Accordingly, for diesel engine pistons, a large expansion length is advantageous for the load carrying capacity of the bowl rim.

2.1.3 Top land

In the piston ring zone, the distance between the edge of the piston crown and the upper flank of the top ring groove is called the top land. Its dimensions are a compromise between the following requirements: low piston mass and minimal dead volume for reducing fuel consumption and exhaust gas emissions, on one side; on the other side, the first piston ring, which is a compression ring, requires a temperature range that is still compatible with its function. This, in turn, depends greatly on the combustion process, the material, and the geometry of the first piston ring and its piston ring groove, as well as the location of the water jacket on the cylinder.

For gasoline engines, the top land width is 4 to 10% of the piston diameter, tending to decrease in order to further reduce the hydrocarbon emissions caused by gaps.

For passenger car diesel engines with direct injection, this value is 8–15% of the piston diameter.

For commercial vehicle diesel engines with direct injection, it is 8 to 13% of the piston diameter for aluminum pistons, and 6 to 10% for steel pistons.

2.1.4 Ring grooves and ring lands

The piston ring zone, in general, consists of three ring grooves that hold the piston rings. The piston rings seal off the combustion chamber and control lubricating oil consumption. Their surface must therefore be of the highest quality. Poor sealing leads to blow-by of combustion gases into the crankcase, to heating due to the contact of the hot gas flow on the surfaces, and to destruction of the critical oil film on the running surfaces of the sliding and sealing partners. The piston ring must not impact the groove root diameter of the piston when it is pressed into the groove so that it is flush with the outer diameter of the piston, that is, it requires radial clearance.

Current lubricating oils permit groove temperatures of over 200°C in gasoline engine pistons, and up to 280°C in diesel engine pistons, without the piston rings binding as a result of residue buildup in the top ring groove.

For diesel engine pistons, which develop significantly greater combustion pressures than gasoline engines, the top ring groove is made much more wear-resistant by casting in a ring carrier. Ring carriers are typically made of Ni-resist, an austenitic cast iron with a thermal expansion coefficient that is approximately the same as aluminum. The ring carrier forms a permanent metallic bond with the piston through the established process of Al-fin composite casting. This process also enables better heat transfer.

The ring land is the part of the ring belt of a piston that is located between two piston ring grooves. In particular, the first ring land, which is severely loaded by the gas pressure, must be sized sufficiently to prevent fracture of the ring land. Its height depends on the maximum gas pressure of the engine and the land temperature. For gasoline engine pistons, the ring land width is 4 to 6% of the piston diameter, for turbocharged passenger car diesel engines it is 5.5 to 10%, and for commercial vehicle pistons about 10%. The second or remaining ring lands can be smaller, on account of the lower pressures they are subjected to.

2.1.5 Total height

The total height GL of the piston, relative to the piston diameter, depends on the compression height and the guide length on the skirt. For small, high-speed engines in particular, the total height is kept as low as possible in order to obtain low piston mass.

2.1.6 Pin bore

2.1.6.1 Surface roughness

The pin bore/piston pin sliding system must be in perfect condition in order to ensure reliable engine operation. If the surface roughness is too low, particularly when starting, this can cause galling of the pin bore. Therefore, depending on the pin bore diameter, a surface roughness of $R_a = 0.63{-}1.0$ µm is desired for the pin bore. Pistons with piston pins that move only in the piston (shrink-fit connecting rods) generally have slightly greater surface roughness values, in order to increase oil retention, particularly under less than ideal running conditions.

Other detailed measures are often necessary in order to ensure lubrication under all operating conditions. These include oil pockets (slots) or circumferential oil grooves for improved lubrication in the pin bore.

2.1.6.2 Installation clearance

The clearance of the piston pin in the piston pin boss is important for smooth running and low wear of the bearing surfaces. As the materials of the piston and piston pin have different thermal expansion, the running clearances in a warm engine are greater than the installation clearances in a cold engine. This difference can be approximated as:

$$\text{increase in clearance} = 0.001 \times \text{pin diameter [mm]}$$

The increase in clearance for a 30 mm diameter piston pin is therefore approximately 30 µm.

Previously, very tight clearances were typical, so that the piston pin could be inserted only in a preheated piston. Today, the clearance is considerably greater, and the piston pin is inserted into the pin bore at room temperature. This prevents deformation of the skirt due to shrinkage stresses and potential galling of the piston pin in the piston when starting at low temperatures.

When designing the minimum clearance, **Table 2.2**, in gasoline engines, differentiation must be made between a floating piston pin or a piston pin with a shrink fit in the small end bore.

Table 2.2: Minimum pin clearance for gasoline engines [mm]–not suitable for motorsport engines

Piston pin with floating design	Shrink fit piston pin
0.002–0.005	0.006–0.012

The floating piston pin is the standard design and is the variant that can be specifically most highly loaded in the piston pin boss.

With shrink-fit connecting rods, the piston pin is seated in the small end bore with some interference. This makes the automatic assembly of the piston, piston pin, and connecting rod easier, because no special piston pin circlip is needed. The shrink-fit connecting rod design is not suitable for modern diesel engines and gasoline engines with turbocharging.

2.1.6.3 Tolerances

Similar considerations apply to matching the piston pin and piston as for the piston and cylinder. In order to facilitate assembly–aided by lower production tolerances for the pin bore and piston pin–only one defining group is typically used. The tolerance for piston pins is 4 to 8 µm, depending on the pin diameter. The pin bore tolerance is about 1 µm greater in each case.

2.1.6.4 Piston pin offset

The kinematics of the crank mechanism of a reciprocating piston engine leads to multiple contact alteration of the piston on the cylinder wall during a working cycle. After the top dead center point, the gas pressure presses one side of the piston skirt against the cylinder wall. This zone is known as the thrust side, and the opposite side of the skirt is the antithrust side.

An offset of the piston pin axis relative to the piston axis (piston pin offset) causes a change in the contact behavior of the piston as the side changes, and decisively affects the lateral forces and impacts. By calculating the piston motion, the location and amount of offset from the piston axis can be optimized, thus drastically reducing piston running noise and the risk of cavitation on the cylinder liner.

2.1.7 Piston skirt

The piston skirt, as the lower part of the piston, guides the piston in the cylinder. It can fulfill this task only if it has suitable clearance to the cylinder. Sufficient skirt length and tight guidance keep the tipping of the piston low during contact alteration from one cylinder wall to the other.

For diesel engine pistons, the full-skirt piston was previously dominant, with its closed skirt, interrupted only in the area of the pin bore. This construction is still sometimes used for pistons in two-stroke gasoline engines. Aluminum diesel engine pistons for commercial vehicle engines still feature a full-skirt design at times, with only a slight setback in the area of the piston pin boss, but window-type pistons are used across the board in passenger cars.

Gasoline engine pistons have a wide range of designs for the piston skirt; **Figure 2.4**. In order to keep inertia forces low, they now have only relatively narrow skirt surfaces, which led to the box-type piston, sometimes with different running surface widths (asymmetrical duct pistons) and/or inclined box walls (including EVOTEC® pistons).

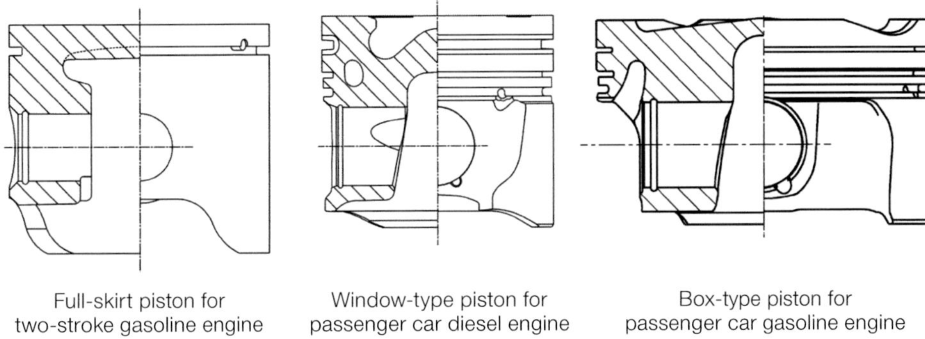

Figure 2.4: Skirt profiles

The piston skirt must meet a few requirements related to its strength. First, it must bear the lateral forces without major deformations, and second, it should elastically adapt to the deformations of the cylinder. The piston crown deflects under the temperature and peak cylinder pressure, and deforms the piston skirt to an oval in the thrust and antithrust direction. This increases the diameter in the direction of the piston pin, and reduces it along the thrust-antithrust axis. Residual skirt collapse due to plastic deformation, however, should be avoided. Remedial measures for at-risk pistons include greater wall thickness, oval interior piston profile, or small circumferential length of the piston skirt.

The lower end of the piston skirt should protrude out of the cylinder only a little or not at all (lower edge of pin bore). The protrusion must be considered appropriately when designing the piston profile.

2.2 Piston profile

2.2.1 Piston clearance

The piston is deformed under the effect of gas temperatures and forces, particularly the gas force. This change in shape must be taken into consideration when designing the shape of the piston in order to ensure that it runs without binding at the operating temperature. To this end, the piston is installed with some clearance in the cold state, which takes the expected deformation and the secondary piston motion into consideration. Its shape, known as the piston profile (or fine piston contour), also deviates from an ideal circular cylinder.

Local clearance in the cold state is made up of the difference of the cylinder diameter and the piston, imagined as a circular cylinder (the installation clearance), as well as the deviation of the piston from this circular cylinder shape. The piston profile deviates from the ideal circular cylinder in the axial direction (conicity, barrel shape) and in the circumferential direction (ovality).

2.2.2 Ovality

Pistons typically have a slightly smaller diameter in the piston pin axis than in the thrust-antithrust axis. The difference is the (diametric) ovality; **Figure 2.5**.

The oval shape of the crown and skirt provides many design opportunities. The skirt ovality creates space for thermal expansion in the piston pin axis direction. The ovality can be varied to generate an even wear pattern with sufficient width. It is typically (diametric) 0.3 to 0.8% of the piston diameter.

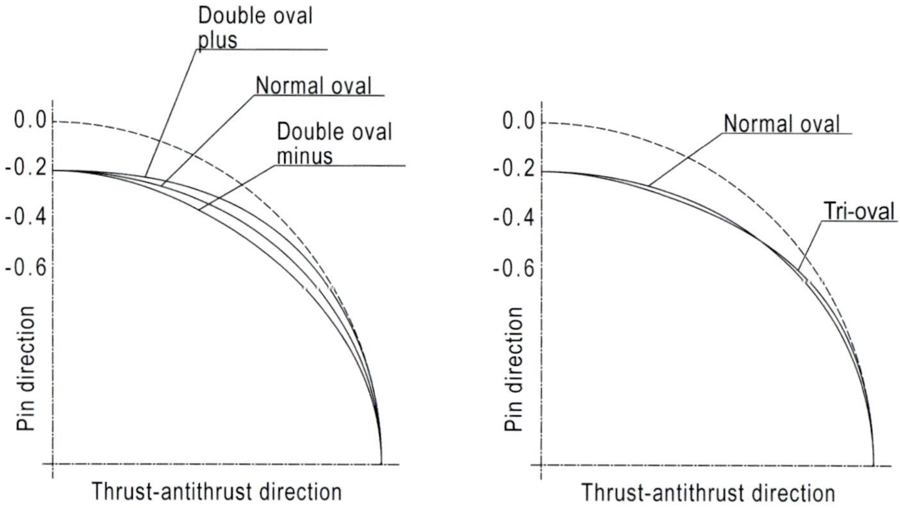

Figure 2.5: Ovality and superposition, double oval (left), tri-oval (right)

In addition to the normal ovality, ovalities with superposition are also possible, such as double or tri-ovality. For double ovality, in the form of a positive (double oval plus) or negative (double oval minus) superposition, the local piston diameter is greater or less than for normal ovality; **Figure 2.5, left**. The positive superposition widens the wear pattern relative to normal ovality, and the negative makes it narrower. Tri-ovality in the form of positive superposition widens the wear pattern, which is limited because of a significantly reduced local piston diameter starting at about 35° from the thrust-antithrust axis; **Figure 2.5, right**.

The resulting running surfaces in the thrust and antithrust direction should not be too narrow, so that the specific pressures between piston and cylinder remain low. In order to prevent hard contact and the risk of galling, the support area should not, however, extend out to the box walls. **Figure 1.7** in Chapter 1.2.4 shows the support area of an advantageous piston profile.

Further opportunities for optimizing the piston profile are provided by different ovalities in the thrust and antithrust direction, as well as ring belt offsets and so-called corrections.

2.2.3 Skirt and ring belt tapering

The piston is tapered slightly at the upper and lower skirt ends, in order to promote the formation of the wedge of lubricating oil that acts as a support element.

The greater taper in the area of the ring belt compensates for the great thermal expansion due to high temperatures in this area and for the deformation due to gas force. It also prevents the piston ring belt from impacting the cylinder owing to secondary piston motion.

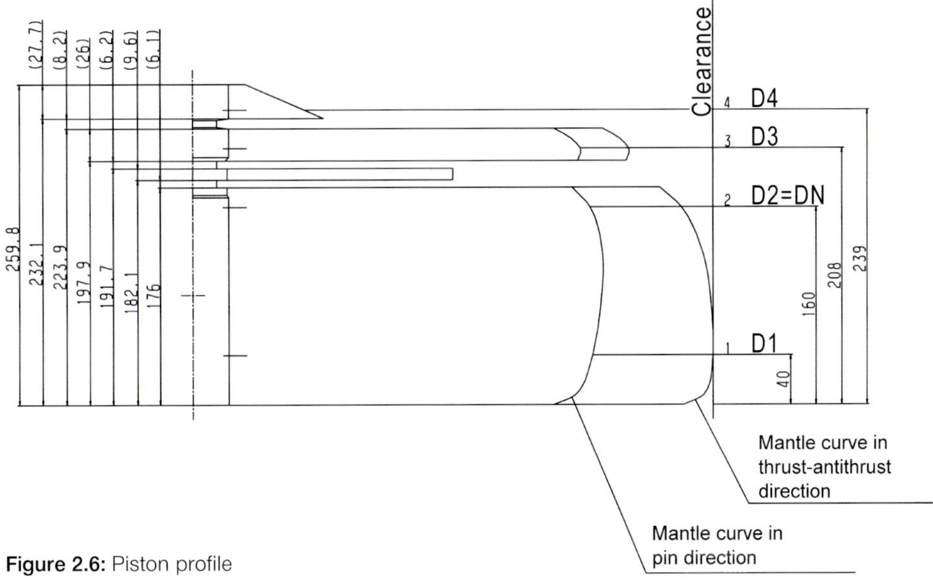

Figure 2.6: Piston profile

For noise-sensitive gasoline engines, in particular, there should be no contact between the ring belt and the cylinder.

All of these aspects require optimized machining forms of the outer surface for the various piston types. The final piston profile can only be verified through extensive simulation and engine testing. **Figure 2.6** shows a detail from a piston profile drawing.

2.2.4 Dimensional and form tolerances

The piston diameter is typically determined absolutely at one of at least three measuring planes. This reference measuring plane is designated as DN. It is preferably located at the point with the tightest clearance between the piston and cylinder (DN = D1) or in an area with a stable shape (DN = D2). The dimensional tolerance (diametric) is 8 to 18 µm, depending on the piston diameter.

The outer contour of the piston is manufactured by NC-controlled precision turning. A funnel-shaped tolerance band results from the elasticity of the piston, as shown schematically in **Figure 2.7**. Deviations from the nominal form are called form tolerances. The form tolerances of the diameters D1, D2, D3, and D4 for passenger car and commercial vehicle pistons are

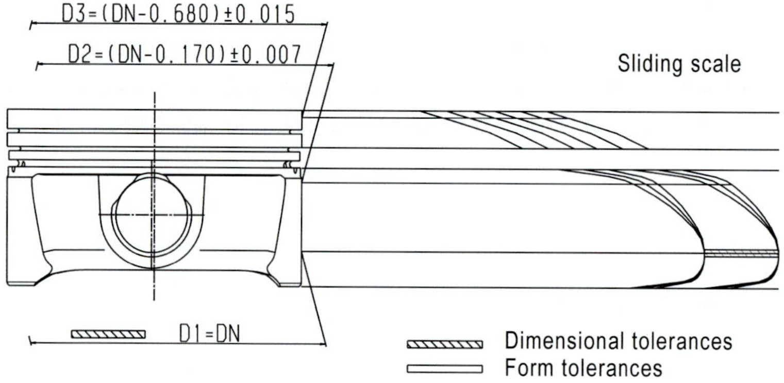

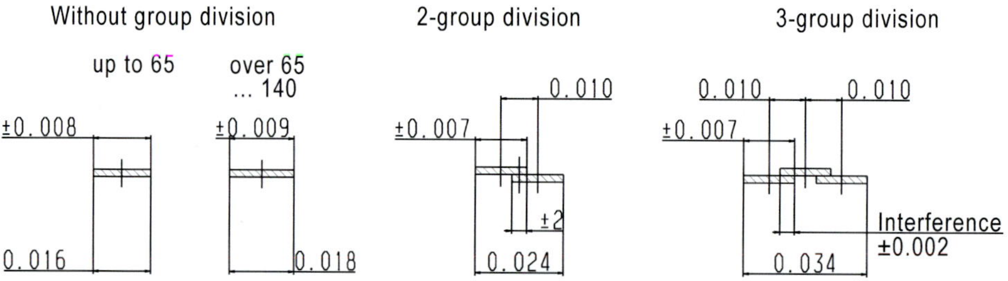

Figure 2.7: Piston profile, dimensional and form tolerances

about 7 μm in the skirt area (diametric) relative to DN, and 10 to 15 μm in the ring belt area (diametric). The principle of the sliding scale applies. The tolerance band for the form tolerances shifts according to the actual diameter in the classification plane.

2.2.5 Installation clearance

The installation clearance is the difference between the cylinder diameter and the largest piston diameter D1. For low friction power loss, the installation clearance must not be too small. However, it also must not be too large, so that consistently smooth running is achieved under all operating conditions. Because of the difference in thermal expansion, these goals are most difficult to achieve for the combination of an aluminum piston and gray cast iron cylinder. Previously, cast-in steel struts were often used to reduce thermal expansion. Table 2.3 gives an overview of the (diametric) clearances at the skirt for various piston types.

The installation clearance decreases with increasing operating temperature, which is caused by the greater heating of the piston relative to the cylinder, and possibly the different thermal expansion of the piston and cylinder materials. At operating temperature, the piston runs in the cylinder with interference. Because of the ovality, the interference is limited to the elastically adaptable area of the skirt.

Table 2.3: Typical installation clearances for light-alloy pistons [‰ of nominal diameter]

	Gasoline engines			Diesel engines
	Two-stroke	Four-stroke (passenger cars)		Four-stroke (passenger cars)
Engine block material	Aluminum-base alloy	Aluminum-base alloy	Gray cast iron	Gray cast iron
installation clearance	0.6–1.3	0.2–0.6	0.4–0.8	0.6–0.9
Clearance at upper skirt end	1.4–4.0*	1.2–1.8	1.7–2.4	1.9–2.4

* Only for single-ring designs and high-performance engines (skirt end close to top land)

2.2.6 Defining group

One defining group for the piston and cylinder makes logistics easier in large-scale production. If the highest priority is the economic efficiency of production, then slightly wider bands must be used for the dimensional tolerances than for division into several groups, e.g., (diametric) 18 μm compared to (diametric) 14 μm for a two-group division; **Figure 2.7**.

When pistons up to 140 mm in diameter are divided into several classes, overlap zones of 2 μm are required at the group boundaries. The pistons in the overlap zones can be assigned to the larger or smaller defining group, as desired. This ensures that the desired quantity can be supplied for each defining group.

2.2.7 Skirt surface

Besides the skirt profile, the surface of the skirt running surface also has a great influence on the sliding behavior of the piston. Too little surface roughness means the piston will not run in properly, while too much increases the friction power losses. Skirt roughness profiles with roughness values of R_a = 1.5–5 µm (R_z = 6–20 µm), generated precisely by diamond turning, provide good results (cf.: Chapter 1.2.4).

Thin metal layers of tin (0.8 to 1.3 µm) or synthetic resin graphite coatings (10 to 40 µm) further improve the boundary lubrication properties, particularly in the critical run-in process or when starting the engine under less than ideal conditions, such as during cold start.

3 Simulation of piston operational strength using FEA

Today's requirements for modern combustion engines can be met only with highly efficient combustion and charging processes. Modern engines can now achieve specific power output and peak pressures in the combustion chamber that were previously unheard of outside of motorsport. This exposes the pistons to extreme loads, but the high requirements for service life and cost efficiency remain unchanged for these components. They are the decisive variables in the current global competitive environment.

One prerequisite in the context of extensive engine testing is simulation, particularly of operating loads, and verification of the operational strength of the piston. Precise, physically based approaches and the use of efficient calculation methods are therefore critical. In the industrial environment, finite element analysis (FEA) has become established as a standard process. The special application of this method to the piston engine component is described in the following.

3.1 Modeling

The basis for FE analysis is modeling, using discretization or meshing, that is, the partitioning of the affected structure into so-called volume elements. All the individual parts of the entire model are meshed, typically in three dimensions, including all significant details, and with only minor simplifications.

Thanks to the processing capacity of current computers, symmetries (the use of half or quarter models) is no longer absolutely necessary. The initial geometry created by design can nearly always be implemented to its full extent. Powerful mesh generators and user interfaces with refined graphics (known as preprocessors) effectively support modeling. Modern software running on multiprocessor computers allows models with a large quantity of volume elements to be processed, or solved.

In the case of piston calculations, the entire model includes not only the piston, but also the piston pin, the small end bore, and the cylinder; **Figure 3.1**. The cylinder and small end bore are fixed to the appropriate interfaces for the calculation. For the piston and piston pin components, which move during engine operation, support is provided solely by contact conditions. Interactions and load transmission can thus be shaped under conditions that are close

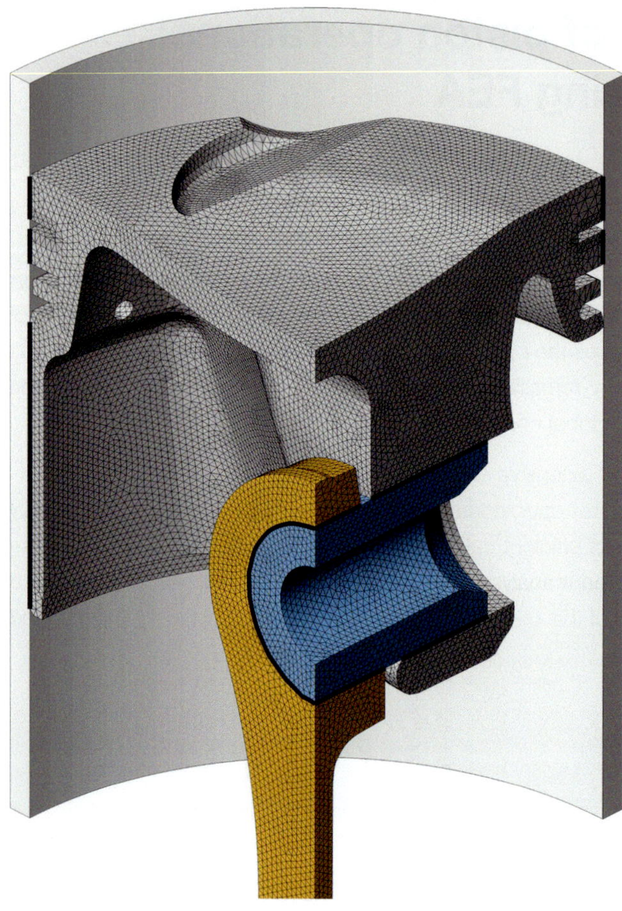

Figure 3.1:
FE model of a piston, with
piston pin and small end bore

to reality. Within the contact definitions, any existing refined profiles of the contact surfaces, such as the piston skirt profile or the pin bore profile, can also be taken into consideration.

The application of the load in the model is depicted in the following Chapters 3.2 and 3.3. The assignment of material fatigue data concludes the modeling process. Depending on the component, the application, and the load, both purely linearly elastic material properties and elastic-plastic properties are used.

3.2 Boundary conditions from engine loading

First, loads are differentiated as either thermal or mechanical; see Chapters 3.2.1 and 3.2.2. This determines whether the load on the piston should be treated as dynamic, or simplified as a static load.

The operating load on the piston is determined by the operating principle of the engine (e.g., four-stroke gasoline or four-stroke diesel process). On account of the cyclic sequence of the strokes for this operating principle, the load is considered to be cyclic. The cyclic loads include

- the temperature from the combustion process;
- the gas force from the cylinder pressure;
- the inertia force; and
- the lateral force.

Loads that do not originate directly in the engine operating principle are either static, such as residual stresses (Chapter 3.3), or they result from changes in engine operation. These could be randomly caused (e.g., road profile for passenger cars), or definitively prescribed (e.g., acceptance run, exhaust gas test cycle).

3.2.1 Thermal load

The thermal load from the gas temperatures in the combustion process is also a cyclic load on the piston. It acts primarily during the expansion stroke on the combustion chamber side of the piston.

In the other strokes, depending on the operating principle, the thermal load on the piston is reduced, interrupted, or even has a cooling effect during gas exchange. In general, heat transfer from the hot combustion gases to the piston occurs primarily by convection, and only a slight portion results from radiation.

In relation to the expansion stroke, the duration over which the thermal load from combustion acts is very short. Therefore, only a very small portion of the component mass of the piston, near the surface on the combustion side, follows the cyclic temperature fluctuations. Nearly the entire mass of the piston, therefore, reaches a quasi-static temperature, which can, however, have significant local variations. A temperature field is established; see Chapter 3.4. In the simulation, the cyclic temperature fluctuations at the surface can be taken into consideration, but this is very difficult. It is generally omitted.

The thermal boundary conditions required to simulate the quasi-static temperature field in the piston are determined using an iterative thermodynamic calculation. They may be supplemented by any temperature measurements from engine tests; see Chapter 7.2. Additional knowledge about heat transfer that are dependent on the contact pressure allows suitable heat transfer coefficients to be defined for all relevant surfaces and contact areas of the piston. In combination with the corresponding ambient temperatures, they can be used to describe the local heat flow.

An alternative to determining the thermal boundary conditions required to simulate the quasi-static temperature field in the piston is the use of the CFD method (Computational Fluid Dynamics). It allows numerical simulation of the combustion process.

Using the CFD method, for example, the consequences of nonhomogeneous mixture formation and the corresponding nonuniform heat release in the combustion chamber can be described. It also allows near-realistic modeling of the complex conditions on the piston-side walls of the combustion chamber, particularly in a direct injection diesel engine. A practical example of the high spatial resolution of the boundary conditions for simulating the temperature field is the detailed consideration of the interaction between the wedge-shaped flame jets and the contour of the piston recess; **Figure 3.2**.

Thanks to the multiport nozzles used in this combustion process, nearly any pattern can occur at the circumference of the piston recess. Without the support of the CFD method, such relationships can be modeled only at a simplified level with reasonable effort.

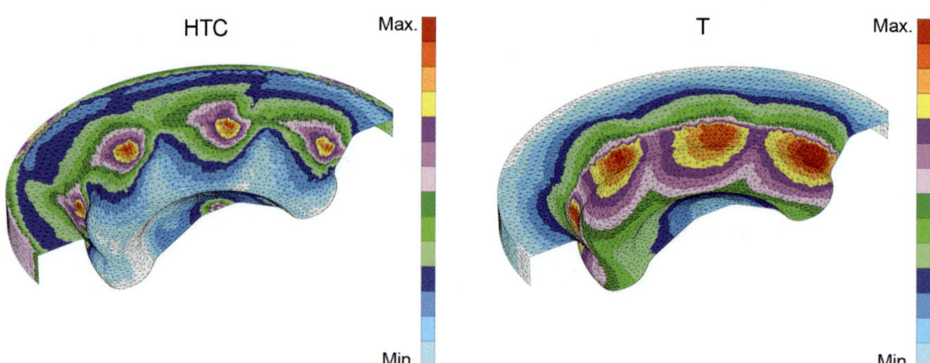

Figure 3.2: Distribution of heat transfer coefficients (HTC, left) and gas temperature (T, right) on the piston crown of a commercial vehicle engine, from a CFD analysis

3.2.2 Mechanical load

The cyclic loading of the piston due to

- the gas force from the combustion pressure;
- the inertia force from the oscillating motion of the piston; and
- the lateral force from the support of the gas force by the inclined connecting rod, and the inertia force of the oscillating connecting rod

determines the mechanical load.

According to the fundamental principles of machine dynamics, the correct superposition of these dynamic forces, as a function of the kinematic relationships in the crank mechanism, can be determined at any given point in time, and can be defined as an external force in the simulation of the operating loads. In the simulation of the operating load on the piston, the external forces acting on the piston are considered to be static at the point in time that is under analysis.

In order to avoid neglecting any combination of these forces–so-called load cases–that may be critical, a practical number of appropriately selected points in time are analyzed. This include primarily the times at which individual components of the mechanical load reach a maximum. For example, the effect of the piston's inertia force is at a maximum at the dead center points of the crank mechanism. Other examples include the maximum of the combustion pressure, or the lateral force, which in turn can act in two different directions. Because the same arrangement of the crank mechanism elements occurs for all of these cases, the time points are defined on the basis of the angular position of the crankshaft. For example, in the four-stroke cycle, two revolutions of the crankshaft are needed for the entire process, which corresponds to 720° of crank (shaft) angle.

3.2.2.1 Gas force

Simulation of the gas force is intended to represent the effect of gas pressure on a piston at operating temperature. The gas pressure is applied to the simulation model over the entire piston crown, down to the lower flank of the top ring groove. The force flow of the longitudinal force generated thus runs through the piston pin to the small end bore.

3.2.2.2 Inertia force

The oscillating motion of the piston in the cylinder generates accelerations that reach a maximum at the top dead center (TDC).

In this context, the length of the connecting rod relative to the crank radius of the crankshaft pin, known as the stroke-connecting rod ratio, plays a decisive role. As the length of the connecting rod increases, the acceleration of the piston and the lateral forces decrease. Maximizing the length of the connecting rod is therefore a recognized design principle. In order to achieve this while minimizing overall dimensions, the compression height of the piston must also be reduced as much as possible. This has corresponding consequences for the piston design; see Chapter 2, **Figure 2.2**. The accelerations are linearly dependent on the stroke-connecting rod ratio of the crank mechanism, and quadratically dependent on the engine speed. This means that the effects of the inertia forces in an engine increase significantly at high speeds.

The resulting acceleration force is applied to the connecting rod by the piston, by means of the piston pin. In the FE model, the acceleration is applied globally to the piston and piston pin. The model is fixed in the axial direction at the small end bore.

3.2.2.3 Lateral force

The conversion of the piston's linear motion into the crankshaft's rotational motion leads to force components in the crank mechanism, which press the piston against the cylinder bore, due to the lateral displacement of the big end bore and the resulting inclined position of the connecting rod. These are known as lateral forces. In a crank mechanism that is not offset (cylinder axis and conrod axis are collinear), the greatest lateral forces occur in the expansion stroke. In high-speed engines, however, they can also occur in other crank angle ranges as a result of inertia forces. In the simulation, the lateral forces of interest are transmitted into the piston via the small end bore and the piston pin. The piston then presses against the cylinder.

As already indicated in the simulation of the gas force, the simulation of lateral forces is a purely static consideration at first. This approach is correct if the piston runs in the cylinder without clearance, but this is not always the case in practice. Disregarding the operating conditions for a cold start, in which the installation clearance is fully present for a short period, additional conditions can occur, depending on the application, in which there is a running clearance between the piston and the cylinder. This includes, in addition to most applications of steel or nodular cast iron as a piston material, all thin-walled lightweight pistons, such as are preferred for use in gasoline engines and especially in motorsport applications. These piston types become deformed under the operating conditions during engine load. Despite the typical interference under full-load conditions, this often leads to dynamic clearance conditions between the piston and cylinder.

These effects can be depicted using appropriate simulation tools (in this case, the structural dynamic simulation of the secondary piston motion) and the resulting additional dynamic loads in the static approaches of the lateral force can be correctly taken into account for the physical FE analysis.

3.3 Boundary conditions due to manufacturing and assembly

Pistons with design features such as pin bore bushings or ring carrier inserts, as well as composite pistons, are made up of several parts that are connected to one another using various methods: casting, joining or pressing a shrink fit, and bolting. These joining and composite methods cause residual stresses, which must be considered in a precise strength analysis. Such loads act statically on the piston and must be included in relaxation processes and for problems dealing with creep under load.

3.3.1 Casting process/solidification

Residual stresses in the component are an unavoidable characteristic of casting. They arise as a result of differences in wall thicknesses and different local cooling conditions; they are therefore present in the piston as well. Heat treatment after the casting process, however, reduces these stresses. Both processes–the casting process and the heat treatment–can be modeled using a numerical casting simulation, and the resulting residual stresses can be considered as well.

3.3.2 Inserts

Aluminum pistons for diesel engines usually have a cast-in ring carrier made of austenitic cast iron with lamellar graphite (Ni-resist), which is intended to reduce ring-side wear in the top ring groove; see Chapter 1.3.3. Cast-in and composite stresses arise during the casting of this ring carrier. The aluminum alloy shrinks onto the ring carrier during solidification and final cooling, thus generating primarily compressive stresses in the ring carrier.

3.3.3 Pressed-in components

The combustion pressures in a passenger car diesel engine currently reach about 200 bar, while they reach about 250 bar in commercial vehicle diesel engines and are even greater than 250 bar in large-bore engines. This leads to high stresses in the piston pin boss. Particularly for pistons made of aluminum or composite pistons with aluminum skirts, therefore, it may be necessary to increase the load carrying capacity of the piston pin boss by shrinking in a pin bore bushing made of a suitable material, such as bronze or special brass. The shrink-in process generates compressive stresses in the pin bore bushing, and tensile stresses in the surrounding aluminum alloy, primarily at the bottom of the pin boss (nadir).

3.3.4 Screw joints

Screw joints in composite pistons must be analyzed under the described thermal and mechanical loads. Analysis of the assembled state at room temperature, as well as friction at the various contact surfaces, is also required.

Depending on the process used for tightening the bolts (hydraulic, torque, or rotation angle tightening) the pretensioning force of the bolts is calculated and adjusted at room temperature for the simulation of the screw joint in the FE model. The pretensioning force is then superimposed on the thermal and mechanical loads (gas force, inertia force, lateral force).

With this approach, it is possible to establish

- whether the bolt pretensioning force increases or decreases in the piston at operating temperature;
- how clearances and inclinations on the contact surfaces affect the distribution of the pretensioning force;
- whether the upper part lifts off from the piston skirt in the area of the bolt; and
- how the clearances affect the stress amplitudes from the mechanical loads.

3.4   Temperature field and heat flow due to temperature loading

The extent of the temperature field and the resulting temperature gradients are determined by the cooling of the piston. Pistons are generally categorized as having no cooling, oil-spray cooling, or cooling-gallery/cooling-cavity cooling. The partially filled cooling galleries and cooling cavities, together with the oscillating piston motion, lead to extensive turbulence in the engine oil cooling medium. This turbulence generates high relative velocities of the cooling medium at the gallery wall, which in turn improves heat transfer at the wall. This condition is referred to, in a simplified manner, as the "shaker effect." The piston is then said to have "shaker cooling."

With regard to component strength, the temperatures at the piston crown (gasoline engine pistons), at the bowl rim (diesel engine pistons), in the support area (transition from the boss to the piston crown), and in the pin bore are of interest. The temperatures in the top ring groove and the cooling gallery are also significant in terms of oil carbonization. Typical temperature values for passenger car engines are

- center of piston crown (gasoline engine, port fuel injection)   270–310°C;
- piston crown bowl (gasoline engine, direct injection)   270–350°C;
- bowl rim (diesel engine, direct injection)   350–400°C;
- support area   200–250°C;

- pin bore (zenith)   200–250°C;
- top ring groove (spray jet cooling, salt-core cooling gallery)   200–280°C;
- top ring groove (cooled ring carrier)   180–230°C;
- cooling gallery (zenith)   250–300°C.

Numerical simulation of the temperature field allows detailed analysis of the heat flows in the piston. Depending on the piston cooling method used in each case, typical values for the percentual distribution of the heat flow through the piston crown are shown in **Table 3.1** for various areas of the piston. In the uncooled piston, the heat flow to the cooled cylinder bore (through the ring grooves, the piston rings, and the piston skirt) dominates the heat flow pattern, while the heat flow into the engine oil by convection is predominant in the cooled piston types.

Table 3.1: Distribution of heat flow for various piston types

Piston type	Uncooled piston	Oil-spray cooling	Oil-spray cooling	Salt-core cooling gallery	Cooled ring carrier	MONO-THERM® ring gallery	2-chamber cooling cavity
Operating principle	Gasoline	Gasoline	Diesel	Diesel	Diesel	Diesel	Diesel
Heat flow [%] Cooling gallery	0	0	0	40–50	50–60	75–90	90–100
Ring belt	50–60	15–25	50–55	25–45	10–30	0–10	0–5
Skirt	10–15	5–10	10–15	5–10	5–10	0	0
Inner shape	10–20	50–60	20–30	5–15	5–15	0–10	0–5
Window/ undercut	5–10	0–5	0–5	0–5	0–5	0	0
Boss	5–10	0–5	0–15	0–10	0–10	0	0

The typical temperature distribution in a gasoline engine piston, with a temperature gradient from the piston crown to the skirt, generates a thermal deformation as shown in **Figure 3.3**. This deformation characteristic must be taken into consideration when fine machining is done on the outer diameter, and defines the barrel shape used for the piston skirt. The prominent inclined position at the zenith of the pin bore is typical for these deformation patterns. This wedge-shaped gap between the pin boss zenith and the piston pin is completely closed under the gas force load.

Pistons with cooled ring carriers are used in passenger car diesel engines, in order to reduce the temperature at the bowl rim and in the top ring groove. **Figure 3.4** shows a comparison of the cooling effect of a cooled ring carrier (left) and a so-called salt-core cooling gallery (right).

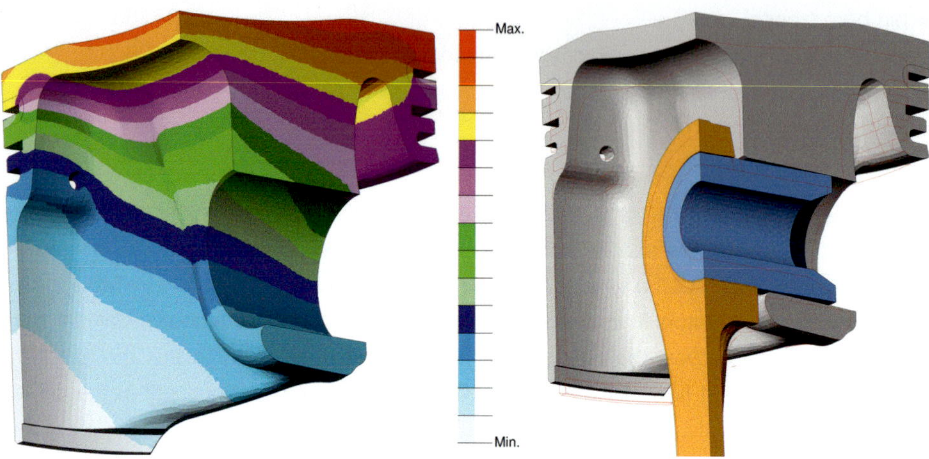

Figure 3.3: Temperature distribution (left) and thermal deformation (right) of a piston in a passenger car gasoline engine with spray jet cooling (deformation shown magnified)

The advantageous influence of the cooled ring carrier on the bowl rim temperature, and especially on the temperature in the top ring groove, is evident.

For commercial vehicles, MONOTHERM® pistons are also used in engines with the highest power ratings. It is a single-piece forged steel piston. It has a skirt that is directly bonded to the piston pin boss, and which can be open or closed at the top. **Figure 3.5** shows the

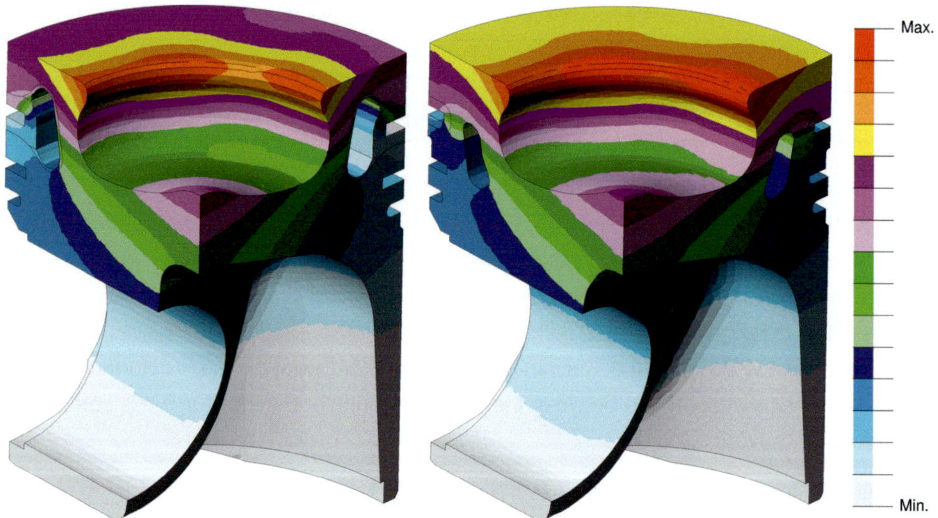

Figure 3.4: Comparison of cooling effect for cooled ring carriers (left) and salt-core cooling galleries (right) on a passenger car diesel piston

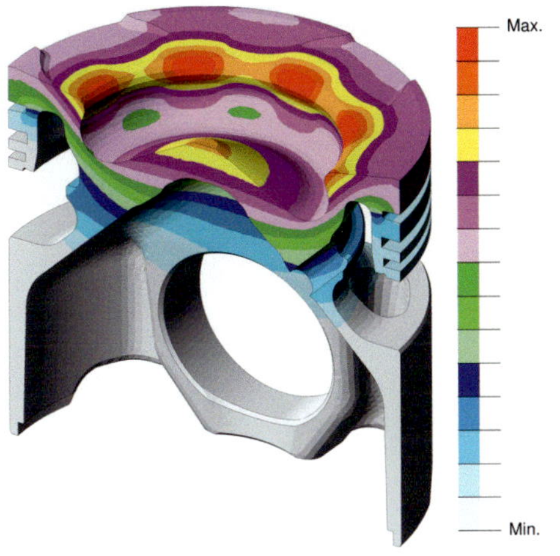

Figure 3.5:
Temperature distribution in a MONOTHERM®
piston with a closed skirt

temperature distribution in a MONOTHERM® piston with a closed skirt. The greatly reduced heat flow in the skirt, and the resulting low skirt temperatures, is typical for this type of piston.

For large-bore engines, a significantly longer service life is required, and the combustion pressures reach levels greater than 250 bar. Under such conditions, piston cooling is decisively significant. Therefore, large-bore pistons are nearly always designed as composite pistons, with an upper part made of heat-resistant steel, and a skirt made of either aluminum, nodular cast iron, or forged steel. With this design, it is possible to make the cooling cavities large, and accordingly effective for cooling. This is accomplished, for example, by using concentric 2-chamber systems, which cover nearly the entire crown of the piston. The heat balances in **Table 3.1** show that the cooling in this piston is very effective. The heat flowing into the piston as a result of combustion can thus be dissipated nearly 100% by the cooling oil.

The thermal deformations caused by the severe temperature gradients in the piston crown are shown in **Figure 3.6** for a piston with shaker cooling. The temperature in the screw joint affects the relaxation behavior in the high-strength screw material used here. If temperatures greater than 180 to 200°C occur over large areas of the bolt, then relaxation must be anticipated, leading to a reduction in bolt pretensioning force during operation.

The skirts of composite pistons are hardly affected by the temperature load–with the exception of the area of the screw joint's countersunk holes. The temperature here can increase the stresses in the radii of the countersunk holes, if the axial tolerance are poorly designed. For screw joints in composite pistons, the distribution of the bolt pretensioning force at the internal contact surface must not drop below 20 to 25% of the pretensioning force under temperature loading. This can be achieved by suitably designing the axial tolerance at the external contact surface.

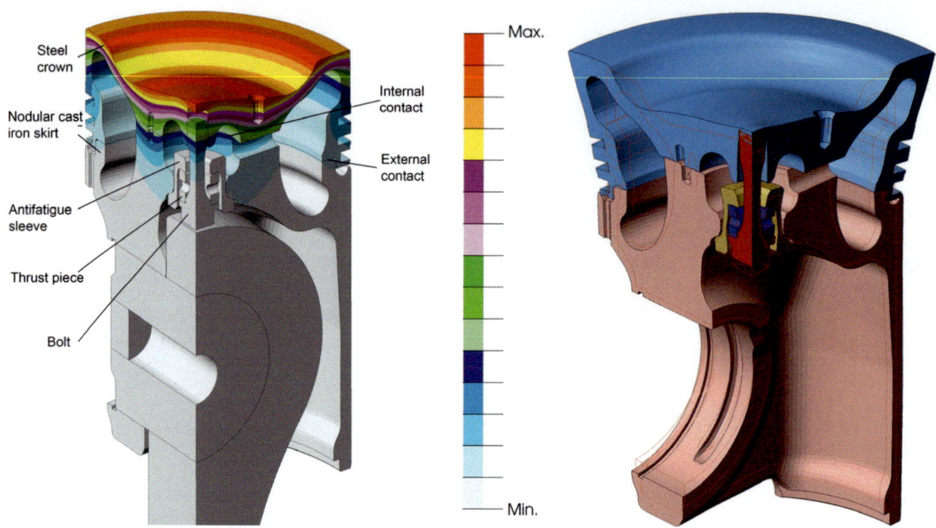

Figure 3.6: Temperature distribution (left) and thermal deformation (right) of a composite piston with shaker cooling (deformation shown magnified)

3.5 Stress behavior

3.5.1 Stresses due to temperature loading

The temperature distributions shown in Chapter 3.4 for various pistons show significant temperature gradients. This is especially true at the piston crown and the bowl rim. The piston expands greatly in the hot areas, whereas expansion is prevented in the cold areas. The thermal stresses thus induced–primarily compressive stresses–are greatest at the crown and bowl area, and can exceed the yield limit of the piston material there.

In order to consider the fact that the yield limit has been exceeded in the FEA, complex analysis using the nonlinear elastic-plastic behavior of the material is necessary. Exceeding the yield limit, however, affects only limited, localized areas of the piston. These areas also have very low stiffness, on account of the high temperatures that are typically present there. The impact on the global stress and deformation behavior of the entire piston is therefore correspondingly small.

The numerical complexity can be reduced by determining the thermal stresses under linear-elastic conditions, considering the temperature-dependent physical material data. The stresses thus determined are then converted into an elastic-plastic behavior at the highly stressed locations of interest, and are analyzed for strength and service life according to the methods described in Chapter 3.6.

Figure 3.7 shows the distribution of the thermally induced stresses in a piston for a gasoline engine with port fuel injection. The temperatures on nearly level piston crowns are approximately 280 to 300°C and generate compressive stresses in the internal crown area.

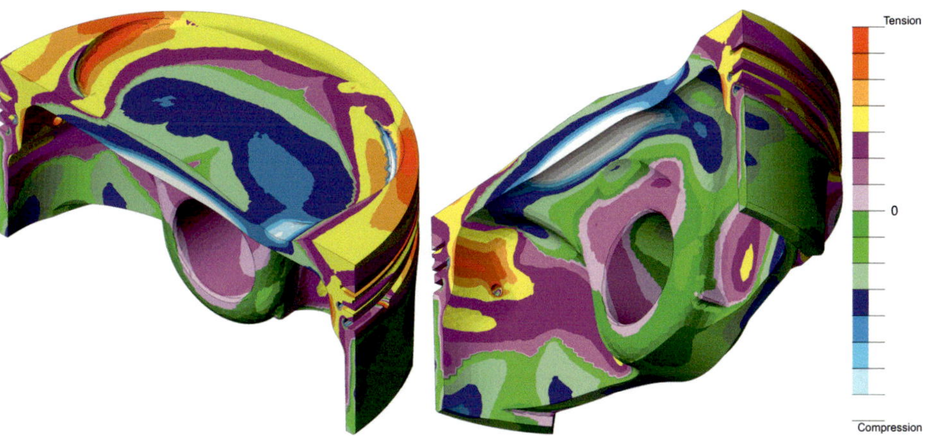

Figure 3.7: Thermally induced stresses in a piston for a passenger car gasoline engine with port fuel injection

Figure 3.8 shows the thermally induced stresses in an aluminum piston with a cooled ring carrier for a passenger car diesel engine with direct injection. The undercut bowl shape reveals a significant concentration of compressive stresses at the bowl rim. The level of the bowl rim stresses, together with the high temperatures (up to 400°C) lead to the yield limit at the bowl rim being exceeded.

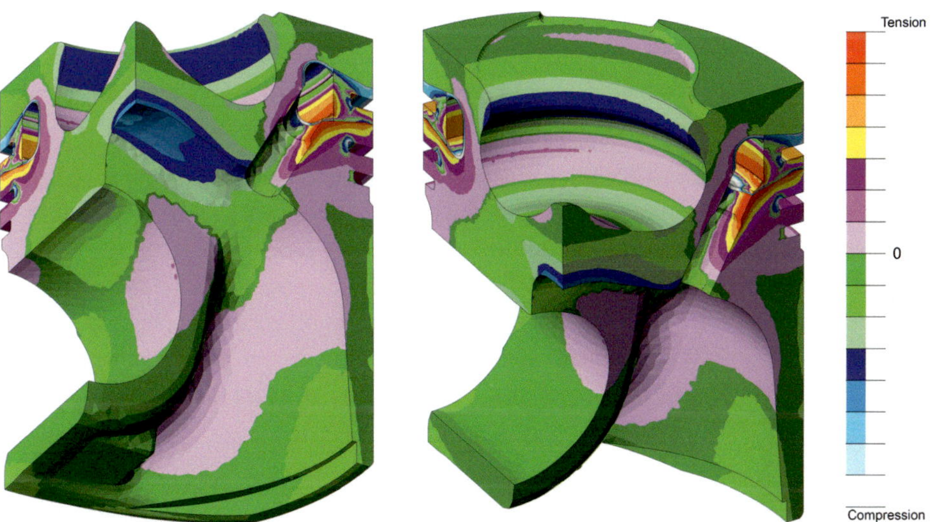

Figure 3.8: Thermally induced stresses in a piston with cooled ring carrier

In general, bowl rims have better strength and temperature performance when the bowl rim radius is greater. The same applies to a small undercut for the bowl as well as to obtuse angles and flat bowl shapes, which are preferably used for commercial vehicle pistons.

MONOTHERM® pistons (made of steel) also have high thermally induced compressive stresses at the bowl rim and in the cooling gallery area. Compared with pistons made of aluminum, however, this has little influence on the strength, as long as the temperatures in these areas are below the scaling limit. Long-term temperatures greater than 450 to 500°C cause scaling, and therefore damage the surface. Such defect locations can be the starting point for cracks in the bowl rim of a steel piston.

Composite pistons with shaker cooling generally have the greatest thermally induced stresses in the steel crown. The stress maxima are located in the wall of the outer cooling cavity, the top ring groove, and at the bowl rim.

3.5.2 Stresses due to mechanical loading

For all mechanical loads, the "temperature" load case, assumed to be quasi-static, is superimposed in order to calculate the correct resulting stresses in the piston analysis.

As a reaction to the gas force load, the piston presses against the piston pin and the cylinder. For aluminum pistons, the rigid piston pin dominates the deformation behavior of the system, because the piston has a lower stiffness on account of its highly temperature-dependent material fatigue data. The structure bends; the piston is deformed "around the piston pin." This deformation generates stresses induced by the gas force in addition to the temperature stresses, and causes a "saddle-shaped" curve in the piston ring grooves.

Figure 3.9 shows the distribution of the stress amplitudes, as calculated from the cyclic gas force loads on a piston. This general stress distribution from the gas force can be explained by the "bending around the piston pin" and applies, in the broadest sense, to all pistons: under gas force load, circumferential tensile stresses arise at the bowl rim along the pin axis, while compressive stresses predominate in the thrust-antithrust direction.

Under gas force load, for all piston variants, the boss and support area and the transition of the support into the piston crown are clearly loaded by compressive stresses; **Figure 3.10**. In aluminum pistons, in particular, the load limit of the bosses can be exceeded as a result of high local pressures and temperatures. This limit can be increased, to a certain degree, by the use of special pin bore profiles that deviate from the round, cylindrical pin bore by 10 to 100 μm. The design of the pin bore profile, however, also affects the stress at other areas on the piston, such as the bowl rim, bowl base, cooling gallery, or support.

In composite pistons, high compressive stresses due to gas force also arise in the support area. The force flow of the gas force, from the combustion chamber, through the crown and the contact surfaces, into the skirt and then into the piston pin, can be significantly affected by the radial location of the contact surfaces. Contact pressures and displacements (relative

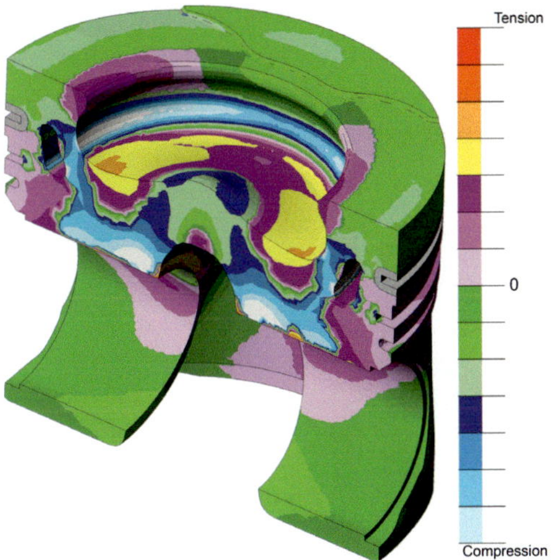

Figure 3.9:
Stress amplitudes due to gas force load on
a commercial vehicle piston with salt-core
cooling gallery

motions) arise under gas force loads at the contact surfaces in each expansion stroke. These
are caused by the different deformations of the piston's crown and skirt. This can lead to
wear at the contact surfaces.

The screw joint is particularly important for composite pistons in order to minimize the amplitudes from the gas force, particularly in the skirt's countersunk holes. This is done by appropriately designing the bolted parts (tappet, optimized countersunk depth) and by placing the
bores outside of the force flow of the gas force as much as possible (central screw joint, twin
screw joint).

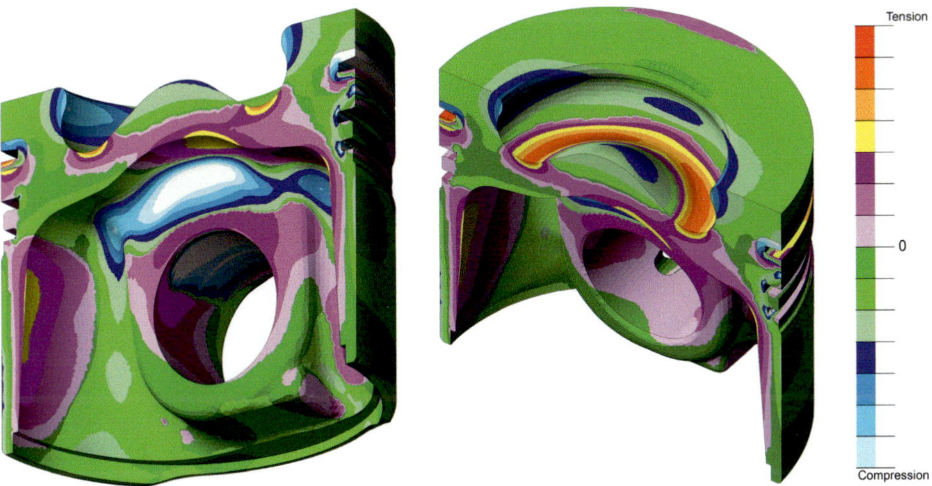

Figure 3.10: Stresses under temperature and gas force in a passenger car diesel piston with spray jet
cooling

At the TDC fired, the effect of inertia force contributes to relieving the contact load due to gas force in the piston pin boss and the small end bore. **Figure 3.11** shows the stress distribution due to the temperature and inertia forces at the TDC nonfired for a passenger car gasoline engine piston, which is primarily concentrated at the side of the pin bore.

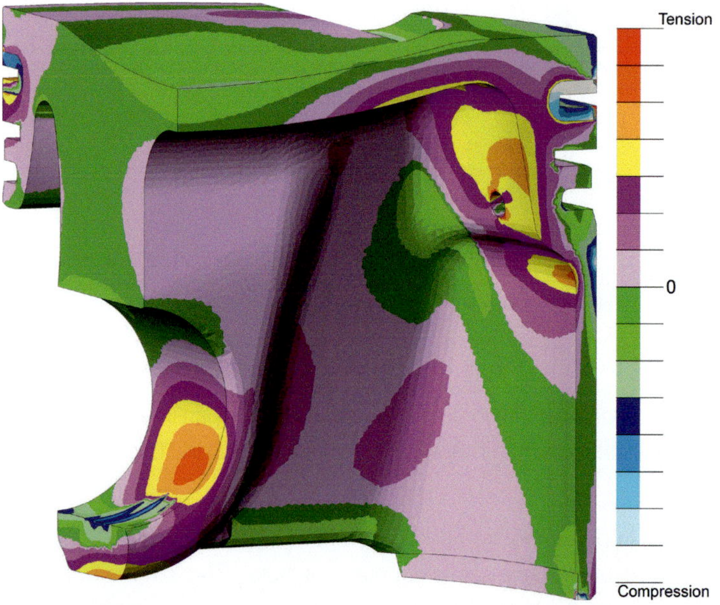

Figure 3.11: Stresses from temperature and inertia force loads in a gasoline engine piston at TDC nonfired

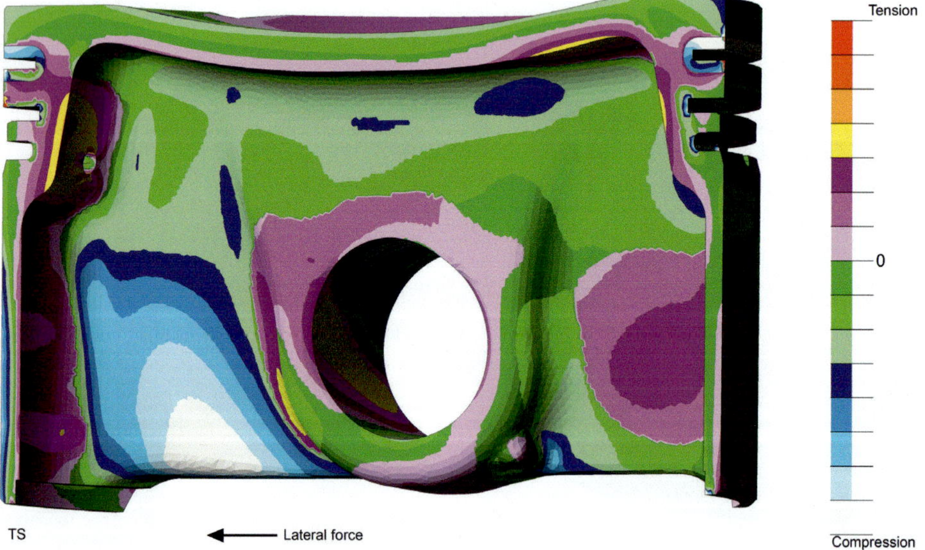

Figure 3.12: Stresses due to temperature, gas, and lateral force loads on a gasoline engine piston

Figure 3.12 shows the distribution of stresses under the influence of temperature, gas force, and the maximum lateral force on a passenger car gasoline engine piston. The location with the greatest stress amplitudes can be easily identified at the transition from the piston skirt to the pin boss. The level of the maximum compressive stresses is greatly influenced by the degree of skirt ovality. Reducing the free deformation of the piston prior to contact with the cylinder, by using a piston skirt with less ovality, significantly reduces the stress amplitudes without requiring increased wall thickness in the skirt area.

3.5.3 Stresses due to manufacturing and assembly

Figure 3.13 shows the stress distribution from the joining process of a shrink-fit pin bore bushing, using the example of a piston for a passenger car diesel engine with direct injection. The bushing is not shown in the illustration. The joining process generates circumferential tensile stresses across broad areas of the pin boss circumference, with maxima at the equator and the nadir (exterior). These must be taken into consideration when designing the boss shape and wall thicknesses.

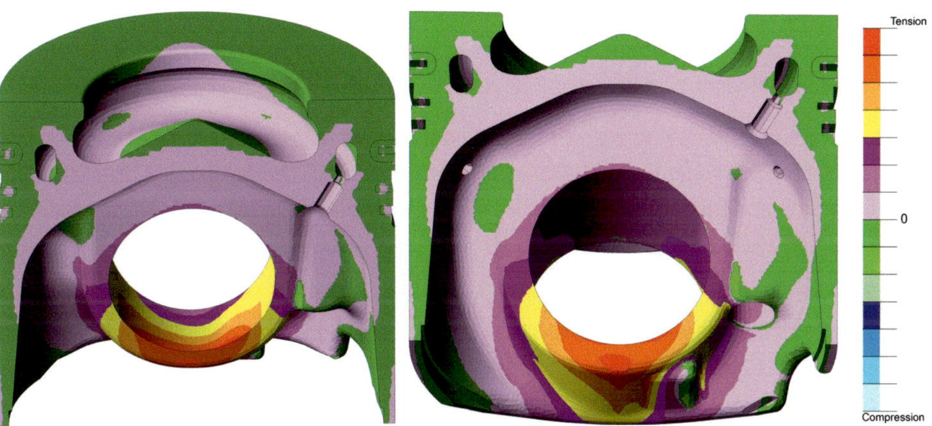

Figure 3.13: Cold shrink fit of the pin boss, stress distribution

3.6   Numerical verification of operational strength

While steel and cast iron materials have pronounced fatigue resistance, aluminum materials lose fatigue strength with increasing numbers of load cycles. This material behavior needs to also be taken into consideration in the strength assessment. For steel pistons, the equivalent stresses are determined in a linear-elastic FEA under full-load conditions, considering all load case combinations, and are analyzed for fatigue resistance as a function of the temperature and surface condition (stress-life method).

For today's highly stressed light-alloy pistons, the evaluation based on a defined number of load cycles (limiting number of cycles: 50 million) is no longer possible considering a minimum safety factor because of the lack of fatigue resistance, as indicated above, at the highly stressed component locations. The alternating plasticization that occurs primarily in the highly thermally stressed bowl area must be taken into consideration, and this requires a different approach.

Numerical verification of the operational strength is an important instrument for reducing the time, and therefore the cost, required for development. It is possible to develop the product close to its final contours at an early stage by considering variants used in series production, and thus reducing the number of engine tests. Service life analysis provides, among others, the following opportunities for analysis:

- Assessment of geometric influences (e.g., bowl rim radius, cooling gallery location, pin bore profiles)
- Effects of load-specific changes (e.g., power output, torque, and engine speed)
- Comparison of customer release procedures
- Correlation between release engine tests and field applications
- Establishing inspection intervals
- Parameter studies for developing reasonable engine test programs

Verification of operational strength is performed using the "local strain approach," analyzing local, transient stresses and deformations (strain-life method); see overview in **Figure 3.14**.

The load-versus-time sequences are typically defined for customer-specific release engine tests, or by the load spectrum of the in-house engine test lab. These are primarily alternating load programs and thermal shock tests, which are intended to cover nearly all conditions of the series application within less time under more severe conditions.

The stress distribution in a piston under engine operating loads is known for a static temperature field from the finite element analysis. In addition to the full-load state, partial-load states can also be calculated as needed, and analyzed for critical levels of equivalent stresses (mean stress and deflection stress). Residual stresses are taken into account in the service life analysis, as are the existing knowledge gained from engine testing, field findings, and strain gauge measurements on identical or similar components in the laboratory.

With regard to materials, cyclic stress-strain curves and total strain controlled Wöhler lines as a function of temperature are necessary. The isothermal material fatigue data for the temperature range from RT (room temperature) to 400°C are based on zero mean strain tests (strain ratio $R_\varepsilon = -1$) and are applicable to the saturated material state.

In order to describe the local stress-strain situation, starting with the initial loading curve and continuing after load reversal with the hysteresis branches (doubling the stress-strain curve, according to Masing), the "true" (elastic-plastic) stresses and strains are iteratively calculated from the existing elastic stresses at the highly loaded component locations, using the Neuber approximation. The mathematical formulation of the cyclic stress-strain curve follows the

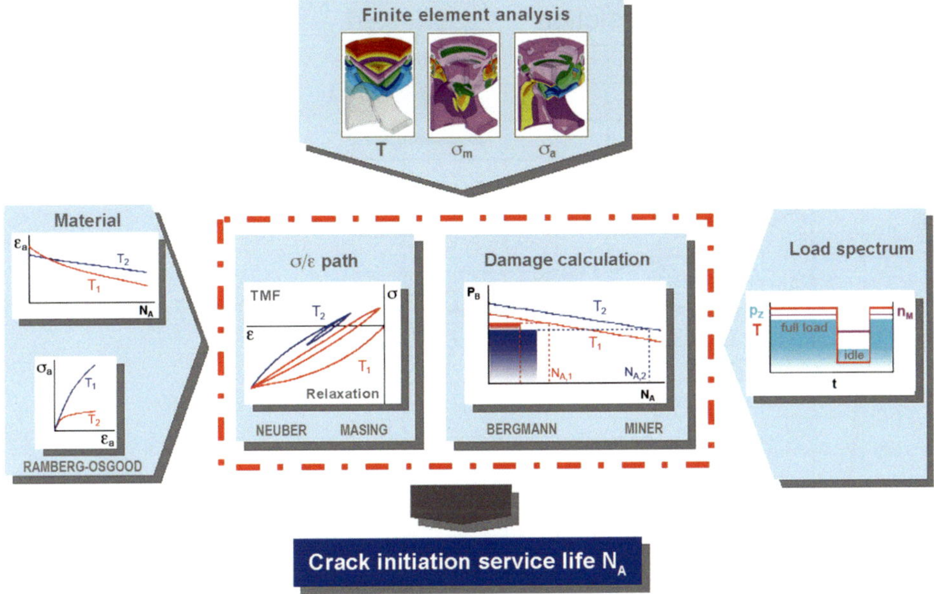

Figure 3.14: Service life concept (schematic)
Definitions: T_1: temperature 1
    T_2: temperature 2
    σ_m: mean stress
    σ_a: stress amplitude
    ε_a: strain amplitude
    p_z: maximum gas force in expansion stroke
    n_M: engine speed
    P_B: damage parameter (Bergmann)
    N_A: number of cycles to crack initiation
    t: time
Remark: The symbols apply only to the images above.

Ramberg-Osgood material law and the hyperbolic curve according to Neuber. The effect of mean stress is described by a damage parameter, according to Bergmann, for example.

Here, nonzero mean stresses are transformed into zero mean stresses. In addition to converting the stress into P_B values (damage parameter according to Bergmann), the temperature-dependent ε/N lines are also described as P-Wöhler lines. The influence of the load spectrum is analyzed using a damage calculation based on the linear damage hypothesis according to Palmgren-Miner, which states that damage is accumulated starting at the first load cycle. Failure occurs at a defined sum of damage. The application of the operational strength assessment is performed at MAHLE using the proprietary MAFAT computer program. The result is the crack initiation service life. The qualitative effects of gas force and temperature on the service life are shown, using the example of a passenger car diesel piston, in **Figure 3.15**: 10% reduction in gas pressure leads to nearly 3 times the service life, while a 10°C lower component temperature approximately doubles the service life.

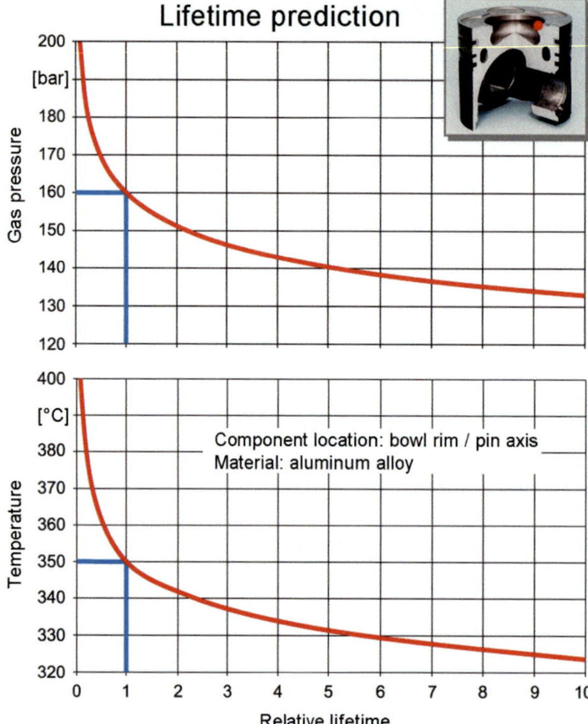

Figure 3.15:
Effect of gas pressure and component temperature on the service life of a passenger car diesel piston

Figure 3.16 shows the close agreement between the analysis and engine test results, based on the example of a passenger car gasoline engine piston. The failure locations are situated in the center of the crown and at the valve pocket intersection, and coincide with the lowest service life estimates on this piston.

Another comparison is shown in Figure 3.17, using the example of a heavy-duty diesel engine piston. The piston failure, with cracks occurring in the bowl base, again matches the service life prediction.

In the damage calculation, the high-cycle stress due to the maximum gas force load acting for a large number of load cycles, and the low-cycle damage due to temperature changes under operating conditions with relatively few events are separated (HC or HCF: high-cycle fatigue, and LC or LCF: low-cycle fatigue). When the operating conditions change, a transient temperature and a change in material properties occurs. At present, the LCF stress is considered as isothermal by using the higher temperature of the two load cases to define the LC stress range. According to the latest TMF (thermomechanical fatigue) fundamental research, under thermal-mechanical loading, this can lead to a nonconservative service life estimate. Because of the temperature and stress gradients present in the piston, the out-of-phase processes are of primary interest: while compressive stresses buildup during heating because of the constraint of the surrounding material, tensile stresses during cooling resulting from cyclic plasticization may lead to crack initiation.

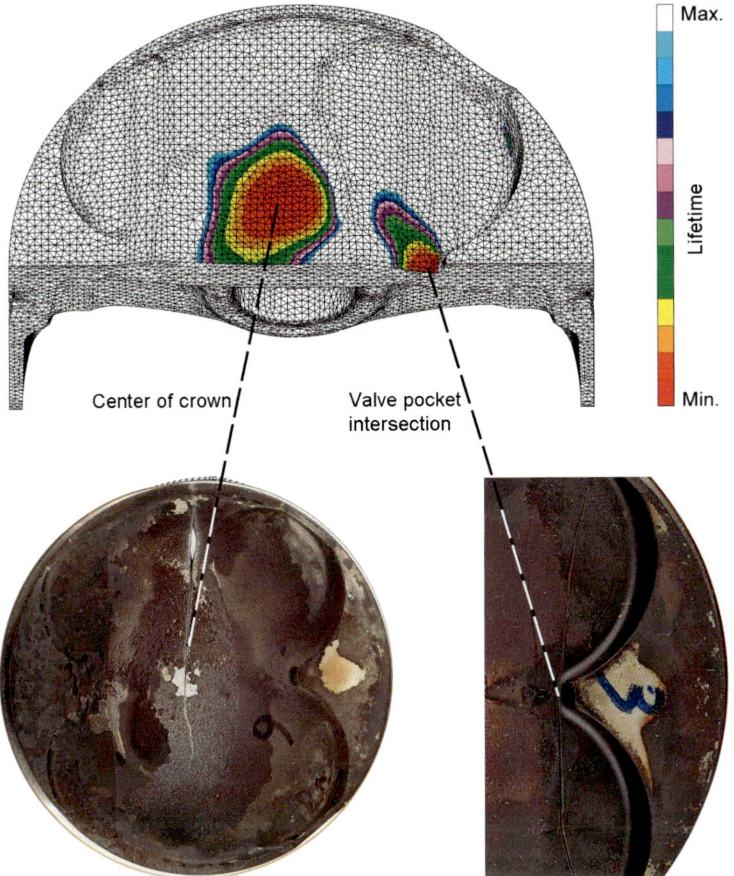

Figure 3.16: Comparison of numerical simulation and engine test (passenger car gasoline engine piston)

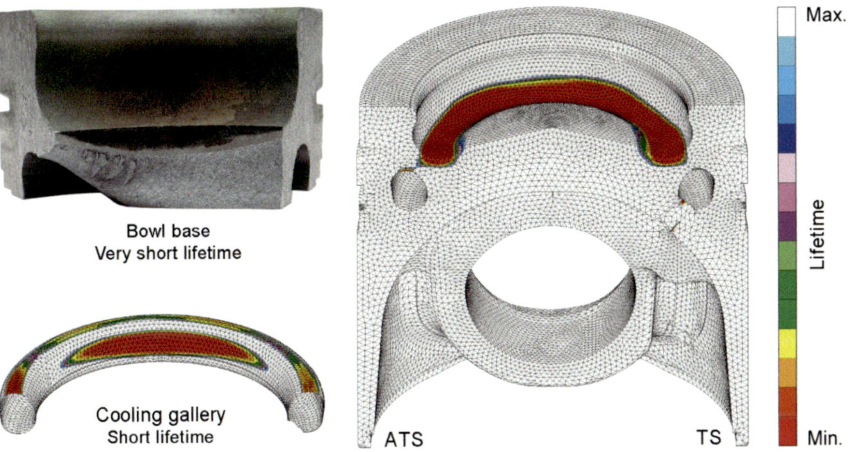

Figure 3.17: Comparison of numerical simulation and engine test (heavy-duty diesel engine piston)

Current research and development activities focus on the simulation of near-engine conditions. Baseline material tests are performed on in-house TMF test benches. Because the transient LC stress–with the exception of the shutdown or stopping process–is simultaneously superimposed on the mechanical loads (HC stress), this must also be taken into consideration in the numerical simulation. One potential method is to investigate the microstructural damage process under various load applications.

One practical approach uses models for cyclic plasticity (Chaboche, Jiang, etc.) that describe material-dependent deformation behavior for monotone and cyclic stresses, creep, relaxation, and cyclic creep. Changes in the deformation model for LCF stresses can be used to include the viscoplastic material law to capture the influence of HCF. The service life of components under cyclic load is often determined by the formation and growth of microcracks. A cycle-dependent damage parameter can be developed on the basis of elastic-plastic fracture mechanics of microcracks, and includes the effects of isothermal and thermomechanical stresses, dwell times, creep effects, etc., resulting in a prediction of the service life.

4 Piston materials

4.1 Requirements for piston materials

The function of the piston and the loads that act on it present a special requirements profile for the piston material.

If the goal is low piston weight, then a low-density material is preferred. Besides the design, the strength of the material is crucial for the load carrying capacity of the piston. The change in loads over time requires good static as well as dynamic strength. Temperature resistance is also important in view of the thermal loads.

The thermal conductivity of the material significantly affects the temperature level. As a rule, high thermal conductivity is advantageous, because it leads to a uniform temperature distribution within the piston. Low temperatures not only allow greater material loading but, in the piston crown, they also have a beneficial effect on process parameters such as filling ratio and knock limit.

Static and dynamic strength values describe the behavior of the material under isothermal conditions. Pistons are sometimes exposed to heavily fluctuating temperatures. The transient thermal stresses that this causes place a cyclic load on the material, sometimes beyond its elastic limit. The materials must also be resistant to this stress. Because of the motions and forces at sliding and sealing surfaces, piston materials must also meet high requirements for seizure resistance, low friction coefficients, and wear resistance.

Material pairings between the piston and its mating parts, as well as lubrication conditions, are of particular significance. They must be considered as a tribological system. Special surface treatments or coatings improve the properties of the base material.

The requirements for the thermal expansion behavior of the piston material depend on the material pairing between the cylinder and the piston pin. In view of changes to clearance between the cold and hot states, the thermal expansion coefficients should differ by as little as possible.

A material with good machining properties ensures cost-effective production of large quantities. The production of the raw forging should allow a shape that is as close as possible to the final contour, as well as high material quality. Suitable processes include permanent mold casting and forging. Sliding and sealing surfaces demand high-precision finishing, which requires the material to have suitable machining properties.

4.2 Aluminum materials

As a light alloy with high thermal conductivity, aluminum is ideal for use as a piston material. Unalloyed, however, its strength and wear resistance are too low. The discovery of precipitation hardening by Wilm in 1906 made aluminum alloys suitable for technical purposes.

Metals have a mutual solubility that depends on temperature. At low temperatures and in the solid state, this value is very low for some metals [1]. Phase diagrams, derived from cooling curves from the liquid state, provide a good visual representation of these conditions.

For example, **Figure 4.1** shows the phase diagram of the 2-material system (binary system) for aluminum/silicon [2]. Drawing a section through the diagram at about 7% silicon, the liquidus line is intersected at about 620°C when cooling down. Below this line, a mixture of primarily precipitated α solid solution and melt is present.

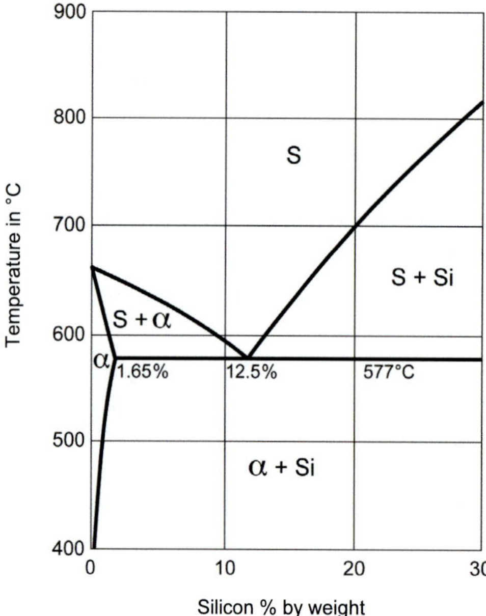

Figure 4.1:
Phase diagram of the 2-material alloy AlSi

The solid solution consists of crystals of the main alloying element (aluminum), with the foreign atoms of the second element (silicon) randomly distributed throughout the matrix. As the temperature drops, the melt becomes more and more rich in silicon, until finally the residual melt solidifies at the so-called eutectic point of 577°C and a content of about 12.5% silicon, forming a eutectic mixture consisting entirely of aluminum α solid solution and silicon crystals. As the diagram shows, the maximum solubility (1.65%) of the silicon in the aluminum solid solution occurs at 577°C, and drops at lower temperatures. At 200°C, the solubility is only 0.01%.

The microstructure of the solidified alloy consists of the primarily precipitated α solid solution and the AlSi eutectic. The diagram also shows that the percentage of eutectic becomes greater with increasing silicon content, until finally the melt transitions from the liquid to solid eutectic state with no solidification interval at the eutectic point of 12.5% silicon and 577°C.

Alloys with a silicon content of <12.5% are hypoeutectic, and those with a silicon content of >12.5% are hypereutectic. It can also be seen in the diagram that the microstructure in the hypereutectic range consists of primarily precipitated silicon crystals and the AlSi eutectic.

Crystallization from the melt is enhanced by nuclei. The number of nuclei and the cooling rate, among other factors, determine the fineness of the microstructure grains. The crystallites growing from the melt restrict the proportion of liquid metal more and more, and form grain boundaries where their edges run into each other. Contaminants and intermetallic phases arising from the melt are concentrated at these boundaries.

4.2.1 Heat treatment

The temperature function of the solubility of solid solutions is utilized for precipitation hardening. With rapid cooling from the melt or from annealing temperatures of around 500°C (solution annealing), supersaturated solid solution is present at low temperatures. Alloys suitable for pistons primarily have aluminum-copper-magnesium and aluminum-magnesium-silicon solid solutions. These alloys, in a supersaturated solid solution, are relatively soft and have minimal volume. Even aging at room temperature causes the supersaturated portions to attempt to reach the state of equilibrium that corresponds to the temperature, which causes stresses in the atomic matrix. No microscopically verifiable changes to the microstructure occur.

This process, known as natural aging, leads to substantially greater hardness and strength and a slight increase in volume after hours or days. Natural aging, however, can be reversed at relatively low temperatures. This makes it unsuitable for thermally loaded parts.

In the range of temperatures between 100°C and 300°C, artificial aging occurs and precipitation can be microscopically verified. The considerable increases in hardness and strength can lead to undesired increases in the volume of precision parts if the heat treatment is performed incorrectly.

The artificial aging curve depends on the exposure time and the temperature level; **Figure 4.2**.

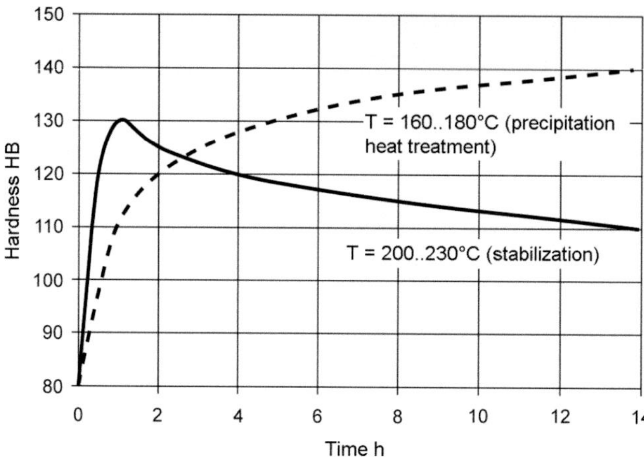

Figure 4.2:
Hardness-time isotherms for artificial aging of aluminum-silicon piston alloys. Initial state: solution-annealed and rapidly cooled

Depending on the temperature used for artificial aging, the hardness-time isotherms pass through a maximum. In the stabilizing process of artificial aging, this maximum is exceeded, which always causes some loss of strength, but establishes sufficient volume stability. The temperature level and the exposure time affect not only the hardness and thus the strength at room temperature, but also the hardness and strength at elevated temperatures, which are of great significance to the operating behavior of the piston material.

Figure 4.3 shows the residual hardness curve for an aluminum-silicon piston alloy, measured on a cooled sample after corresponding temperature exposure. The curves are used to evaluate the suitability of heat-resistant alloys. If the hardness of used pistons is measured, the curves can be used to draw conclusions about the average temperature to which the piston has been exposed.

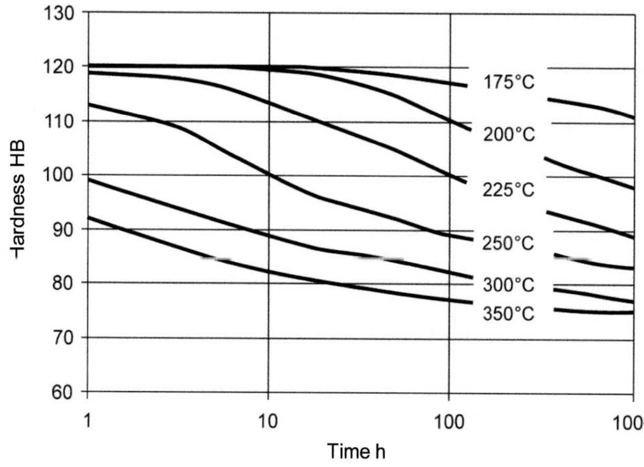

Figure 4.3:
Residual hardness curve of a heat-treated aluminum-silicon piston alloy, by temperature exposure

4.2.2 Piston alloys

Pistons are almost exclusively made of aluminum-silicon alloys of eutectic, and partly hypereutectic composition, which can be cast easily and nearly always can be forged as well. **Table 4.1** shows an overview of the chemical composition of piston alloys used by MAHLE.

The eutectic alloy M124 is the "classic" piston alloy, and has been the basis for the vast majority of pistons in recent decades. Even today it is still a very significant, universally applicable alloy. Pistons made of hypereutectic alloys exhibit even greater wear resistance. From this group, the alloys M138 and M244 are preferred for two-stroke engine pistons, while the M126 alloy is used in the USA for passenger car gasoline engines.

The alloys M142, M145, and M174+ have been developed only recently. Their common characteristic is the relatively high proportion of the elements copper and nickel. This gives the alloys particularly high strength at elevated temperatures and thermal stability. Despite increasing requirements in terms of casting technology and slight disadvantages due to somewhat greater density and lower thermal conductivity, their high strength at elevated temperatures and thermal stability have led to great market penetration for these alloys in high-performance passenger car and commercial vehicle engines. The eutectic alloy M142 is primarily used in gasoline engines, and the alloy M174+, which is likewise eutectic, is used increasingly in diesel engines. The hypereutectic alloy M145 is used in several gasoline engines.

Aluminum-silicon piston alloys are employed mainly for cast pistons. They can also be forged for special purposes, which leads to somewhat different microstructures and properties. The designation "P" is added to the alloy symbol in order to highlight this distinction.

The alloy M-SP25 is a silicon-free, high-strength aluminum alloy used exclusively for forged pistons and piston components, primarily for racing pistons, but also for skirts of composite pistons in large-bore engines.

The **Figures 4.4** show examples of characteristic material structures. The eutectic alloys M124, M142, and M174+ exhibit a granular, heterogeneous, eutectic aluminum-silicon mixture with embedded intermetallic phases. In the hypereutectic alloys M126, M138, M145, and M244, the increasing silicon content results in a clearly visible increase in the proportion of primarily precipitated silicon crystals. Low levels of phosphorus in the aluminum melt act as nucleating agents for the silicon, producing a silicon crystal formation that is beneficial for machinability. Their edge lengths should not be greater than 100 µm.

Table 4.1: Chemical composition of MAHLE aluminum piston alloys [percent by weight]

	M124	M126	M138	M244
	AlSi12CuMgNi	AlSi16CuMgNi	AlSi18CuMgNi	AlSi25CuMgNi
Si	11.0–13.0	14.8–18.0	17.0–19.0	23.0–26.0
Cu	0.8–1.5	0.8–1.5	0.8–1.5	0.8–1.5
Mg	0.8–1.3	0.8–1.3	0.8–1.3	0.8–1.3
Ni	0.8–1.3	0.8–1.3	0.8–1.3	0.8–1.3
Fe	Max. 0.7	Max. 0.7	Max. 0.7	Max. 0.7
Mn	Max. 0.3	Max. 0.2	Max. 0.2	Max. 0.2
Ti	Max. 0.2	Max. 0.2	Max. 0.2	Max. 0.2
Zn	Max. 0.3	Max. 0.3	Max. 0.3	Max. 0.2
Cr	Max. 0.05	Max. 0.05	Max. 0.05	Max. 0.6
Al	Remainder	Remainder	Remainder	Remainder

	M142	M145	M174+	M-SP25
	AlSi12Cu3Ni2Mg	AlSi15Cu3Ni2Mg	AlSi12Cu4Ni2Mg	AlCu2,5Mg1,5FeNi
Si	11.0–13.0	14.0–16.0	11.0–13.0	Max. 0.25
Cu	2.5–4.0	2.5–4.0	3.0–5.0	1.8–2.7
Mg	0.5–1.2	0.5–1.2	0.5–1.2	1.2–1.8
Ni	1.75–3.0	1.75–3.0	1.0–3.0	0.8–1.4
Fe	Max. 0.7	Max. 0.7	Max. 0.7	0.9–1.4
Mn	Max. 0.3	Max. 0.3	Max. 0.3	Max. 0.2
Ti	Max. 0.2	Max. 0.2	Max. 0.2	Max. 0.2
Zn	Max. 0.3	Max. 0.3	Max. 0.3	Max. 0.1
Zr	Max. 0.2	Max. 0.2	Max. 0.2	–
V	Max. 0.18	Max. 0.18	Max. 0.18	–
Cr	Max. 0.05	Max. 0.05	Max. 0.05	–
Al	Remainder	Remainder	Remainder	Remainder

The eutectic piston alloy M124 is also used in a refined state (M124V). The eutectic between AlSi solid solutions, which develop as dendrites, is made particularly fine by the addition of small amounts of sodium or strontium. Machinability is improved, but resistance to wear is worse than for the granular eutectic structure.

The microstructures of the forged alloys M124P, M124VP, and the AlCu alloy M-SP25 are also illustrated. The flow structure can be seen in the line-shaped formation of the microstructure.

The temperature-dependent physical and mechanical material fatigue data for cast pistons are summarized in **Table 4.2**. Along with the static parameters of tensile strength and yield

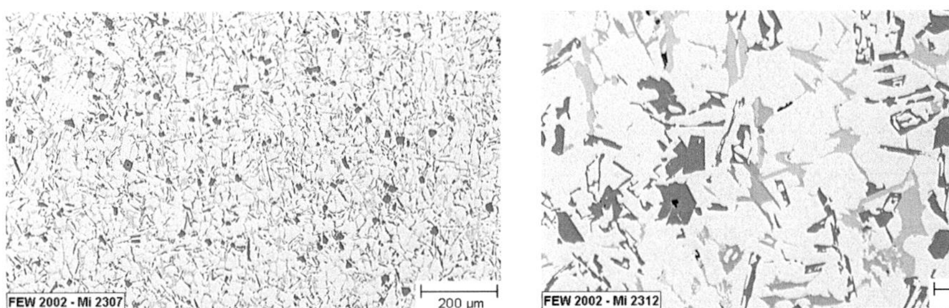

a) Eutectic AlSi alloy M142 (representative for M124 and M174+ as well), casting state with granular structure

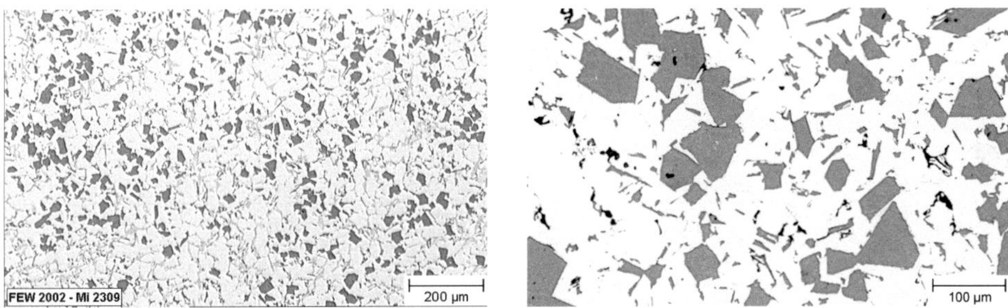

b) Hypereutectic AlSi alloys, casting state with granular structure, M145 (14–16% Si), left, M244 (23–26% Si), right

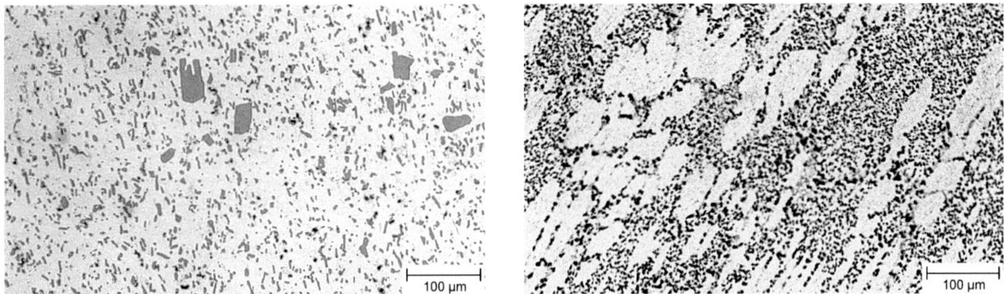

c) Aluminum-base alloys for forged pistons, M124P with granular structure, left, M124VP with modified structure, right

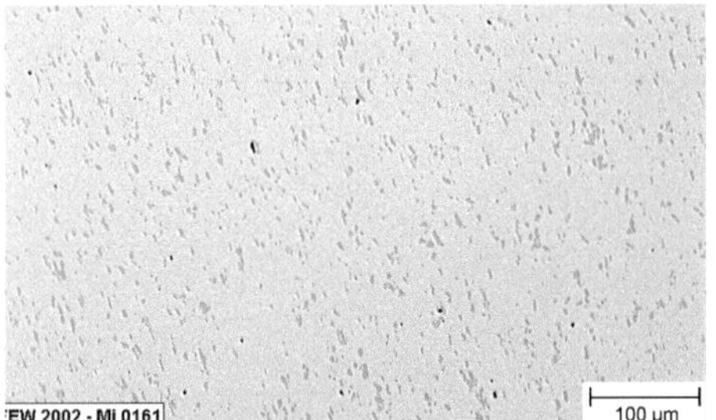

Figures 4.4:
Microstructure of MAHLE aluminum piston alloys

d) Silicon-free alloy M-SP25

Table 4.2: Physical and mechanical properties of cast MAHLE aluminum piston alloys

Description		M124	M126, M138	M142, M145, M174+	M244
Hardness HB10	20°C	90–130	90–130	100–140	90–130
Tensile strength R_m [MPa]	20°C	200–250	180–220	200–280	170–210
	150°C	180–200	170–200	180–240	160–180
	250°C	90–110	80–110	100–120	70–100
	350°C	35–55	35–55	45–65	35–55
Yield strength $R_{p0,2}$ [MPa]	20°C	190–230	170–200	190–260	170–200
	150°C	170–210	150–180	170–220	130–180
	250°C	70–100	70–100	80–110	70–100
	350°C	20–30	20–40	35–60	30–50
Elongation at fracture A_5 [%]	20°C	<1	1	<1	0.1
	150°C	1	1	<1	0.4
	250°C	3	1.5	1.5–2	0.5
	350°C	10	5	7–9	2
Fatigue strength under reversed bending stress σ_{bw} [MPa]	20°C	90–110	80–100	100–110	70–90
	150°C	75–85	60–75	80–90	55–70
	250°C	45–50	40–50	50–55	40–50
	350°C	20–25	15–25	35–40	15–25
Young's modulus [MPa]	20°C	80,000	84,000	84,000–85,000	90,000
	150°C	77,000	80,000	79,000–80,000	85,000
	250°C	72,000	75,000	75,000–76,000	81,000
	350°C	65,000	71,000	70,000–71,000	76,000
Thermal conductivity λ [W/mK]	20°C	145	140	130–135	135
	350°C	155	150	140–145	145
Thermal expansion α [10^{-6} m/mK]	20–100°C	19.6	18.6	18.5–19.5	18.3
	20–200°C	20.6	19.5	19.5–20.5	19.3
	20–300°C	21.4	20.2	20.5–21.2	20.0
	20–400°C	22.1	20.8	21.0–21.8	20.7
Density ρ [g/cm³]	20°C	2.68	2.67	2.75–2.79	2.65
Relative wear rate		1	0.8	0.85–0,9	0.6

Table 4.3: Physical and mechanical properties of forged MAHLE aluminum piston alloys

Description		M124P	M142P	M-SP25
Hardness HB10	20°C	100–125	100–140	120–150
Tensile strength R_m [MPa]	20°C	300–370	300–370	350–450
	150°C	250–300	270–310	350–400
	250°C	80–140	100–140	130–240
	300°C	50–100	60–100	75–150
Yield strength $R_{p0,2}$ [MPa]	20°C	280–340	280–340	320–400
	150°C	220–280	230–280	280–340
	250°C	60–120	70–120	90–230
	300°C	30–70	45–70	50–90
Elongation at fracture A_5 [%]	20°C	<1	1	8
	150°C	4	2	9
	250°C	20	6	12
	300°C	30	20	12
Fatigue strength under reversed bending stress σ_{bw} [MPa]	20°C	110–140	110–140	120–150
	150°C	90–120	100–125	110–135
	250°C	45–55	50–60	55–75
	300°C	30–40	40–50	40–60
Young's modulus [MPa]	20°C	80,000	84,000	73,500
	150°C	77,000	79,000	68,500
	250°C	72,000	75,000	64,000
	300°C	69,000	73,000	62,000
Thermal conductivity λ [W/mK]	20°C	155	140	140
	150°C			155
	250°C			165
	300°C	165	150	170
Thermal expansion α [10^{-6} m/mK]	20–100°C	19.6	19.2	22.4
	20–200°C	20.6	20.5	24
	20–300°C	21.4	21.1	24.9
Density ρ [g/cm^3]	20°C	2.68	2.77	2.77
Relative wear rate		1	0.9	1.3

strength, the dynamic parameter of alternate strength is also provided for evaluating fatigue resistance [3]. In order to determine alternate strength, the load cycle limit has been defined as 50×10^6.

The strength values given here apply to test bars taken from pistons. Prior to testing at elevated temperatures, they were artificially aged at the test temperature for a long period of time. This procedure allows for the current service life of pistons in engine operation, which is significantly longer than the load times that can be produced in materials testing.

The material fatigue data for forged pistons are shown in **Table 4.3**. Compared with the cast state, the material in the forged state exhibits greater strength and greater plastic deformability (greater elongation at fracture). The strength advantage of the forged material structure is greatest in the lower and middle temperature ranges, up to about 250°C, and drops off at high temperatures.

The strength values are based on heat-treated samples taken from pistons and artificially aged at the test temperature.

The wider scatter bands for the strength values compared with cast alloys, because of differences in heat treatment, account for the use of forged pistons for different service life requirements. The lower limits are representative for series vehicles and large-bore engines, and the higher limits for racing and sports vehicles.

The wear rates shown–relative to the material M124–are relative values, obtained using a variant of the wear machine according to E. Koch [4]. The practical applicability of values obtained on wear machines, however, is often questionable. Experience has shown that the data determined by MAHLE allow at least a qualitative evaluation of the materials.

4.2.3 Fiber reinforcement

Ceramic fibers significantly increase the thermal and mechanical load limits of aluminum piston materials. Using pressure-aided casting processes, fiber preforms–short Al_2O_3 fibers with an average fiber diameter of 3–4 μm and average fiber length of 50–200 μm–are infiltrated with the molten alloy [5]. The structure images in **Figure 4.5** show the distribution and orientation of the ceramic fibers in the metallic matrix. As a result of the fiber preform manufacturing process, most fibers are randomly oriented within a plane.

Slight disadvantages in the form of somewhat greater density and lower thermal conductivity are offset by the many advantages of fiber-reinforced composite materials, such as an increase in Young's modulus and a lower thermal expansion coefficient.

The thermal and mechanical alternate strength increases particularly sustainably. For this reason, composite materials are especially well-suited for the local reinforcement of particularly highly stressed zones, such as the combustion bowl in diesel engine pistons. In

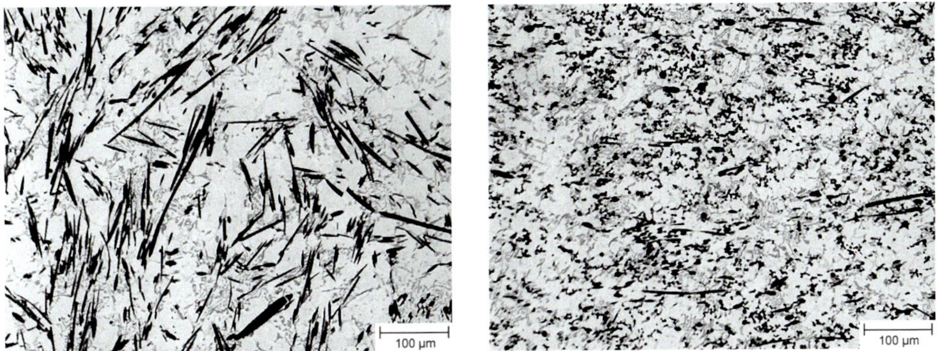

Figure 4.5: Microstructure of the Al piston alloy M124, reinforced with about 15 percent Al_2O_3 fibers by volume, section in the plane, left, and perpendicular to the preferred fiber orientation, right

commercial vehicles, pistons with locally fiber-reinforced combustion chamber bowls have proved their functionality and reliability for many years.

4.3 Ferrous materials

If the strength or wear resistance of aluminum alloys is not sufficient to meet the loads, then ferrous materials are employed. This can begin with local reinforcement (e.g., ring carriers), and extend to parts of composite pistons (e.g., piston crown, bolts), all the way to pistons constructed entirely of cast iron or forged steel.

Carbon is the most important alloying element for iron. The iron-carbon diagram, **Figure 4.6**, enables a thorough assessment of these materials [6]. This diagram differentiates two systems: the metastable or carbidic system, consisting of iron and metastable iron carbide Fe_3C (solid lines), and the stable or graphite system (dashed lines). In the latter case, the majority of the carbon is embedded in the iron as graphite. In the metastable system, the iron can bind the carbon only as cementite (Fe_3C), up to 6.7%. **Figure 4.6** therefore presents a complete overview (0–100% Fe_3C). Depending on the temperature, carbon atoms can be embedded up to a certain degree in the iron lattice on so-called interstitials. The solubility of carbon in the cubic body-centered α solid solution–also known as α iron or ferrite–is significantly lower (max. 0.02 percent C by weight) than in the cubic face-centered γ solid solution–also known as γ iron or austenite (max. 2.0 percent C by weight).

For carbon contents of less than 4.3%, austenite is primarily precipitated from the melt. At 4.3% C, the melt solidifies eutectically as ledeburite. At greater than 4.3% C, cementite is primarily precipitated from the melt. Below 2.0% C, austenite forms first, and its range of

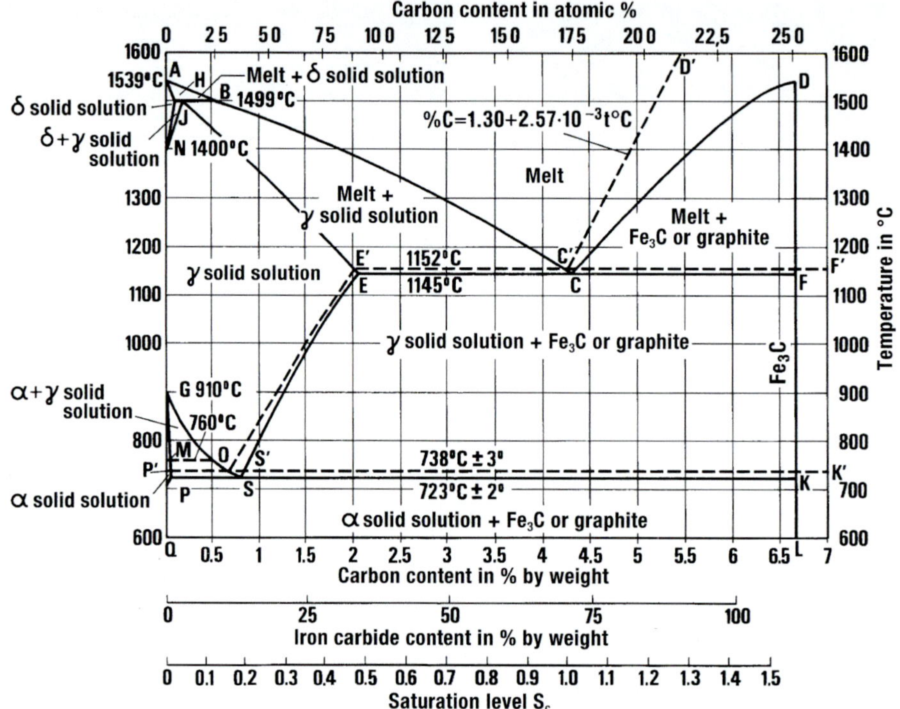

Figure 4.6: Iron-carbon diagram

existence drops off with falling temperature to lower concentrations of C, until finally the high-carbon austenite degrades into low-carbon ferrite and cementite at 723°C. In the range of 0.8% C, the eutectoid perlite is formed, which is a laminar arrangement of ferrite and cementite.

In the stable system, the processes shown with dashed lines occur at normal cooling rates only in the presence of additional silicon (>0.5%), because silicon promotes the formation of free carbon (graphite). Fundamentally, the processes are similar to those in the metastable system, except that graphite is formed instead of cementite.

4.3.1 Cast iron materials

Cast iron materials generally have a carbon content of >2%. In these materials, the brittle cementite or graphite can no longer be brought into solution by subsequent heat treatment. They are therefore not suitable for radical hot forming, but their castability can be optimized.

MAHLE uses high-quality cast iron varieties with lamellar and spherolithic graphite for its products. **Table 4.4** contains the composition, physical properties, and strength values of common alloys. The microstructures of these materials are shown in **Figures 4.7a to c.**

Table 4.4: MAHLE cast iron materials–chemical composition, mechanical and physical properties (guidelines for separately cast sample bars)

Description		Austenitic cast iron for ring carriers		Cast iron with nodular graphite for pistons and piston skirts
		M-H (lamellar)	M-K (spherolithic)	M-S70 (EN GJS 700-2)
Alloying elements [Percent by weight]	C	2.4–2.8	2.4–2.8	3.5–4.1
	Si	1.8–2.4	2.9–3.1	2.0–2.4
	Mn	1.0–1.4	0.6–0.8	0.3–0.5
	Ni	13.5–17.0	19.5–20.5	0.6–0.8
	Cr	1.0–1.6	0.9–1.1	–
	Cu	5.0–7.0		<0.1
	Mo			
	Mg		0.03–0.05	0.04–0.06
Tensile strength R_m [MPa]	20°C	190	380	700
	100°C	170		640
	200°C	160		600
	300°C	160		590
	400°C	150		530
Yield strength $R_{p0,2}$ [MPa]	20°C	150	210	420
	100°C	150		390
	200°C	140		360
	300°C	140		350
	400°C	130		340
Elongation at fracture A_5 [%]	20°C	2	8	2
Brinell hardness HBW 30		120–150	140–180	240–300
Fatigue strength under reversed bending stress σ_{bw} [MPa]	20°C	84		250
Young's modulus [MPa]	20°C	100,000	120,000	177,000
	200°C			171,000
Thermal conductivity λ [W/mK]	20°C	32	13	27
Thermal expansion α [10^{-6} m/mK]	20–200°C	18	18	12
Density ρ [g/cm^3]	20°C	7.45	7.4	7.2

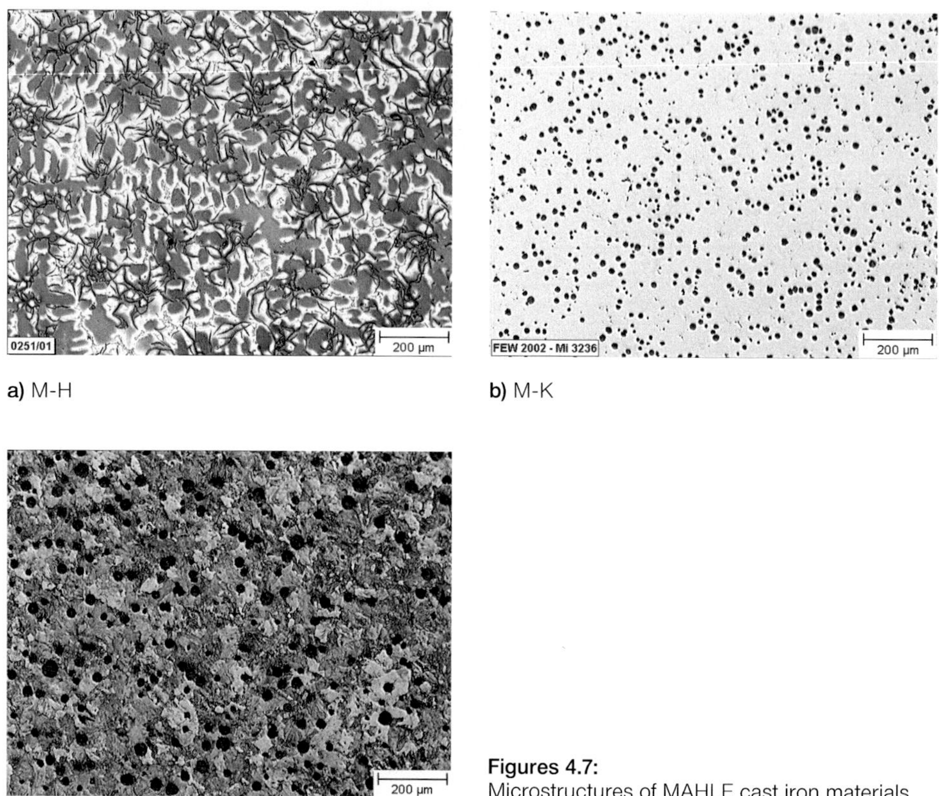

a) M-H

b) M-K

c) M-S70

Figures 4.7:
Microstructures of MAHLE cast iron materials

Thanks to their relatively high thermal expansion in comparison to cast irons with perlitic or ferritic basic structures, austenitic cast iron materials are of great significance for the production of ring carrier pistons [7]. The alloy M-H, for example, has a coefficient of thermal expansion of $\alpha \approx 17.5 \times 10^{-6}$ m/mK, which, compared with normal cast iron at $\alpha \approx 10 \times 10^{-6}$ m/mK, is close to the aluminum piston alloy M124 ($\alpha \approx 21 \times 10^{-6}$ m/mK). For the solidified cast aluminum piston, this means that critical stresses between the piston body and a ring carrier made of austenitic cast iron are much lower than for a ring carrier made of normal cast iron. Most ring carriers are made of austenitic cast iron with lamellar graphite formation M-H. In special cases, the higher-strength austenitic cast iron with spherolithic graphite formation M-K is used. Ring carriers are machined from centrifugal-cast tubes. Centrifugal casting produces a dense, consistent casting structure.

For the materials used in piston casting, the basic material of the structure is largely perlitic, on account of its good strength and wear properties. Pistons in highly stressed diesel engines and other components subjected to heavy loads in engines and the mechanical engineering industry are predominantly made of M-S70 spherolithic cast iron. This material is used, e.g., for single-piece pistons and piston skirts in composite pistons.

4.3.2 Steels

The iron alloys designated as steels generally have a carbon content of less than 2%. When heated, they transform completely into malleable (suitable for forging) austenite. The ferrous alloys are therefore excellent for hot forming, such as roll-burnishing or forging.

Steels used for components generally have a carbon content of less than 0.8%. If they cool slowly after casting or hot forming, then a ferritic-perlitic structure is formed; **Figure 4.8b**. In this state, the steel typically has insufficient strength and hardness. The strength of the material is therefore increased by various means.

One technically significant process is tempering, where the steel is cooled rapidly from a temperature of more than 850°C (quenching). The conversion of austenite to perlite and ferrite, as shown in the iron-carbon diagram in **Figure 4.6**, no longer takes place, because of the suddenly limited freedom of motion of the carbon atoms in the iron matrix. The carbon remains forcibly dissolved in the crystal lattice, although it does not have sufficient solubility under equilibrium conditions. This leads to great distortions in the lattice, which are expressed macroscopically as high hardness and strength, but also brittleness. This hardened structure with a typically acicular appearance is called martensite. Subsequent heating of the material, or tempering, relieves these stresses somewhat and forms a tempered structure; **Figure 4.8a**. Hardness and strength are reduced somewhat, but toughness is increased [8].

When quenching, the cooling rate decreases from the edge to the core of a component and is ultimately less than the critical cooling rate of the steel, so that the austenite is no longer completely converted to martensite in this area. The material no longer fully hardens. Elements such as manganese, chromium, nickel, or molybdenum increase the hardenability of the alloy by reducing its critical cooling rate. This is particularly important for components with large heat-treatment cross sections, because it can limit the loss of strength toward the core.

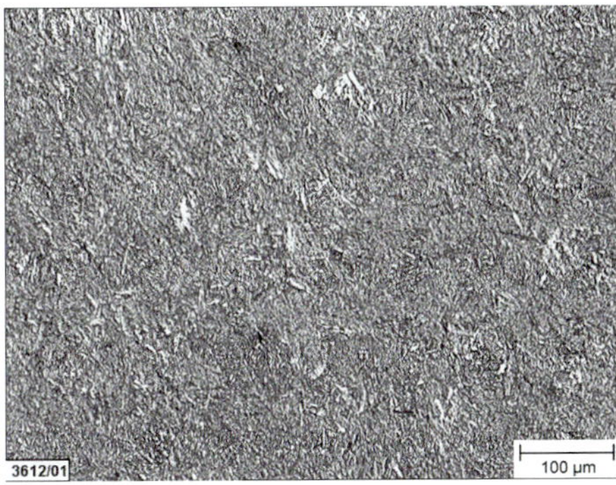

Tempered structure
a) Steel 42CrMo4

Ferritic-perlitic structure
b) Precipitation-hardened ferritic-perlitic steel 38MnVS6

Figures 4.8: Microstructures of steels for pistons

Table 4.5: Steels for pistons–chemical composition, mechanical and physical properties

Description		42CrMo4	38MnVS6
State of heat treatment		**Heat-treated**	**Controlled cooling from heat of deformation**
Alloying elements [Percent by weight]	C	0.38–0.45	0.34–0.41
	Si	Max. 0.40	0.15–0.80
	Mn	0.60–0.90	1.20–1.60
	Cr	0.90–1.20	Max. 0.30
	Mo	0.15–0.30	Max. 0.08
	P	Max. 0.035	Max. 0.025
	S	Max. 0.035	0.020–0.060
	V		0.08–0.020
	N		0.010–0.020
Brinell hardness HBW 30		265–330	240–310
Tensile strength R_m [MPa]	20°C	920–980	910
	130°C	870–960	860
	300°C	850–930	840
	450°C	630–690	610
Yield strength $R_{p0,2}$ [MPa]	20°C	740–860	610
	130°C	700–800	570
	300°C	680–750	540
	450°C	520–580	450
Elongation at fracture A_5 [%]	20°C	12–15	14
	130°C	8–13	9
	300°C	10–13	11
	450°C	15–16	15
Fatigue strength under reversed bending stress σ_{bw} [MPa]	20°C	370–440	370
	130°C	350–410	350
	300°C	340–400	320
	450°C	280–340	290
Young's modulus [MPa]	20°C	212,000	208,000
	130°C	203,000	201,000
	300°C	193,000	189,000
	450°C	180,000	176,000
Thermal conductivity λ [W/mK]	20°C	44	38
	130°C	43	39
	300°C	40	39
	450°C	37	37
Average linear thermal expansion α [10^{-6} m/mK]	20–300°C	13.2	13.1
	20–450°C	13.7	13.7
Density ρ [g/cm^3]	20°C	7.80	7.78

For very highly stressed pistons and piston components, the chromium-molybdenum alloy of heat-treated steel 42CrMo4 is used. In addition to improved full hardenability, both alloying elements promote carbide formation, and molybdenum also increases strength at elevated temperatures. However, even for this steel, a decrease in strength toward the core area must be expected for very large heat-treatment cross sections or changes in cross section. The scattering ranges for strength values shown in **Table 4.5** highlight this fact.

Bolts that connect the piston crown of a composite piston to the piston skirt are also generally made of 42CrMo4 heat-treated steel. They must comply with the highest DIN 267 strength classification of 10.9. In special cases, 34CrNiMo6 heat-treated steel is used, which has even greater full hardenability thanks to the addition of nickel.

Another significant technology for increasing the strength of metallic materials is precipitation hardening. Precipitation-hardened ferritic-perlitic steels exhibit small amounts of vanadium or niobium added (about 0.1 percent by weight). They are therefore referred to as microalloyed steels [9]. When the material is heated to forging temperatures, these microalloyed elements dissolve completely in the γ solid solution. The forged part is allowed to cool in air at a controlled rate immediately after hot forming. As the austenite converts to ferrite and perlite, the carbides and carbonitrides of these microalloyed elements precipitate in a very fine distribution in the ferrite and increase the strength, particularly the yield limit, by preventing dislocation mobility.

Figure 4.8b shows a typical AFP structure, using 38MnVS6 as an example. This material is preferably used in steel pistons for commercial vehicle engines and for forged steel skirts in composite pistons. The advantages of this group of materials, compared with heat-treated steels, are improved machinability of the ferritic-perlitic structure and the elimination of costly subsequent heat treatment.

Both steel grades used for manufacturing pistons today, 42CrMo4 heat-treated steel and 38MnVS6 precipitation-hardened ferritic-perlitic steel, are suitable for use at temperatures of up to 450°C with regard to strength at elevated temperatures and oxidation resistance. When testing new engines, it is sometimes common to test the service life under overloaded conditions. This can easily involve steel piston temperatures of 500–550°C, particularly at the edge of the combustion chamber bowl. In this temperature range, the iron reacts with the excess oxygen of the burning fuel-air mixture, and appreciable scaling occurs. If future engine concepts are going to have similar thermal loads even under normal operating conditions, either antioxidation coatings or other heat-resistant and oxidation-resistant steel grades may be used.

4.4 Copper materials for pin bore bushings

Piston pin bosses of diesel engine pistons are highly specifically loaded and sometimes have pin bore bushings made of copper materials. The bushings are inserted with interference, i.e., using a shrink fit connection. For aluminum pistons, the pin bore bushing increases fatigue strength of the piston pin boss. For steel pistons, seizure resistance of the piston pin bearing is the primary emphasis. Besides high strength and good sliding properties, bushing materials should feature these additional properties:

- Similar thermal expansion behavior as the piston material (for constant interference under hot and cold conditions)
- Corrosion resistance against hot, acidified lubricating oil

MAHLE prefers to use solid piston pin bore bushings. They are machined from drawn tubing, which gets its strength from the deformation process or heat treatment. The range of applications indicated above is covered by two material grades.

Special brass CuZn31Si1 (material 2.1831) per DIN/ISO 4382-2
A nonhardenable material, which obtains its strength from cold forming. It is very resistant to corrosion due to oils and has good sliding properties. The microstructure consists of an α matrix with some proportion of β phase. Alloying silicon improves the wear and corrosion resistance.

Aluminum bronze CuAl10Ni5Fe4
A precipitation-hardened material that is hardened during hot forming. It is characterized by high strength and good corrosion resistance against oils. The microstructure consists of an α matrix with precipitated iron, nickel, and manganese (κ phase). Alloyed aluminum improves the mechanical properties. Nickel increases strength at elevated temperatures, and iron improves fine-graining and strength. The addition of manganese increases corrosion resistance and ductility.

Figures 4.9 display the structures, and **Table 4.6** shows the composition and properties of the two materials. Because of its very high strength, aluminum bronze CuAl10Ni5Fe4 is preferred for bushings in steel pistons, while the special brass CuZn31Si1 is the preferred pin bore bushing material for aluminum pistons, on account of its greater thermal expansion coefficient.

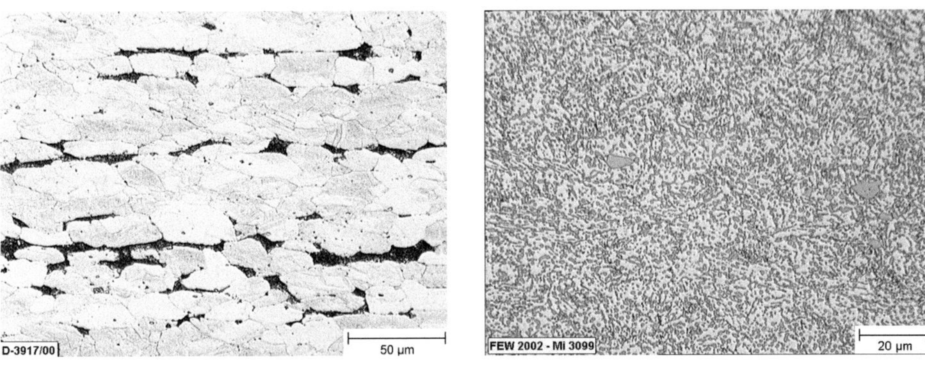

a) CuZn31Si1   **b)** CuAl10Ni5Fe4

Figures 4.9: Microstructures of pin bore bushing materials

Table 4.6: Chemical composition and properties of pin bore bushing materials

		CuZn31Si1	CuAl10Ni5Fe4
Composition [Percent by weight]			
Reference values	Al	–	8.5–11
	Ni	Max. 0.5	4–6
	Fe	Max. 0.4	2–5
	Mn	–	0.5–1.5
	Si	0.7–1.3	–
	Pb	0.1–0.3	Max. 0.05
	Zn	28.5–33.3	Max. 0.5
	Others, total	Max. 0.5	Max. 0.3
	Cu	Remainder	Remainder
Physical properties			
Density	[g/cm^3]	8.4	7.6
Thermal expansion, RT–300°C	[10^{-6} m/mK]	19.2	16.6
Thermal conductivity	[W/mK]	71	63
Young's modulus at room temperature	[MPa]	108,000	120,000
Mechanical properties at room temperature, minimum values			
Hardness	[HB 10]	150	170
Tensile strength R_m	[MPa]	540	700
Yield strength $R_{p0,2}$	[MPa]	430	460
Elongation at fracture A_5	[%]	10	10

4.5 Coatings

4.5.1 Piston coatings

Coating of the piston skirt is primarily intended to prevent local welding between the piston and the cylinder, or piston seizing. A protective coating against wear is generally not necessary here.

Under normal, moderate operating conditions, a piston does not require a coating on the skirt if it is carefully and correctly dimensioned. Risk of seizing does exist, however, under extreme operating conditions:

- Lack of local clearance, caused by mechanical and/or thermal deformation of the cylinder
- Insufficient oil supply, such as during a cold start
- Insufficient lubricity of the engine oil, caused by fuel contamination in the oil, extremely high operating temperature, or excessive aging of the oil
- In brand-new condition, when the piston and cylinder have not yet been run in

A coating on the piston skirt provides protection in such extreme situations. It is important that the skirt coating be tribologically matched to the cylinder bore (cast iron or aluminum).

4.5.1.1 GRAFAL® 255 and EvoGlide

The standard coating for the piston skirt is GRAFAL®, for pistons of all sizes and types that are paired with cast iron cylinders.

GRAFAL® is an approximately 20 μm thick sliding lacquer coating with fine graphite particles embedded in a polymer matrix. It withstands temperatures of up to 250°C that can occur at the piston skirt and is resistant to oils and fuels. The film-forming polymer matrix supports the action of the solid graphite lubricant during dry running, with advantageous tribological properties. This provides great seizure resistance for very low clearances and a lack of oil. Under normal stress, the coating does not wear. Under extreme stress, particularly in case of high local contact pressure, it can be partially worn off locally. The resulting greater clearance reduces the tendency to seize. As a result of this adaptability and the self-lubricating properties of GRAFAL®, pistons can exhibit very low clearances, which produces favorable acoustic properties with low friction.

Because the lateral forces on the piston skirt are increased when implementing a downsizing concept, the sliding lacquer layer must demonstrate improved wear resistance in the piston/cylinder bore system over the service life of the engine. EvoGlide was developed for this purpose. The addition of certain additives makes the resin matrix more wear-resistant. The ability to wear off locally under extreme stress, however, is retained.

4.5.1.2 Tin

Tin is a soft, malleable metal that is also suitable as a skirt coating, helping with run-in and improving boundary lubrication properties, especially for a cold start. The tin coating is 1–2 μm thick. The potential for preventing seizing is thus somewhat less than that of a GRAFAL® layer. Tin-plated pistons are paired with gray cast iron or NIKASIL® cylinders, and are preferably installed in passenger car gasoline engines.

4.5.1.3 Ferrostan/FerroTec®

The skirt coatings that have been proven in gray cast iron cylinders, such as tin and GRAFAL®, are not tribologically suited as partners for aluminum running surfaces, because they would lead to rapid seizing and high wear. The Ferrostan coating system and FerroTec® have proved to be excellent for uncoated AlSi cylinders. Both are characterized by a high level of running reliability and low wear. Ferrostan consists of a 10–13 μm thick iron coating that is topped with a thin coating of tin as a run-in aid. FerroTec® consists of a pure iron coating that can optionally be tin-plated as well. Ferrostan/FerroTec® pistons are used exclusively in passenger car gasoline engines, because diesel engines typically do not feature AlSi running surfaces.

4.5.1.4 FERROPRINT®

An alternative to the Ferrostan coating is the FERROPRINT® coating. It also serves to protect the running surfaces for pistons that are paired with high-silicon aluminum cylinders. Similar to GRAFAL®, this layer consists of a highly temperature-resistant polymer in which harder stainless steel particles are embedded as a reinforcement. The layer thickness is about 20 μm. The properties are similar to GRAFAL®.

4.5.1.5 Hard oxide in the top ring groove

While the top ring groove in a diesel engine piston, or the compression ring groove, is traditionally well protected against wear by a ring carrier, gasoline engine pistons have not previously required this protection. Recently, however, damage has occurred in the form of disruption and material wear at the groove flanks, particularly at the lower groove flank (microwelding). It presumably occurs as a result of local microwelds between the piston ring and the groove flank, and particularly on pistons with short top lands, which thus exhibit higher groove temperatures. These higher temperatures presumably arise in conjunction with ring packs that are tuned for extremely low oil consumption. On account of the minimal oil quantity that is available for lubricating the piston rings, microwelds can occur.

One effective means of protection against microwelding is localized hard-anodizing of the groove with a coating thickness of about 15 μm. The quasi-ceramic structure of the hard oxide layer eliminates metallic contact between the piston ring groove and the piston ring, and thus prevents microwelding. Although the hard oxide layer is significantly rougher than the finely machined groove flank, there is no problem sealing the piston rings, because the coating surface is quickly smoothed out by the micromotion of the piston rings.

4.5.1.6 Hard oxide on the crown

After long running times in diesel engines for passenger cars and commercial vehicles, thermal and mechanical overloads can cause cracks in the bowl rim and piston crown. In order to prevent such cracks in the bowl rim, hard oxides are used once more, albeit with a layer thickness of 50–80 µm in this case.

The advantageous effect of this coating on the cracking behavior of the piston material can be attributed to bond stresses that reduce the compressive stresses as a result of thermal and mechanical loading of the piston crown. Because the layer is not applied from the outside, but arises from the conversion of the base material at the surface, the bond forces between the layer and the base material are strong enough to withstand the substantial local stresses.

4.5.1.7 Phosphate

Metal phosphates form crystalline or amorphous coatings that stand out for their good oil bonding, high adhesive strength, and good deformability. Phosphate coatings can provide effective protection against seizing and scuffing between sliding pairs, especially in run-in phases. This effect can also be exploited for pistons, particularly for diesel engine pistons to protect the boss.

Relatively thick layers (averaging 5 µm) of manganese-iron mixed phosphates are deposited on steel pistons (MONOTHERM®, FERROTHERM®, MonoWeld®). They enable direct pairing with hardened steel pins, without the use of pin bore bushings.

Thin (<0.5 µm) aluminum phosphate layers are deposited on aluminum pistons as conversion layers. They also provide limited antiscuffing protection in the piston pin boss.

4.5.1.8 GRAFAL® 210

Run-in of steel pistons in heavy-duty engines of commercial vehicles is a critical phase with regard to the pin bore/pin tribological system. This requires greater protection, such as is provided by GRAFAL® 210. The layer consists of a highly temperature-resistant polymer matrix, in which graphite particles and molybdenum sulfide pigments–added as a pressure-resistant component–are embedded. The layer thickness is about 8 µm and provides longer antiscuffing/antiseizing protection in the pin bore during the run-in phase.

4.5.1.9 Chromium contact surfaces

For composite pistons in large diesel engines, relative motion occurs at the surfaces where the upper part contacts the piston skirt as a result of cyclic deformation of the components. Although these relative motions normally amount to less than 100 µm, they can lead to undesired frictional wear at the contact surfaces. For the pairing of an aluminum skirt and steel upper part, the harder steel upper part wears more severely. Dimensional chrome plating of the contact surfaces of the steel upper part with a layer thickness of about 20 µm provides an effective solution.

4.5.1.10 Chromium ring grooves

Heavy fuel oil, used as a fuel for large diesel engines, can contain large quantities of small, hard particles, which can cause severe abrasive wear to the piston rings and piston ring grooves if they enter the combustion chamber. Heavy fuel oil also incidentally contains relatively high concentrations of sulfur compounds, which produce acidic gases during combustion. These condense in the region of the piston ring groove and then cause corrosion damage to the piston rings and piston ring grooves.

Hard chrome plating of the side faces of the compression ring groove, paired with a piston ring with chrome-plated sides, provides highly effective protection against these effects. In order to ensure the very long service life required for these engines, the layer must be appropriately thick, with the standard being at least 200 µm.

4.5.2 Application table

Table 4.7: Application table for coatings

	Pistons for passenger cars	Pistons for passenger cars	Pistons for commercial vehicles	Pistons for commercial vehicles	Large-bore pistons
	Gasoline	Diesel	Steel	Aluminum	
GRAFAL®/EvoGlide	X	X	X	X	X
Tin	X			X[1]	
Ferrostan/FerroTec®	X				
FERROPRINT®	X				
Hard oxide compression ring groove	X				
Hard oxide crown		X		X	
Phosphate on Al	X	X		X	X[2]
Phosphate on Fe			X		X[3]
GRAFAL® 210			X		
Hard chrome					X

[1] only for old types, [2] aluminum piston skirts, [3] gray cast iron piston skirts

5 Piston cooling

As specific power output increases in modern combustion engines, the pistons are subjected to increasing thermal loads. In order to ensure operating safety, therefore, they must be cooled as effectively as possible.

5.1 Thermal loads

The chemical energy stored in the fuel is converted into heat in the cylinder during combustion. The piston, as a moving wall of the combustion chamber, converts part of this heat into mechanical work and drives the crankshaft through the connecting rod. The heat that is not converted into mechanical work is partially dissipated with the exhaust gas. The remainder is transferred, through convection and radiation, to the parts of the engine that are adjacent to the combustion chamber. In addition to mechanical stresses, this leads to thermal loading of these components, which can reach very high levels locally.

5.2 Combustion and flame jets

In gasoline engines, the spark plug normally initiates combustion of the fuel-air mixture. The flame front expands quite evenly in all directions, so that all the surfaces enclosing the combustion chamber are subjected to the gas temperature. This results in a temperature gradient in the piston from the combustion chamber side to the crankcase.

For diesel engines with direct fuel injection, combustion starts in the area of the injector holes, because the combustible mixture is first present at that point. After the mixture in this area has autoignited, several flame fronts in the form of combustion lobes run along the injected streams of fuel toward the surfaces enclosing the combustion chamber; **Figure 5.1**. This generates a nonhomogeneous temperature field in the piston, particularly at the rim of the combustion chamber bowl.

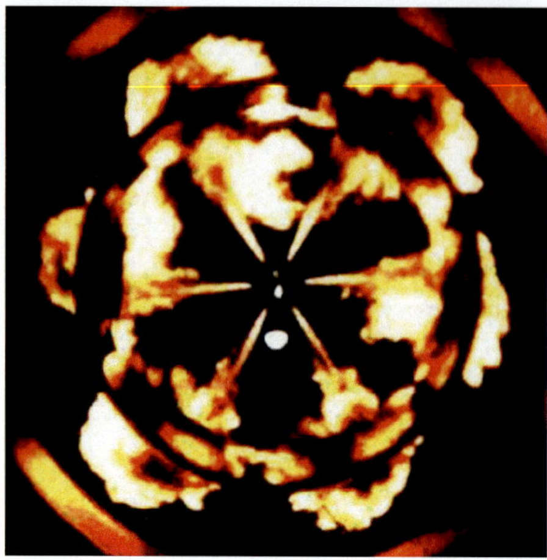

Figure 5.1:
Flame fronts in the diesel process

5.3   Temperature profile at the bowl rim

As a result of the nonuniform introduction of heat through the "combustion lobes" described above, a quasi-stationary, uneven temperature profile is induced at the bowl rim of the diesel engine piston. The temperature profile in **Figure 5.2** shows this temperature condition along the bowl rim for a six-hole injector. The temperature difference between the locations that are located directly in the center of a combustion lobe and the areas in between can exceed 40 K.

The resulting thermal stresses, in conjunction with superimposed loads (mechanical stresses, temperature cycles, oxidation, etc.) can cause cracks in the bowl rim.

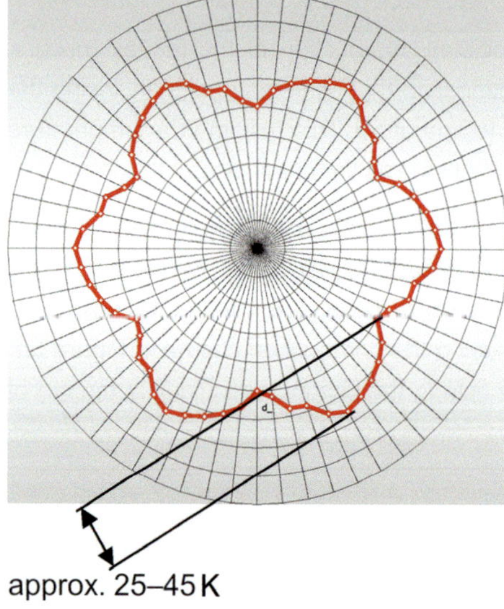

Figure 5.2:
Uneven temperature profile
at the bowl rim

Between the intake and exhaust sides, the temperature profile is shifted toward the exhaust side. Moreover, additional influences must be considered, such as the coolant flow path between the individual cylinders.

5.4 Piston temperature profile

Because of the various combustion processes and the required different piston geometries, diesel and gasoline engine pistons exhibit different temperature profiles; **Figure 5.3**.

Pistons for gasoline engines typically exhibit their maximum temperature in the center of the piston crown, but also at the edge of the bowl in the case of direct fuel injection and associated pronounced combustion bowls. The temperature drops off quite evenly toward the top land. In pistons for diesel engines, however, the maximum temperature occurs at the locations on the bowl rim that are exposed to the center of combustion lobes. The temperature drops off evenly toward the center of the bowl and toward the top land. The temperature profile is largely determined by the number and angle of the injection ports, the injection pressure, the injection timing and duration, and the combustion bowl geometry.

Qualitatively similar temperature profiles develop for diesel and gasoline engine pistons along the piston mantle curve, from the top land, past the piston ring pack, to the piston pin boss and the end of the piston skirt.

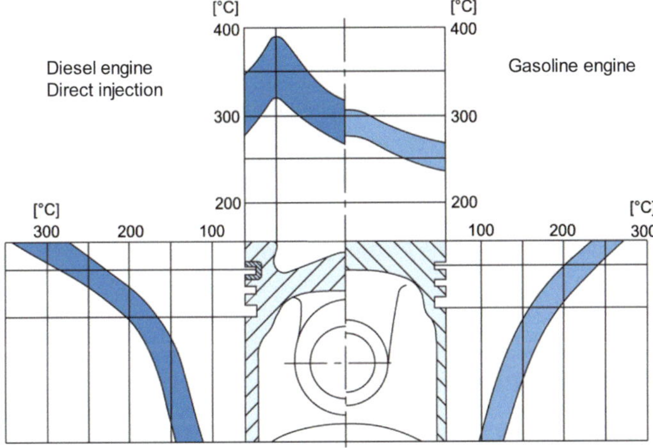

Figure 5.3: Temperature profile for diesel and gasoline engine aluminum pistons

5.5   Effects on the function of the piston

The thermal loads on the piston and the resulting temperature profile within the piston cause temperature-induced deformations. In conjunction with the cyclic thermal loads, these influence the local material stress and strength, as well as the function of the piston. In extreme cases, this can cause the component to fail, resulting in engine damage.

5.5.1 Thermally induced deformation

The deformation of the piston due to the temperature profile must be taken into consideration when designing the shape of the piston. **Figure 5.4** shows this thermal deformation schematically for a piston for a gasoline engine. Pistons in diesel engines show basically comparable deformation behavior.

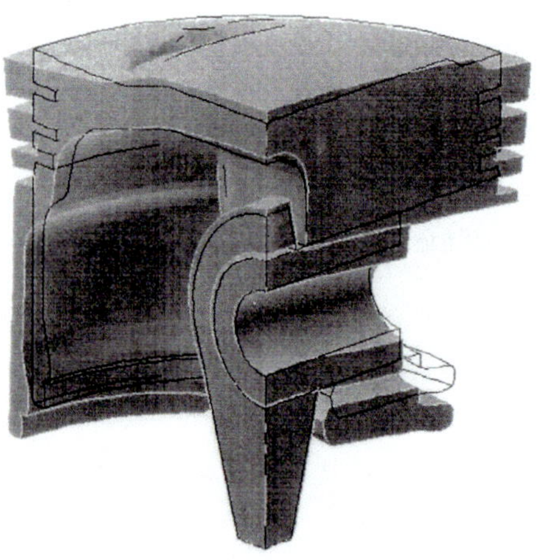

Figure 5.4:
Deformation of a piston in a gasoline engine under thermal load

The magnitude of the deformation depends on the thermal expansion coefficient and the temperature change between the cold state and operating temperature. Radial deformation affects noise, susceptibility to seizure, and friction power loss, and must be compensated for with appropriate clearances, especially in the ring belt. The axial thermal deformation must be considered when designing the clearance at the top dead center point when coordinating the piston position and the valve lift.

5.5.2 Temperature-dependent material fatigue data

As **Figure 5.5** shows, the component temperature has a significant influence on the fatigue resistance of the piston material.

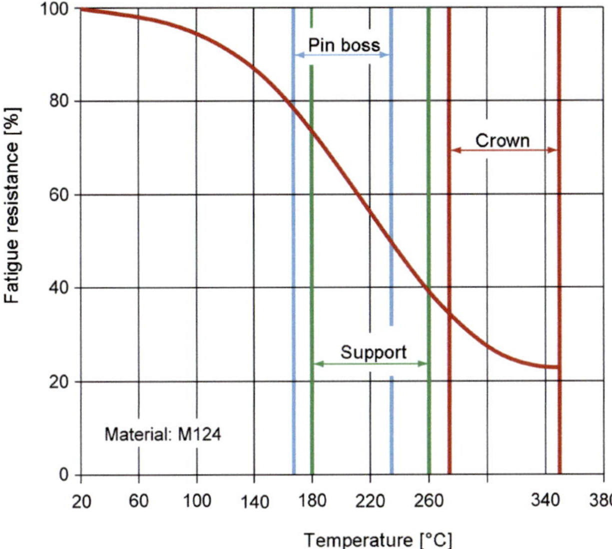

Figure 5.5:
Influence of temperature on the fatigue resistance of M124

For the piston alloy M124, for example, this means that the piston pin boss and support areas, which typically reach a temperature level of 160 to 260°C, have 20 to 60% lower fatigue resistance compared with the value at room temperature. For the piston crown and bowl rim of aluminum diesel engine pistons, which can reach temperatures of 300 to 400°C, the fatigue resistance drops by as much as 80%. Ferrous materials are significantly less sensitive in this temperature range.

5.5.3 Influence of temperature on the piston rings

If the maximum tolerable temperature of the ring belt is exceeded, then plastic deformation and increased wear can occur, particularly in the top ring groove. The chemical decomposition of the lubricating oil brought about by heat can also cause carbon buildup in the groove root. These residues then act as thermal insulators in the ring groove/piston ring system and reduce the heat transport from the piston to the cylinder wall. They also hinder the piston rings from moving freely and can even block them completely. However, the rotational motion of the piston rings in the piston ring grooves, in particular, is essential for their function. This rotation ensures that the ring gap, through which small amounts of combustion gas continuously flow into the crankcase, is not always at the same location. If the piston rings do not undergo any rotational motion for a prolonged period of time, then the hot combustion gases that flow past in the area of the ring gap may damage both the piston ring groove and the cylinder surface. The lubricating oil film in this area is likewise disturbed and can result in ring/piston scoring and engine damage.

5.6   Potential influences on the piston temperature

The influences of engine operating parameters on piston temperature must be considered during piston development. These include the engine speed, load, ignition and injection timing, air ratio, and mean operating temperatures. Table 7.2 (Chapter 7.2) lists reference values for temperature changes in the region of the top ring groove for a gasoline and a diesel engine piston. It is notable that addition of just 50% glycol (antifreeze) to the coolant can increase the temperature in the top ring groove by 5°C.

A comparison of various cooling types–spray jet cooling with a spray jet at the connecting rod end, spray jet cooling with a fixed nozzle on the crankcase, and piston cooling with oil-filled cooling galleries–indicates that cooling galleries result in the greatest reduction in temperature. Depending on the design possibilities, up to 50°C lower temperatures can be achieved at the top ring groove.

5.7   Types of cooling

In order to ensure effective heat dissipation from the piston and to reduce the temperature in critical piston zones, pistons with different types of cooling are used in place of an uncooled piston, depending on the amount of heat to be dissipated and the combustion process (diesel or gasoline). The coolant in all of the piston cooling methods considered here is the engine oil.

5.7.1   Pistons without piston cooling

Pistons without piston cooling are primarily used in gasoline engines with low thermal loading. They are additionally employed in diesel engines with low power output. The temperature profile of these pistons is more than 20°C higher, depending on the point of measurement, than for the same piston with spray jet cooling.

5.7.2   Pistons with spray jet cooling

The simplest way to cool a piston is by means of spray jet cooling; **Figure 5.6**. The interior of the piston is continuously sprayed over as large an area as possible with oil from a cooling oil nozzle. This technology is most common in gasoline engines because the low overall height often leaves no room for cooling galleries. For gasoline engines with direct injection, the use of a cooling gallery is facilitated, on account of the bowl-like shape of the piston crown. Spray jet cooling is also used for diesel engines in the lower power output segment.

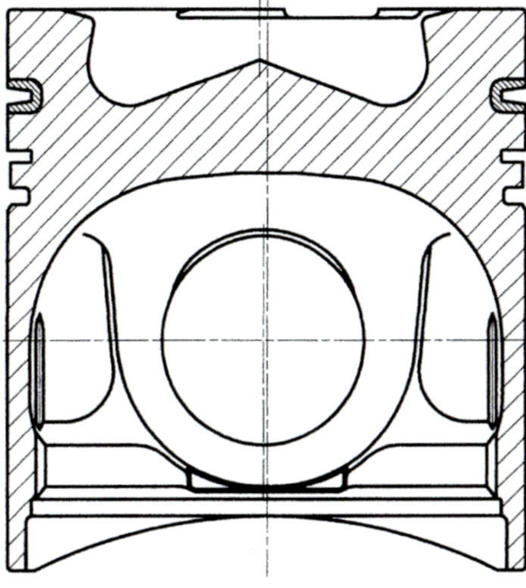

Figure 5.6:
Diesel engine piston without or with spray jet cooling

5.7.3 Pistons with cooling galleries

If spray jet cooling does not reduce the piston temperature sufficiently, then pistons with a cooling gallery must be used. Cooling oil is continuously fed to the cooling gallery during operation. One, or occasionally more, oil spray nozzles feed the cooling oil through inlet openings into the cooling gallery. The cooling oil leaves the cooling gallery through one or more drain holes.

Under engine operating conditions, only partial filling of the cooling gallery is possible, and the cooling oil moves up and down in the cooling gallery synchronous to the oscillating motion of the piston. This leads to a highly turbulent flow in the cooling gallery, and thus a high heat transfer coefficient.

Differentiation is made between salt core cooling galleries, cooled ring carriers, and machined cooling galleries, on the basis of the manufacturing process used.

5.7.3.1 Salt core cooling gallery pistons

For aluminum pistons with a salt core cooling gallery, the cavity in the piston is produced by a salt core that is placed in the mold prior to casting. After the piston blank is removed, the salt core is completely dissolved using water at high pressure.

If the salt core cooling gallery is placed on two support mandrels in the foundry mold, the inlet openings and drain holes are drilled after the salt has been removed; **Figure 5.7**. Inlet openings with pin hole bores are primarily used with vertically oriented cooling oil nozzles. This design also provides the best possible oil supply, and thus the largest cooling effect.

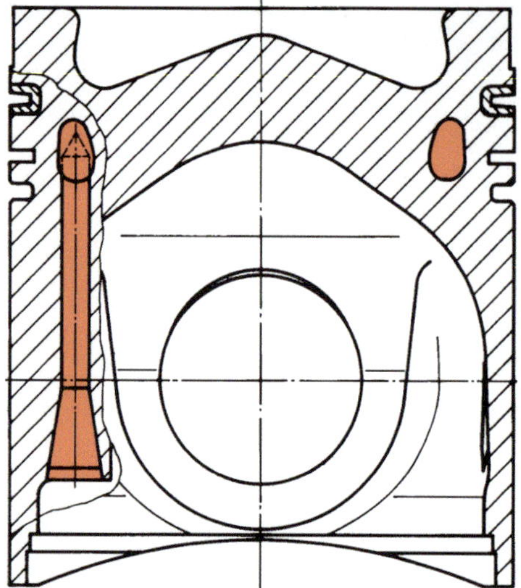 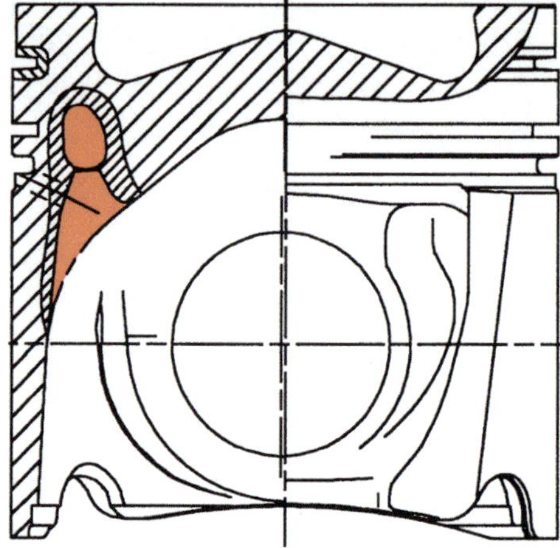

Figure 5.7: Salt core cooling gallery piston with pin hole bore

Figure 5.8: Salt core cooling gallery piston with funnel inlet

For cooling oil nozzles set at an angle, a kidney-shaped inlet opening is typically used in conjunction with a funnel, **Figure 5.8**, in order to compensate for the varying jet impact point on the piston during the piston stroke caused by setting the nozzle at an angle, and to ensure that the oil flow into the cooling gallery is interrupted as little as possible. For such a piston design, typically no or very little machining is needed at the inlet and openings and drain holes.

The salt core cooling gallery should be placed as close to the combustion bowl as possible. Limits are set here, however, by the mechanical load carrying capacity and castability of the piston, as well as the geometric conditions (bowl diameter, location of the top ring groove).

The salt core cooling gallery reduces the temperature at the bowl rim by up to 20°C in comparison with spray jet cooling; **Figure 5.9**. This improved cooling effect amounts to about 10°C at the top land and in the top ring groove.

5.7.3.2 Pistons with cooled ring carrier

For pistons with cooled ring carriers, a sheet metal channel is welded onto the Ni-resist ring carrier, creating a closed cooling gallery; **Figure 5.10**. This component is given an alfin coating prior to casting of the piston, then placed in the mold and cast into the piston blank. The inlet openings and drain holes are produced by drilling.

The same design criteria apply for the inlet and outlet areas as for a piston with salt core cooling gallery, depending on the spray direction of the cooling oil nozzles.

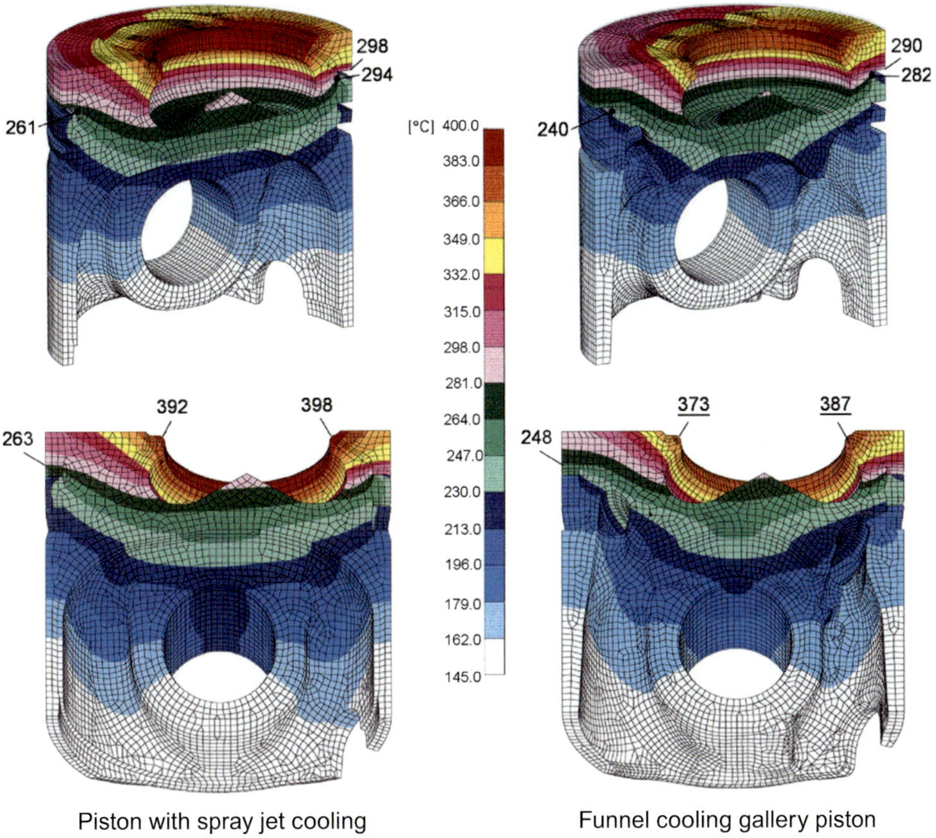

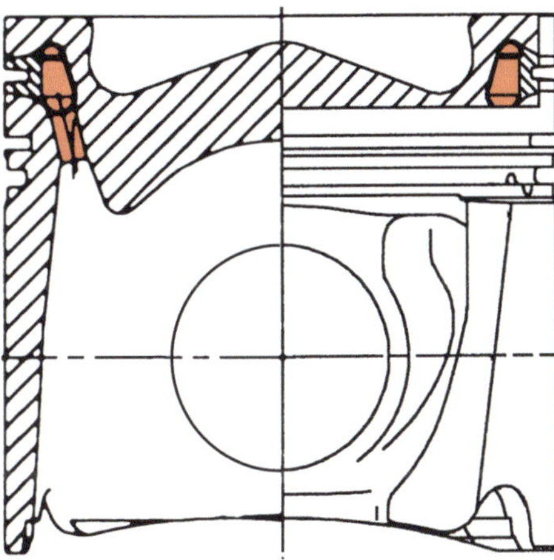

Figure 5.9: Temperature profiles for different types of cooling

Figure 5.10:
Piston with cooled ring carrier

Cooled ring carrier   Salt core cooling gallery

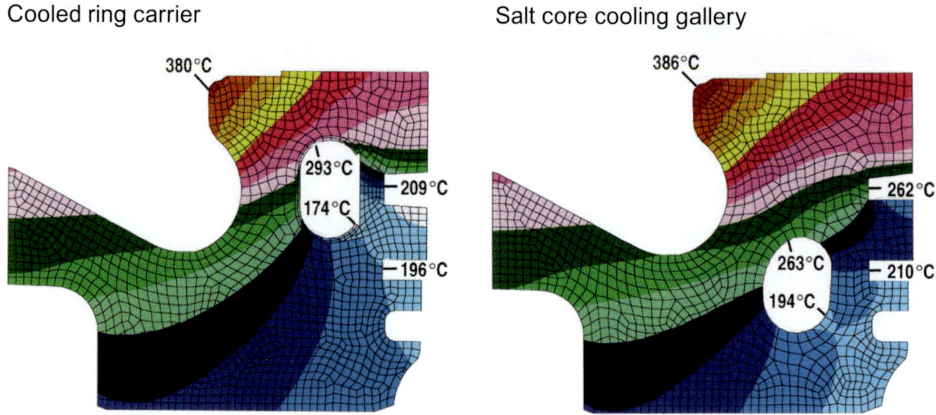

Figure 5.11: Temperature field of the piston with cooled ring carrier and with salt core cooling gallery

A cooled ring carrier reduce the temperature at the bowl rim a bit more in comparison with a salt core cooling gallery; **Figure 5.11**. In the top ring groove a cooled ring carrier brings about a temperature reduction of up to 50°C.

Clearly, the surface temperature at the cooling gallery rises the closer the location of the cooling gallery is to the combustion chamber surface. The maximum surface temperature of about 250°C should not be significantly exceeded. Above this temperature, increased oil aging, thermally induced decomposition of the cooling oil, and deposits of oil carbon can be expected.

5.7.3.3 Machined cooling galleries

For forged steel pistons, the cooling galleries cannot be generated with salt cores or in the form of cooled ring carriers. Therefore, the upper part of the cooling gallery is machined. The lower part of the cooling gallery in the FERROTHERM® piston is formed by shaker pockets in the articulated skirt or a cover plate that closes off the bottom of the cooling gallery. The single-piece design of the MONOTHERM® piston uses the cover plate; **Figure 5.12**. For welded piston designs (MonoWeld®, TopWeld®) the cooling cavity is closed by joining the two components of the piston.

5.7.4 Composite pistons with cooling cavities

The multipiece piston design enables the machining of cooling cavities. Composite pistons also feature an inner cooling cavity behind the ring belt, in addition to the outer cooling cavity. The shape and size of the outer cooling cavity is primarily determined by the combustion bowl geometry and the location of the piston ring pack. The geometry of the inner cooling

Figure 5.12:
MONOTHERM® piston with cover plate
to close the cooling gallery

cavity is influenced by the combustion bowl geometry and the type of screw joint of the piston.

For pistons with central screw joints, the inner cooling cavity is smaller, on account of the screw joint that is located there; **Figure 5.13**. In this design, the oil in the cooling cavity is flowing around the screw.

For pistons with multiple screw joints, the screw joints are located in the contact surface between the cooling cavities. The inner cooling cavity can therefore be significantly larger; **Figure 5.14**. The outer and inner cooling cavity are connected to each other by multiple scavenging ports or bores.

The cooling oil is typically fed into the outer cooling cavity. Drain holes are located both in the outer and in the inner cooling cavity. In exceptional cases, the inlet opening is centered in the inner cooling cavity. Drain holes are then located only in the outer cooling cavity. The cooling oil flow in composite pistons is divided, with the typical flow direction from the outer to the inner cooling cavity, at approximately the ratio of the outer to the inner outflow cross section. One partial oil flow leaves the piston after flowing through the outer cooling cavity, and the second partial oil flow enters the inner cooling cavity through the scavenging bores and flows from there back into the crankcase. Depending on the shape of the outer cooling cavity, composite pistons are divided into shaker cooling and bore cooling types.

5.7.4.1 Shaker cooling

In the case of shaker cooling, the outer cooling cavity has an axisymmetric cross section; **Figure 5.13**. The inner cooling cavity is typically connected to the outer cavity by several milled scavenging ports that are located at about half the height of the outer cooling cavity.

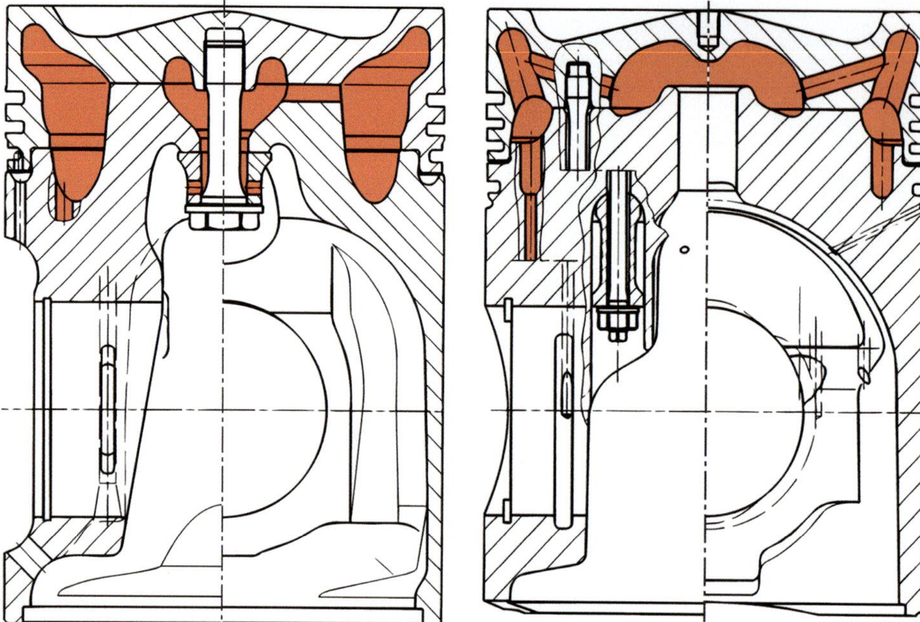

Figure 5.13: Composite piston with central screw joint and shaker cooling

Figure 5.14: Composite piston with multiple screw joint and bore cooling

5.7.4.2 Bore cooling

In the case of bore cooling, the lower part of the outer cooling cavity is likewise axisymmetric. Cooling oil bores run from this side in the direction of the piston crown rim. Some of the cooling oil bores are connected to the inner cooling cavity by means of scavenging bores; **Figure 5.14**.

The advantage of bore cooling, compared with shaker cooling, is the increased stiffness of the piston crown. Material between the cooling oil bores can lead to local temperature maxima on the combustion chamber side. The effort required to machine the cooling cavity is significantly greater.

5.8   Feeding the cooling oil

The cooling medium for piston cooling is the engine oil. The cooling oil delivered by the engine oil pump is distributed via galleries in the crankcase. There are essentially two methods for feeding the cooling oil to the piston. It can be fed to the cooling gallery as through a jet, with nozzles mounted in the crankcase; this solution is typical for passenger car and commercial vehicle engines. Large-bore pistons typically use oil transport from the crankshaft, through the connecting rod and the piston pin, to the piston.

5.8.1 Jet feeding of cooling oil

The cooling oil reaches a nozzle that is fixed to the crankcase through oil galleries, and from there spray nozzles deliver it as a jet to the piston; **Figure 5.15**.

A pressure-regulating valve is typically located between the oil gallery in the crankcase and the nozzle, allowing flow of cooling oil to the nozzle only above a predetermined operating pressure. This ensures that all bearing locations are provided with enough lubricating oil when starting the engine as well as at low speeds and loads. Cooling of the piston is not yet necessary under these operating conditions, because of the low thermal load.

The jet expands as it leaves the nozzle. This expansion depends significantly on the geometry and shape of the cooling oil nozzle.

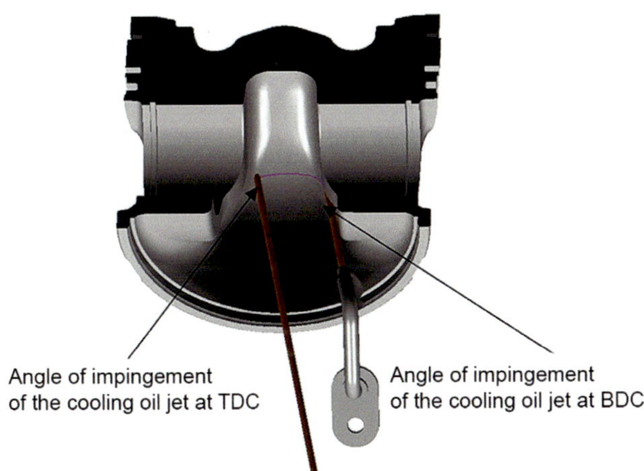

Figure 5.15:
Diesel engine piston with jet feeding of the cooling oil (spray jet cooling), slanted nozzle

5.8.1.1 Nozzle designs for spray jet cooling

Cooling oil nozzles for spray jet cooling should produce a large spread angle α of the oil jet in order to impinge on a large area of the inner contour with the cooling oil; **Figure 5.16**.

A large spread angle α can be achieved by running the cooling oil nozzle at a supercritical point (Reynolds number >2,300). This requires that the ratio of the nozzle diameter d to the length of the flow-calming section l be chosen to be as large as possible, and sharp edges eliminated from the end of the nozzle (e.g., with a chamfer). This jet characteristic can also arise unintentionally with an orifice that has been deburred excessively.

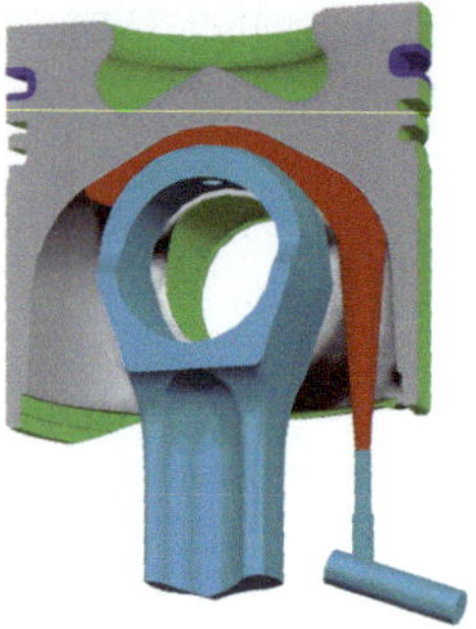

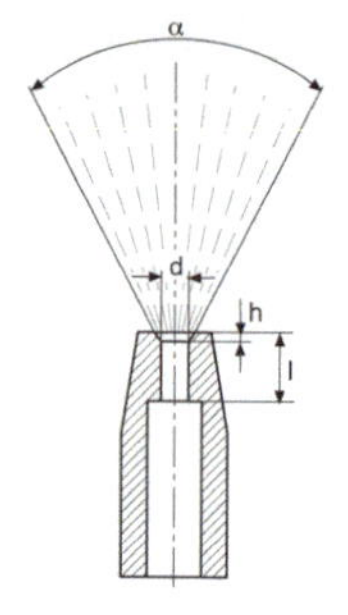

Spray jet cooling

Figure 5.16:
Jet shape for spray jet cooling

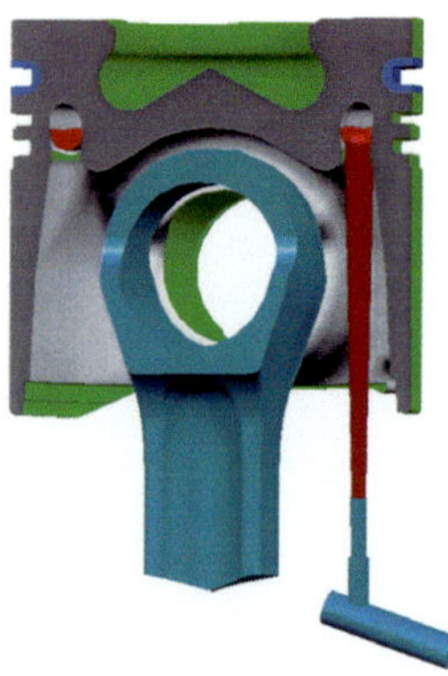

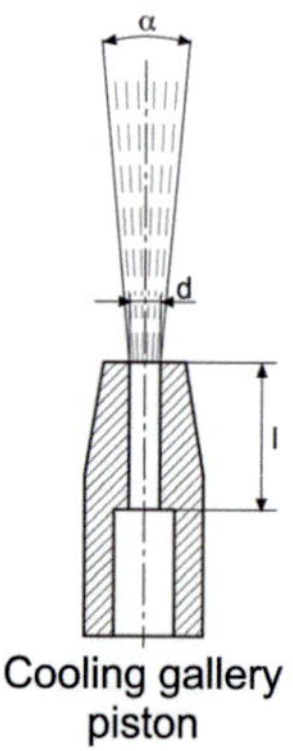

Cooling gallery
piston

Figure 5.17:
Jet shape for cooling gallery supply

5.8.1.2 Nozzle design for supplying cooling galleries and cooling cavities

Cooling oil nozzles for supplying cooling galleries and cooling cavities should have as small a spread angle as possible, so that the jet reliably hits the inlet opening of the cooling gallery and cooling cavity throughout the entire piston stroke; **Figure 5.17**. This bundled jet can be achieved by operating the cooling oil nozzle subcritically (Reynolds number <2,300). This requires that the ratio of the nozzle diameter d to the length of the flow-calming section l be chosen to be as small as possible (l = 4 ... 6 · d), and the nozzle end be sharp edged, but free of burr.

5.8.2 Feeding via crankshaft and connecting rod

For this method of feeding the cooling oil, the oil first flows through the main bearing into bores in the crank web of the crankshaft. From here, the oil flows to the conrod bearing journals, onward through cross-drilled holes in the crankshaft pin, a groove in the lower rod bearing, and a longitudinal bore in the connecting rod, to the small end bore. In this manner, the flowing cooling oil lubricates all the bearing locations of the crankshaft and the rod bearings. There are two main concepts for further transport to the cooling galleries and cooling cavities.

5.8.2.1 Feeding via piston pin and piston pin boss

The cooling oil flows through a circumferential groove in the small end bore and several cross-drilled holes in the piston pin, which has a longitudinal bore and is closed off at the ends. From there, it reaches a groove in the boss of the piston through additional cross-drilled holes of the piston pin. These grooves are connected to the outer cooling cavity through bores.

5.8.2.2 Feeding via slide shoe

In this case, cooling oil is fed through a circumferential groove in the bearing of the small end bore to a bore, and flows from there through a slide shoe mounted on the piston, which feeds it to the inner cooling cavity of the piston. In this case, the cooling oil flows through the piston from the inner to the outer cooling cavity. One advantage of this design is that the piston pin is not weakened by cross-drilled holes.

5.9 Heat flows in the piston

The heat flows in the piston vary according to the type of piston cooling; **Figure 5.18**. Following the principle of conservation of energy, the sum of the heat flows into the piston is equal to the sum of all heat flows out of the piston.

For an aluminum piston with spray jet cooling, about 55 to 65% of the impinging heat is transferred to the cylinder wall through the piston ring pack and the associated ring land surfaces. The remaining 35 to 45% is dissipated through the inner contour and the cooling oil, and partially through the bottom part of the piston skirt, including the piston pin boss. If a cooling gallery is used instead of spray jet cooling, **Figure 5.18, right**, the heat flow dissipated through the piston ring pack is reduced to 30 to 40%, depending on the position of the cooling gallery. The heat flow dissipated via the inner contour and the bottom part of the piston skirt is reduced to 15 to 25%. The remaining heat flow of 35 to 55% is dissipated directly by the cooling oil via the cooling gallery. Salt core cooling galleries, typically positioned at the height of the oil ring groove, reduce the heat flow from the bowl to the piston ring pack and to the inner contour by a slight amount in comparison with the cooled ring carrier. The cooled ring carrier is located directly in the main direction of the heat flow toward the cylinder wall and therefore dissipates significantly more partial heat directly to the oil.

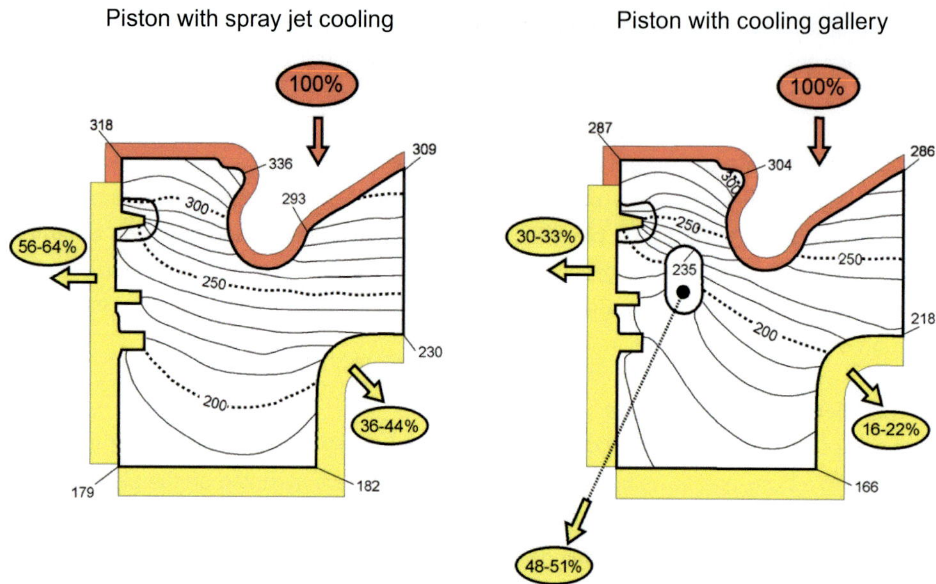

Figure 5.18: Heat flows in the piston as a function of the cooling type

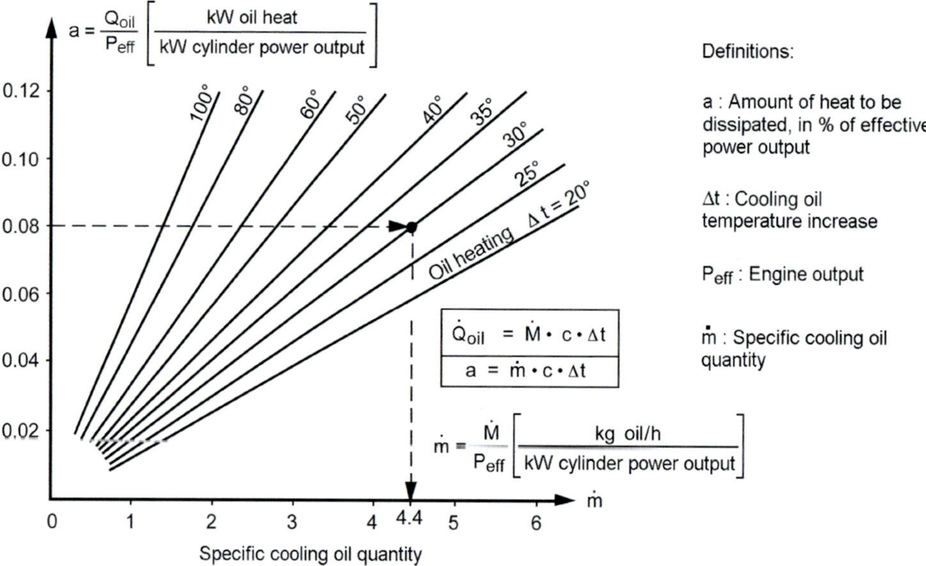

Figure 5.19: Dissipated heat quantity and specific cooling oil quantity

Since the absolute quantity of cooling oil depends on the size and power output of the engine, the basis for comparison is the specific quantity of cooling oil. It is defined as the ratio of the absolute quantity of cooling oil and the engine output, expressed in kg/kWh. Reference values for the specific cooling oil quantity required at the cooling nozzle are:

- For spray jet cooling: 3–5 kg/kWh
- For jet feeding: 5–7 kg/kWh
- For feeding via crankshaft and connecting rod (large-bore engines):   3–5 kg/kWh

Figure 5.19 shows the relationship between the specific cooling oil quantity, the quantity of heat dissipated by the cooling oil, and the difference between the oil inflow and return flow temperatures. For a quantity of heat to be dissipated of, for example, 8% of the effective power output and a permissible temperature increase of the cooling oil of 30°C, a specific cooling oil quantity of 4.4 kg/kWh is required.

5.10 Determining thermal load

Various methods for determining the temperatures at the piston and the effects of piston cooling are in use. They exhibit different levels of effort and precision in the engine.

The methods and their effect on piston temperatures are described in detail in Chapter 7.2.

5.11 Numerical analysis using FE analysis

The initial design of a new piston and comparison of variants typically uses finite element analysis (FEA); **Figure 5.20**. The boundary conditions are varied, on the basis of temperatures measured in the engine, until the calculated and measured temperatures conform to a sufficient degree. The heat flows can then be determined. If no temperature measurements are available, then the boundary conditions from similar designs can be carried over in order to obtain a first estimate of the expected thermal load.

By varying the piston geometry (e.g., bowl geometry, shape and position of the cooling gallery, position of the ring pack), its influence on the piston temperature field is investigated. Optimization efforts are made significantly faster and simpler in this manner than would be possible with engine temperature measurements.

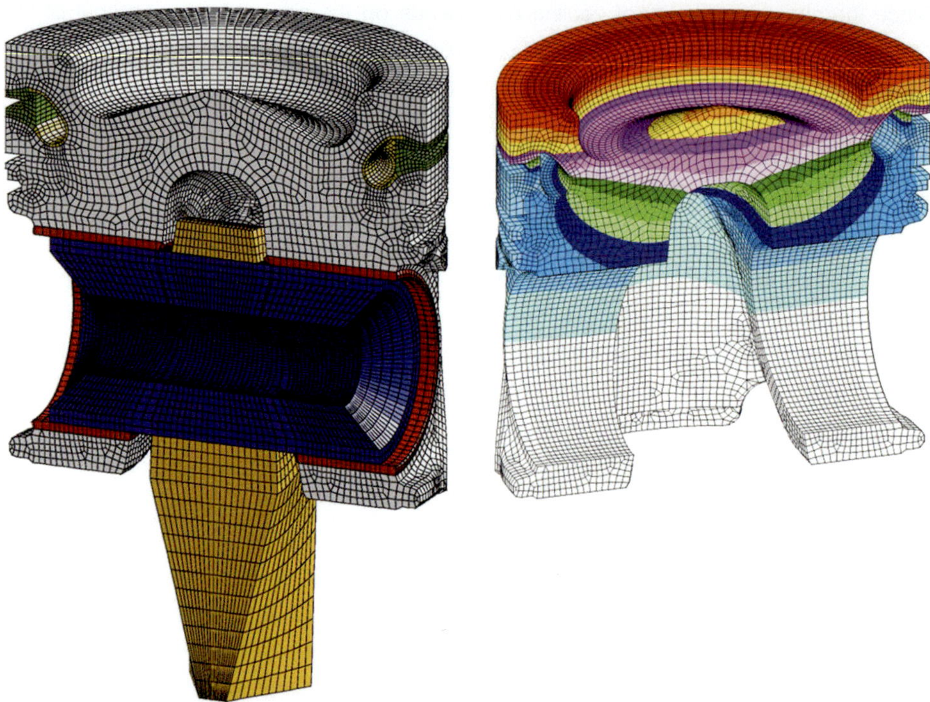

Figure 5.20: FE model and calculated temperature field

The boundary conditions on which the analysis is based are derived from theoretical considerations and experimental test results. The alignment of results of the analysis and the temperature measurements in the engine is regularly checked.

5.12 Laboratory shaker tests

The motion of the cooling oil along the inner contour of the piston and in the cooling galleries and cooling cavities has a decisive influence on the achievable heat dissipation. This oil motion cannot be measured during engine operation. Shaker tests provide a substitute. The kinematic and fluid dynamics conditions of the engine are simulated on appropriate test benches. Piston models made of PMMA allow oil motion in the cooling gallery to be studied.

5.13 Characteristic quantities

In order to evaluate the feeding and throughput of oil through the cooling gallery or cooling cavity when feeding cooling oil through a spray nozzle, the dynamic catching *efficiency* η_F, **Figure 5.21**, is used. It is defined as:

$$\eta_F = \frac{\dot{V}_{out}}{\dot{V}_{in}}$$

where:

$\dot{V}_{out}$ volume flow at the drain hole of the cooling gallery (on a moving piston model)

$\dot{V}_{in}$ volume flow through the cooling oil nozzle

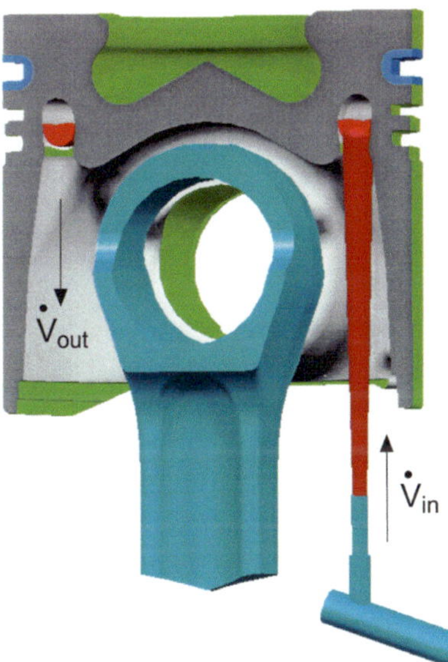

Figure 5.21:
Definition of dynamic catching efficiency

In order to achieve best possible heat dissipation, the cooling oil supplied by the oil nozzle must enter the cooling gallery or cooling cavity as completely as possible, i.e., the highest possible dynamic catching efficiency is required. This dynamic catching efficiency must not be confused with the static catching efficiency used by the manufacturers of cooling oil nozzles, which is based on the jet flow rate through a defined static mask.

The filling ratio ψ_F is a measure of the expected cooling effects due to the shaker effect in a partially filled cooling gallery (**Figure 5.22**), and is defined as:

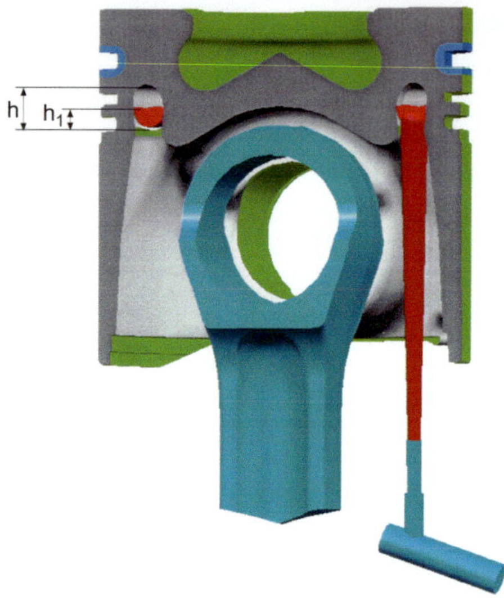

Figure 5.22:
Definition of filling ratio

$$\psi_F = \frac{V_{oil}}{V_{KK}} \quad \text{and} \quad \psi_{FA} = \frac{A_{oil}}{A_{KK}} \quad \text{and} \quad \psi_{Fh} = \frac{h_{oil}}{h_{KK}}$$

where:

V_{oil} volume of the cooling gallery filled with oil

V_{KK} volume of the cooling gallery

A_{oil} cross sectional area of the cooling gallery filled with oil

A_{KK} cross sectional area of the cooling gallery

h_{oil} height of the cooling gallery filled with oil

h_{KK} height of the cooling gallery

Since the cross sectional area is proportional to the volume for axisymmetric cooling galleries, the area-related filling ratio ψ_{FA} can be used for analysis. For a nearly rectangular cross sectional area, the height-related filling ratio ψ_{Fh} can be used to further simplify the calculation.

An intensive convective heat transfer, which can be attributed to the highly turbulent mixing motion that enhances energy exchange, prevails at the cooling cavity surfaces of the moving piston because of the shaker effect. A completely filled cooling gallery or cooling cavity would exhibit a lower heat transfer coefficient as a result of the lower oil velocities. Basic tests with a heated model show that the filling ratio affects the heat transfer *coefficient* α; **Figure 5.23**. The optimum filling ratio under the selected conditions was 30 to 60%. In this case, the heat transfer coefficient drops off again above a filling ratio of 60%, on account of the reduced turbulence in the gallery filled to a higher level.

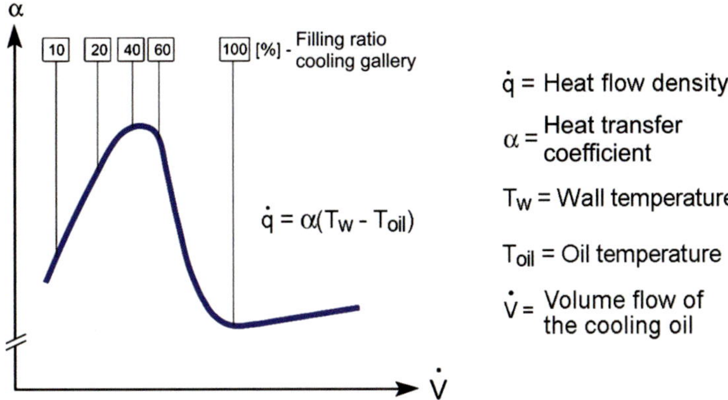

Figure 5.23: Relationship of filling ratio and heat transfer coefficient

The main variables that affect the catching efficiency and the filling ratio are the engine speed, the quantity of oil available, and the design of the inlet and outlet areas; **Figure 5.24**.

Both the catching efficiency and the filling ratio decrease with increasing engine speed. To a small extent, this can be attributed to the change in relative velocity between the piston and the oil jet. During the downward stroke, the velocity vectors of the piston and the cooling oil jet point in opposite directions. This results in a large relative velocity between the piston and

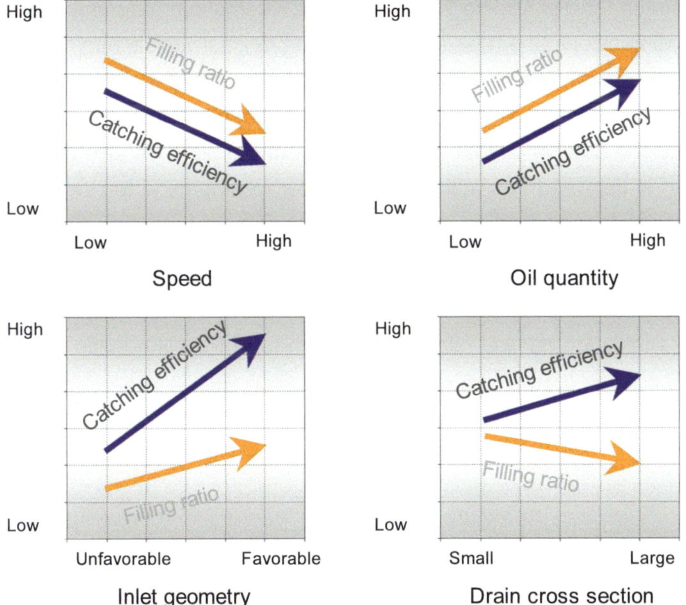

Figure 5.24: Factors affecting the catching efficiency and filling ratio

the cooling oil jet. In this time interval, a great deal of cooling oil enters the cooling gallery or cooling cavity at high velocity. The piston speed is nearly zero in the region around the top and bottom dead centers. At these points in time, the cooling oil enters the cooling gallery or cooling cavity at the speed of the oil jet. The particularly critical range is the upward stroke, during which the velocity vectors of the piston and the cooling oil jet point in the same direction. If the piston speed is greater than the speed of the cooling oil jet, then no fresh cooling oil will enter the cooling gallery for a brief period of time, which may slightly reduce the catching efficiency and filling ratio. This phenomenon can be compensated for in part by a properly designed infeed area. Pin hole bores, for instance, have a positive effect on the catching efficiency as well as the filling ratio.

An increasing quantity of oil for a constant cooling oil nozzle cross section leads to increased catching efficiency and filling ratio. They can be affected significantly by the size of the drain hole.

Another measured variable can be derived from direct measurement of the local heat flow in the shaker test. To this end, heat flow sensors are installed in the piston model to be tested. Their measurement signals are transmitted by a telemetry system.

5.14 Test facilities

The test benches for laboratory shaker testing are based on a kinematic arrangement that corresponds to the major dimensions of a passenger car or commercial vehicle engine and that is driven at an adjustable speed; **Figure 5.25**. The test setup consists of a push rod and a model piston made of PMMA. Cooling oil is fed to the model through PMMA tubes via the original cooling oil nozzles. The cooling oil that exits the outlet of the piston is also guided through PMMA tubes, and is captured in a test device in order to determine the catching efficiency. The filling ratio is determined using computer-aided image processing of video recordings taken at certain crank angles.

A nozzle test stand is also available and can be used to determine the characteristic flow curves and jet shapes for evaluating cooling oil nozzles. These tests form the basis for an optimized nozzle jet shape.

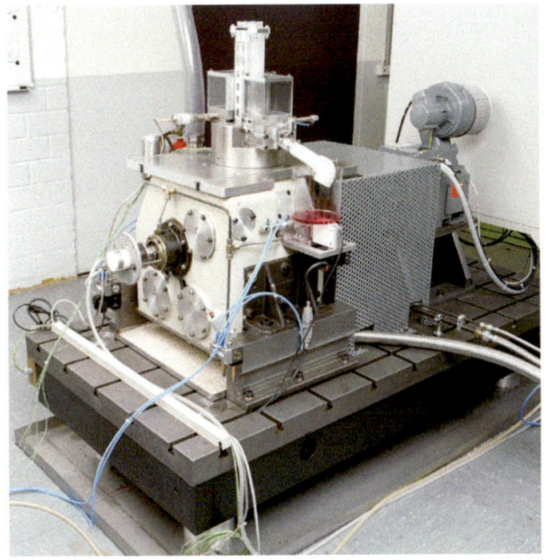 

Figure 5.25: Shaker test bench and test setup

5.15 Simulation of oil motion

Another option for analyzing piston cooling is a CFD (Computational Fluid Dynamics) simulation of the oil motion, including the heat transfer. This is based on modeling approaches from thermal fluid dynamics, which are incorporated into the simulation programs. The theoretical approaches, and their numerical conversion, must be confirmed by experimental measurements.

The ongoing development of commercial computation and simulation programs for describing the oil flow from the nozzle or the conrod to the piston, through the cooling galleries and cooling cavities, and back into the oil sump will make even more realistic simulation and analysis possible in the future. Both the complex flow characteristics (three-dimensional, turbulent, transient, and multiphase) and the heat transfer present great challenges for accurate, detailed representation of the conditions in the simulation model. Near-reality simulation of oil motion and validation of the calculation results with experimental data will still require substantial development work.

6 Component testing

Component strength of the piston can be ensured in various ways. Finite element analysis (FEA) is typically used for this purpose on a regular basis. A strain gauge measurement or pulsator test is used less often. In mathematical component analysis, all relevant thermal and mechanical loads, shrinkage, joining, and residual stresses can be analyzed. The basis of the FEA model is the ideal contour as outlined by the CAD dataset. The computational and strain gauge measurement results are analyzed using temperature-dependent material fatigue data. This is based on statistically confirmed test values, determined using test bars taken from pistons and artificially aged at the test temperature prior to testing. This ensures that the least favorable material parameters are taken as a basis.

Strain gauge measurement and component testing, in comparison with FEA, are based only on individual load cases, but they use the actual component geometries, tolerances, and, for the component-based strength test, incorporate technical influences such as the alloy composition, heat treatment, and casting quality. The results of the three methods thus complement one another. For strain gauge measurement as well as pulsator testing, however, finished components are required, while computational analysis is based on the CAD data of the piston design.

Strain gauges are applied to highly stressed areas. Static or cyclic loading is applied in the laboratory in order to determine local stress amplitudes. Dynamic measurements in the engine can provide additional insight about the stresses.

Several standardized tests have been established for component testing of pistons. Specially adapted loading fixtures are used. It is important that the test devices apply load to the component as realistically as possible. Because of the wide range of load/temperature combinations, an individual test installation must normally be conceived for each loading situation. Correct phasing of the combination of several dynamic mechanical loads, however, is difficult to implement.

Component testing in the lab is carried out primarily when an absolute conclusion about the strength or wear behavior is expected, or in order to compare different design variations. For engine components, purely static loads typically play a secondary role. Therefore, it is preferable to perform durability tests with a high number of load cycles. The time required often does not allow a sufficient number of test parts to be tested in order to obtain statistically reliable results. The staircase method is typically used for evaluation. Additionally, accelerated testing at excessive loads is used, but here the applicability of the results to actual engine conditions must be ensured.

6.1   Static component testing

Stress analysis with strain gauges and tear-off or failure load testing, where the component is statically loaded with a continuously increasing load until fracture, are among the commonly used static test methods; **Figure 6.1**. A comparison with empirical values allows the strength reserves of a component to be derived, such as for loading due to inertia forces at high speeds or the critical load, e.g., in the case of piston seizing.

Strain gauge measurement enables analysis of the geometric influences (overall design, local radii and transitions, elasticity, etc.) on the locally measured stress; **Figure 6.2**. The stresses thus derived can be compared with empirical values, on the basis of the local component temperature. The test results can also be used to confirm computational results, and thus validate the boundary conditions used.

Using the sectioning method, it is possible to determine the local residual stress state in the component, in order to adequately consider these stresses in the strength analysis. For local residual stress analyses, the hole drilling method can also be used.

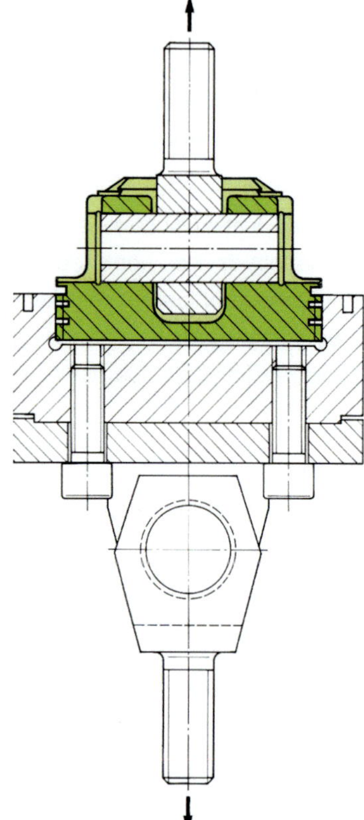

Figure 6.1:
Static tear-off testing of a piston

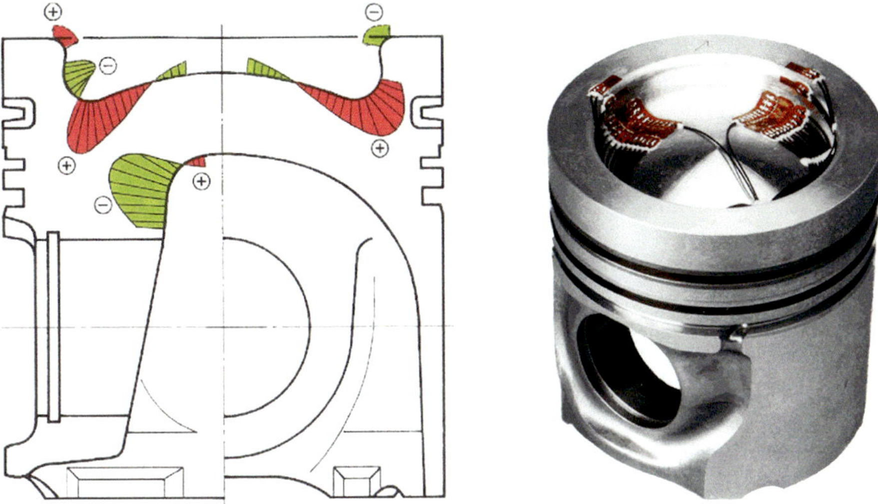

Figure 6.2: Example for strain gauge measurements on a piston

Other static component tests include deformation and stiffness measurements as well as investigation of pressure distribution between loaded surfaces. For deformation measurements, a defined force is applied to the structure. The deformation is measured at selected locations. The stiffness can be derived from the deformation at the point at which the load is applied.

Pressure distribution in a flange connection, for example, or between the piston skirt and the cylinder is measured with a pressure-sensitive film placed between two surfaces that are in contact with each other. Depending on the local compression force, a measurement signal or color change is induced; **Figure 6.3**.

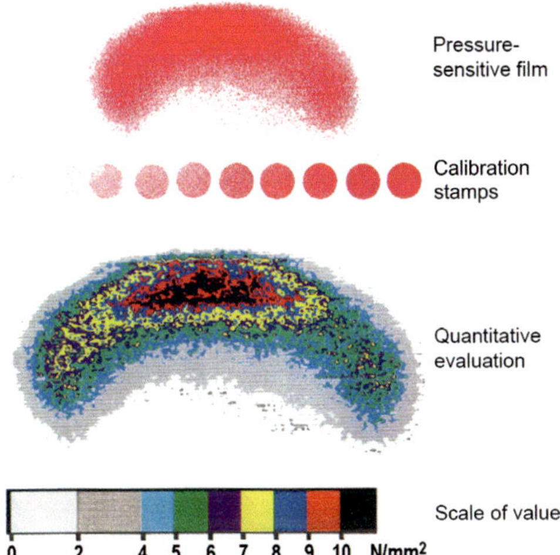

Figure 6.3:
Pressure distribution on piston skirt

Creep tests on components are also part of the static test program. They are used to determine permanent deformations as a function of load, time, and temperature. For bolt joints, the bolt force may experience a decrease due to relaxation. Other components exhibit settling or creep-related permanent deformation.

6.2 Cyclic component testing

Primarily cyclic mechanical loads occur in the engine, in addition to variable thermal stresses. Therefore, in order to analyze the actual component behavior, tests with pulsating or alternating loads are most suitable. The variable mechanical load is typically applied by resonant or hydraulic test machines, which may also include the influence of temperature. The temperature gradients found in the piston during actual engine operation, however, cannot be realistically simulated in the laboratory. Tests at a constant component temperature are used as a substitute. In order to perform a useful analysis of the results of cyclic component testing, it is necessary to have knowledge of the damage scenarios that occur in the engine under specific operating conditions. A correlation between the lab tests and engine loading can then be derived.

Typical examples of cyclic endurance tests on the component include
- crown pulsator test on pistons;
- skirt pulsator test on pistons in order to simulate normal force loads;
- hydropulsator tests to determine the pin bore load carrying capacity and simulation of inertial force; **Figure 6.4**;

Figure 6.4:
Hydropulsator tests
on piston

- tension-compression fatigue tests on connecting rods;
- fatigue resistance tests on the piston pin, piston rings, and bearings;
- thermal fatigue resistance of combustion chamber bowls;
- cyclic tensile tests on blind bore cylinders.

Flame test benches are used to test thermal fatigue resistance, particularly of combustion bowls for diesel engines. The bowl rim is heated rapidly by a gas-powered burner, similar to the conditions in the engine, and is cooled quickly by water spray after reaching the target temperature (e.g., 400°C). The surface is checked regularly for cracks. Typical cycle counts are 500 to 3,000. This test enables the comparison of different coatings and geometries; **Figure 6.5**.

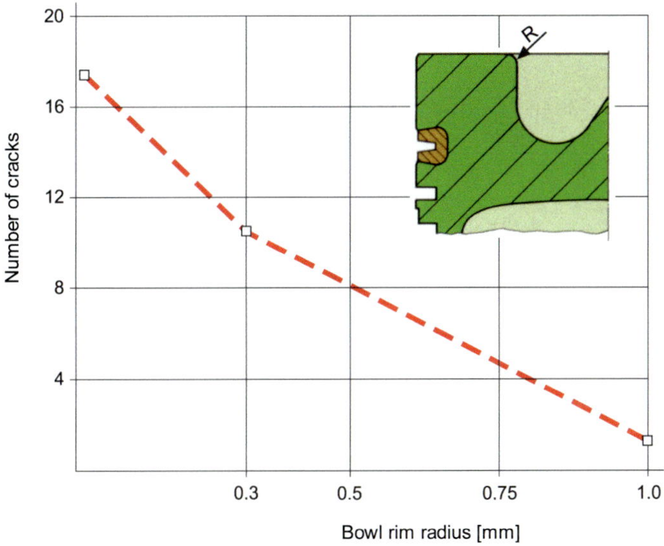

Figure 6.5: Flame bench test, influence of different bowl rim radii

Various test benches are used to analyze the oil motion in cooled pistons; cf. Chapter 5. Tests are performed
- for the amount of oil present in the cooling cavity as a function of the oil inflow and speed;
- in the case of free jet nozzles, for the catching efficiency, i.e., the proportion of oil transported through the cooling cavity relative to the total amount of oil available.

The test benches, **Figure 6.6**, are designed as crank mechanisms or based on modified, externally driven single-cylinder assemblies. They simulate the kinematic conditions of a passenger car or commercial vehicle engine. The models are sometimes made of transparent synthetic resin, in order to be able to determine the instantaneous quantity of oil at a spe-

Figure 6.6:
Shaker test bench

cific location in the cooling cavity, using video recordings. The test results serve to optimize geometries and oil feed cross sections. Comparisons to temperature measurements in the engine are used to derive the ideal parameters.

6.3 Wear testing

Characterization of the wear properties in lab tests is an important means for optimizing surfaces and coatings. Significant factors for wear include the running partners, surface condition, mechanical load, relative velocity, lubricant, and possibly the atmosphere and temperature. Therefore, for near-actual test conditions, lab methods using models such as the pin-on-disc, block-on-ring, SRV (vibrational frictional wear), and the like provide results with only limited applicability, particularly if the parts used do not represent the finish or manufacturing process of the real component. As a result, test concepts in the lab generally use the complete, actual component, or at least parts of it.

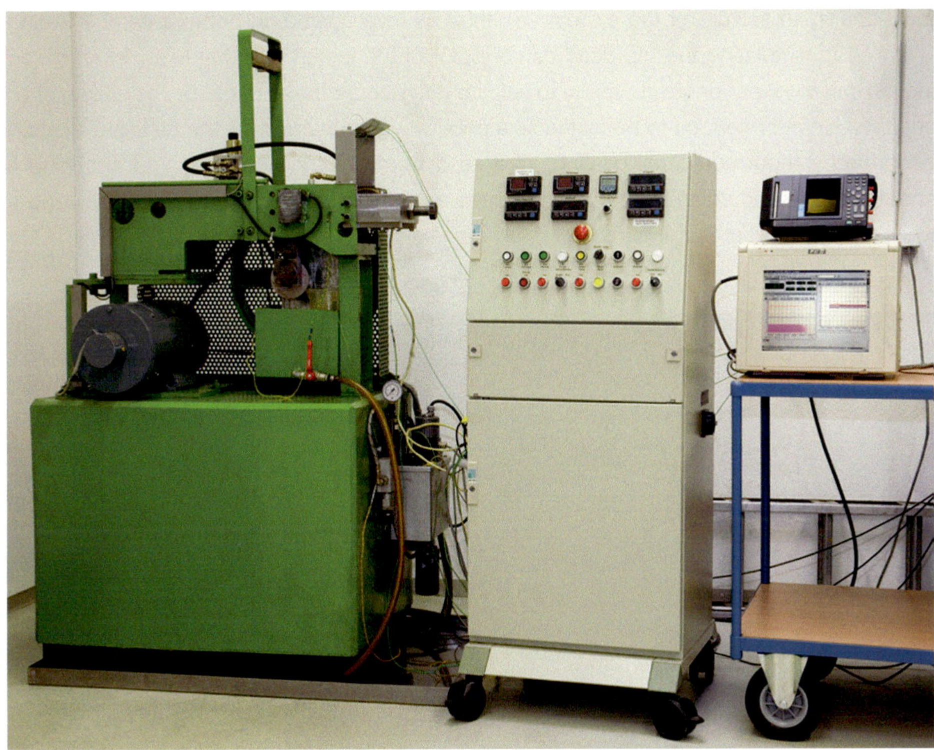

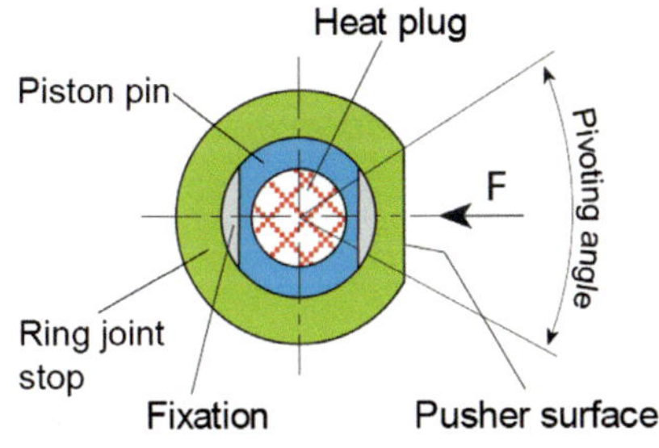

Pin/bore

Load:	Compressive cycling
Temperature:	150°C
Analysis:	Visual (wear marks)
	Change in diameter
	Roughness
	Number of load cycle
	Friction coefficient

Figure 6.7: Pin/boss test machine

In order to characterize the pivoting motion of the connecting rod about the piston pin under pulsating load, MAHLE has conceived a pin/boss test machine; **Figure 6.7**.

A sample boss moves about a defined pivoting angle relative to a fixed, heatable piston pin and is fed with a metered quantity of oil. The axial force drops to nearly zero at the dead center points and reaches a maximum at the time of greatest rotational speed. The surface roughness, dimensional changes, frictional torque, optical appearance, and any signs of seizing are evaluated.

The piston skirt or piston ring and cylinder segments are used to characterize the piston/ring/cylinder contact. The influence of piston ring coatings, ring surface finish, cylinder material, and honing on the wear or susceptibility to seizure of the system can thus be investigated. In both cases, the testing is limited to simulating the top dead center point of the piston ring that is at risk of mixed friction and the wear or susceptibility to seizure that occurs there. Because the effects of oil additives are not intended to be tested as a priority system parameter, the tests are typically performed without additional heating. Wear is determined by measuring and obtaining a surface profile, and sometimes by contact-free analysis of the sample topography using a white-light interferometer; **Figure 6.8**.

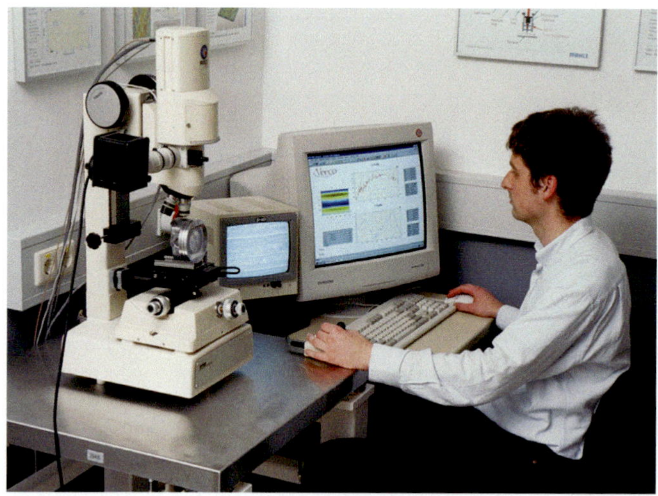

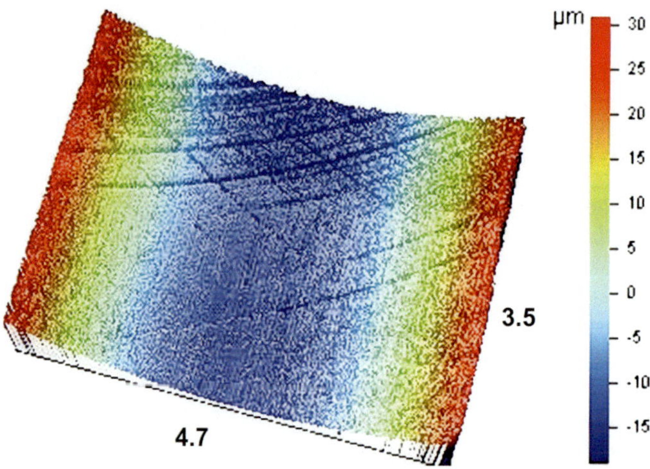

Figure 6.8: White-light interferometer: topography of worn cylinder surface

7 Engine testing

In the past decade, numerical calculation has increased in significance for development. The importance of engine testing, however, has by no means been relegated to the sidelines. It is no longer used just for direct component development, but is also employed for validating new simulation programs and systematically establishing design specifications.

In order to reduce expensive running times, extensive special measurement techniques are also used increasingly today in the development of engine mechanics. Without them, mechanisms and background circumstances can no longer be sufficiently well understood. Individual engine problems are analyzed precisely, corresponding critical boundary conditions and running programs are defined, and complex measurement and evaluation methods are developed and–wherever possible–automated.

Systematic parameter studies ultimately lead to extensive action plans that allow design engineers and developers to select initial values in the early design phase that are as close to series production as possible. These eventually lead to a reduction in the number of experimental iterations on the test bench during the subsequent development phase, thus helping to reduce development time and expenses.

In the course of these process developments, new specific measuring instruments, running programs, and measurement and analysis methods for effective engine testing are continuously created at MAHLE. Some selected subjects are addressed in this volume.

7.1 Test run programs with examples of test results

In order to test engine components, MAHLE uses test run programs in its engine testing, some of which are described as examples below. They include

- full-load curves;
- development runs;
- durability tests (standard for every development program).

Special test programs allow ongoing development of components, particularly for obtaining specific characteristics. They include

- cold start tests for testing cold friction resistance of piston coatings;
- burning mark tests for evaluating scuffing resistance of piston ring running surfaces.

MAHLE engine test laboratories are thus able to perform all of the tests required by customers for series production development. In addition, in the course of product development, new in-house inspection procedures and facilities that are specifically adapted to the new task at hand are continuously being created and refined.

7.1.1 Standard test run programs

7.1.1.1 Full-load curve

The performance settings of the engine are checked in the full-load curve. To this end, the relevant associated operating values under full load at various engine speeds are determined. Depending on the engine type, these may include

- performance;
- torque;
- fuel consumption;
- lambda value;
- smoke value;
- intake air temperature;
- charge air temperature;
- charge air pressure;
- exhaust gas temperature;
- exhaust back pressure;
- peak cylinder pressure;
- blow-by value.

Figure 7.1 shows examples of selected operating values of an internal combustion engine under full load as a function of engine speed.

7.1.1.2 Blow-by behavior

Figure 7.2 shows the behavior of the blow-by value as a function of load and engine speed. This is the amount of gas that flows out of the combustion chamber into the crankcase via the piston, piston rings, and cylinder. Oil is removed from these gases and then they are fed back into the combustion process via the intake section. It is common to measure the amount of gas at defined points across the entire speed range of the engine, such as under no load, partial load, and full load.

7.1.1.3 Seizure test

The seizure test is used for testing seizure resistance of pistons, piston rings, and cylinders under extreme engine conditions; **Figure 7.3**.

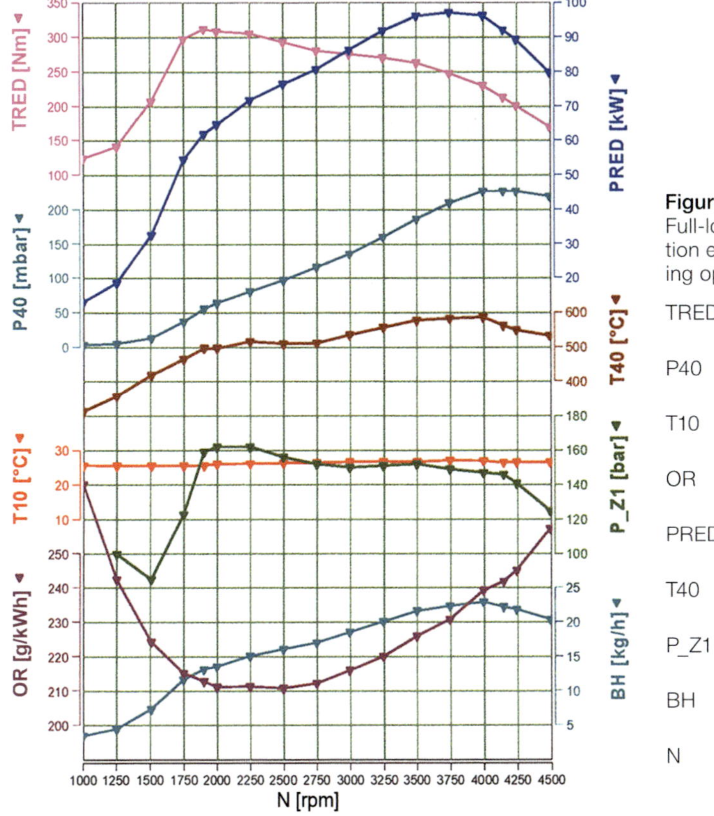

Figure 7.1:
Full-load curve for a combustion engine, with corresponding operating values

TRED	Torque reduced to standard state
P40	Exhaust back pressure after TC
T10	Intake air temperature
OR	Specific fuel consumption
PRED	Power reduced to standard state
T40	Exhaust gas temperature after TC
P_Z1	Peak cylinder pressure cylinder 1
BH	Fuel consumption per hour
N	Engine speed

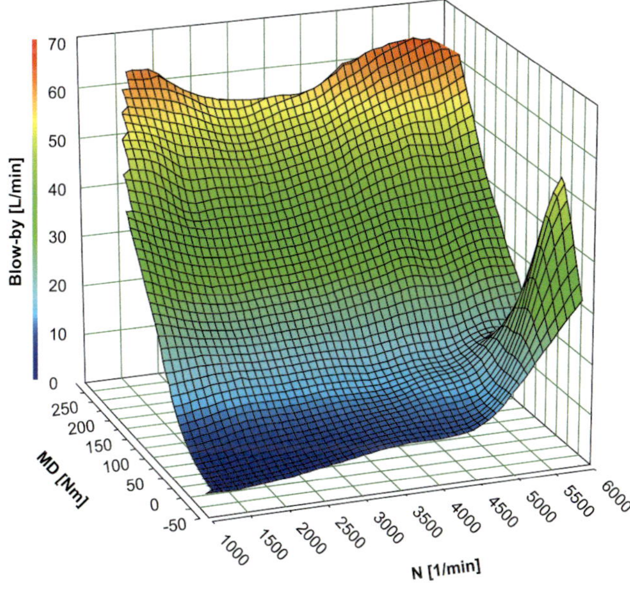

Figure 7.2:
Blow-by behavior of a gasoline engine

MD	Torque
Blow-by	Blow-by value
N	Engine speed

Figure 7.3:
Visible seizure damage on a
cylinder bore

To this end, the engine is operated at full load and nominal speed with increased coolant
and oil temperatures. Increasingly, the piston installation clearance can also be reduced and
a run-in phase can be eliminated. The engine is run under full load at nominal speed immediately after starting.

The running times for seizure tests are typically between 0.5 and 5 hours.

7.1.1.4 Development test

In the course of development, the piston, piston rings, and cylinder are tested for the following behavior:

- Blow-by
- Oil consumption

Further key development areas are

- wear patterns on pistons and piston rings;
- carbon buildup on the top land and in the region of the ring belt, and therefore potential
  cylinder polishing.

The running times of development tests are typically 50 to 100 h.

The programs typically contain a high proportion of full load points at nominal speed and
points of maximum torque. Depending on the development goal, other points, such as partial
load and no load points, can also be run; **Figure 7.4.**

In order to achieve the test objectives indicated previously, primarily geometric changes are
made to the components. These substantially affect fine optimization of shapes and clearances. For the piston, this means changes to clearances on the piston crown and skirt, as
well as adjustment of chamfers in the ring belt area. The axial groove clearances must be
tuned along with the piston rings, because it is important both to ensure sufficient sealing

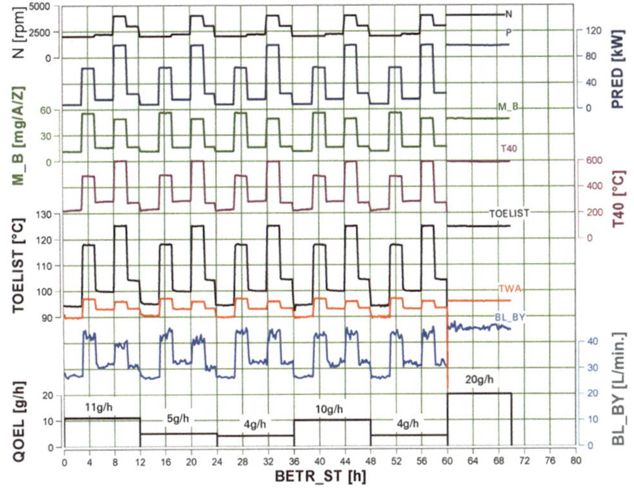

Figure 7.4:
Test run diagram for an example of a development test

M_B   Fuel consumption for each operating cycle and cylinder

TOELIST Current oil temperature

QOEL   Oil consumption

BETR_ST Operating hours

against blow-by and oil pumping, and to provide resistance against piston ring sticking due to carbon buildup–particularly in diesel engine pistons. Further optimization can be performed at the oil drain points in the oil ring groove.

For piston rings, the potential influences include the various parameters of ring geometry, including ring width and depth, running surface shape, and internal bevels.

The design of the cylinder and engine block can influence the distortions and thus the blow-by and oil consumption behavior. Oil consumption can also be influenced by the type and roughness of honing.

7.1.2 Long-term test run programs

7.1.2.1 Standard endurance test

On account of the long running times of up to 1,000 hours for passenger car engines and up to 3,000 hours for commercial vehicle applications, and the associated high costs, endurance tests are performed only after the development of the components involved has been completed. They document the effectiveness of design measures developed in previous, shorter development tests.

The purpose of endurance tests is to verify the

- unrestricted long-term functionality of engine components with respect to low oil consumption, blow-by, and wear, for example; and
- durability of the developed components.

Successful endurance tests are often the basis for series production approval for the tested components.

7.1.2.2 Cold-warm endurance test

One variant of the standard endurance test is a cold-warm endurance test. As a result of thermal expansion, extreme temperature cycles of the coolant cause significant changes in stress in the material of the affected components, such as the pistons, piston rings, cylinder head, and cylinder head gasket. In the worst case, cracks will form, as seen here on the running surface of a piston ring; **Figure 7.5**.

Figure 7.6 shows an example of a cold-warm cycle of this kind with a cycle time of about 10 min.

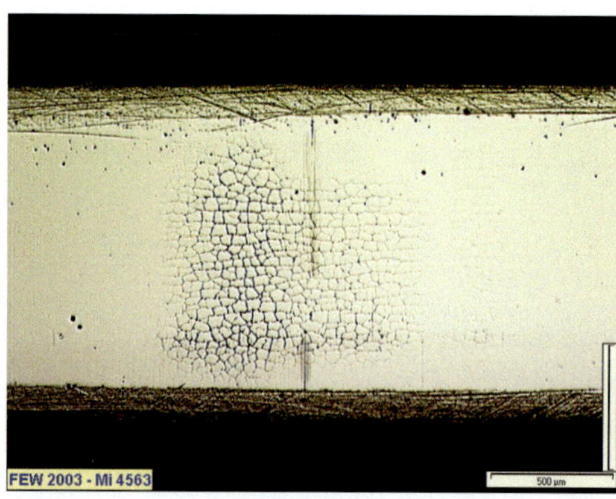

Figure 7.5:
Crack network in the piston ring running surface

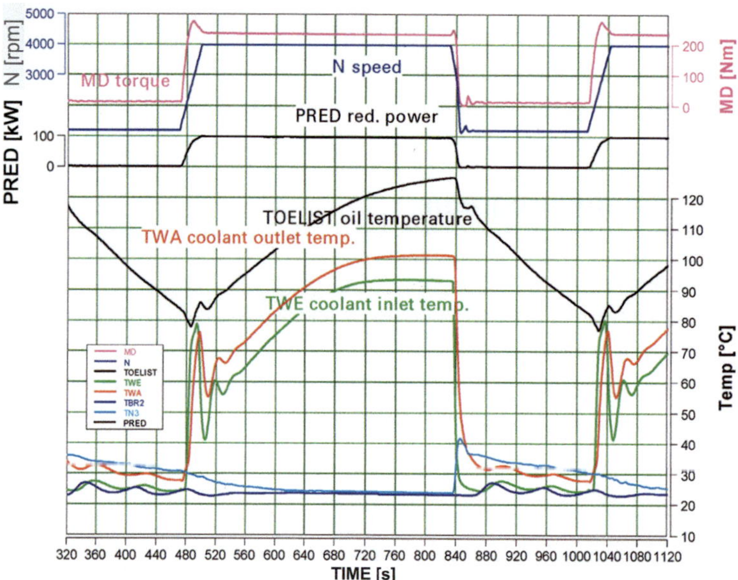

Figure 7.6: Temperature cycle in cold-warm endurance test

N	Engine speed	TOELIST	Current oil temperature
MD	Torque	TWA	Coolant temperature at engine outlet
PRED	Power reduced to a standard state	TWE	Coolant temperature at engine inlet

7.1.3 Specialized test run programs

7.1.3.1 Cold-start test

The cold-start test is used in trials of piston coatings to prevent "cold start scuffing." This occurs in gasoline engines when increased fuel injection washes away the lubricating film from the cylinder bore in a cold engine. Uncoated pistons that exhibit cold start scuffing are initially used as the basis in the cold-start test. Various coatings are then tested in direct comparison.

The engine can be run without load for the cold-start test. Prior to actual testing, the engine runs at room temperature, in order to ensure uniform distribution of the oil in the engine. The engine, including the oil and water, is then chilled down to −25°C in the cold chamber. **Figure 7.7** shows a detailed example of a cycle of this kind.

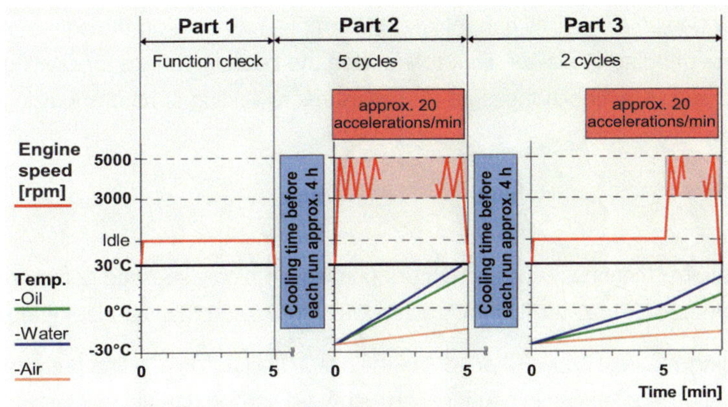

Figure 7.7:
Speed and temperature profile in cold-start testing

7.1.3.2 Microwelding test

Microwelding, or groove flank damage, refers to a local process of destruction on the lower flank of the piston ring groove in an aluminum piston; **Figure 7.8**. It occurs primarily in the top ring groove of gasoline engine pistons. Sporadic cases of microwelding in the middle piston ring groove of aluminum pistons for diesel engines are also known, because this ring groove is typically not protected against wear by a ring carrier, as the top ring groove is.

Microwelding occurs as follows: material is locally torn out of the loaded flank of the top ring groove, microwelded onto the first piston ring, and then pressed back into the groove flank at another location.

Because microwelding can be identified after just a brief running time–e.g., after run-in or audit test runs on the engine–the running time for the test program developed specifically for this problem is only 23 hours. This includes run-in, full-load curves, and the actual microwelding cycle, which lasts three hours and is repeated several times. In this cycle, the maxi-

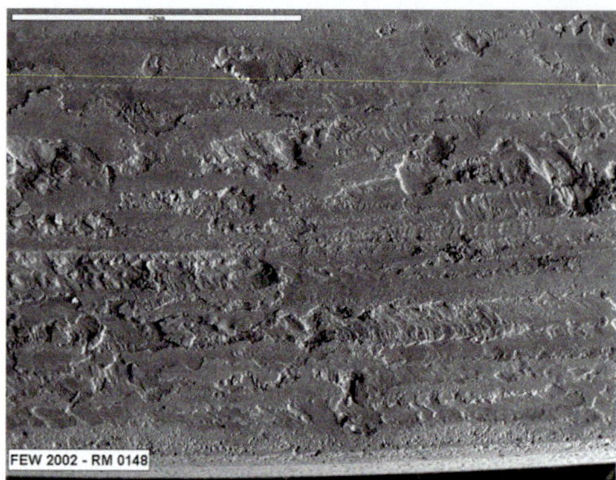

Figure 7.8:
Microwelding damage in the lower flank of the top ring groove of a piston without a ring carrier

mum torque, rated power, and governed speed are run in an alternating sequence. Visual assessment of the damage is carried out after the test. For less severe damage, optimization of the piston ring may be sufficient. This can mean, for example, rounding off the edges of the ring (tumbling), fine grinding the lower flank to increase the bearing surface, or coating the lower flank of the ring. A remedial measure for severe microwelding is hard anodizing the top ring groove.

7.1.3.3 Fretting test

Fretting refers to vibrational frictional wear and occurs primarily in highly stressed commercial vehicle diesel engines, on the lower flank of the top ring groove and the first piston ring.

Similarly to microwelding, material is locally torn out of the piston and piston ring and pressed back in at another location, sometimes in combination with hard carbon deposit residues. In the beginning, only slight surface damage of just a few µm is evident, but during the course of later runs, local cavities of up to 1/10 mm can arise; **Figures 7.9** and **7.10**.

A full-load program is used to test surface modifications or coatings on the piston groove or the ring side face. The running time is 100 hours. This allows a good initial assessment

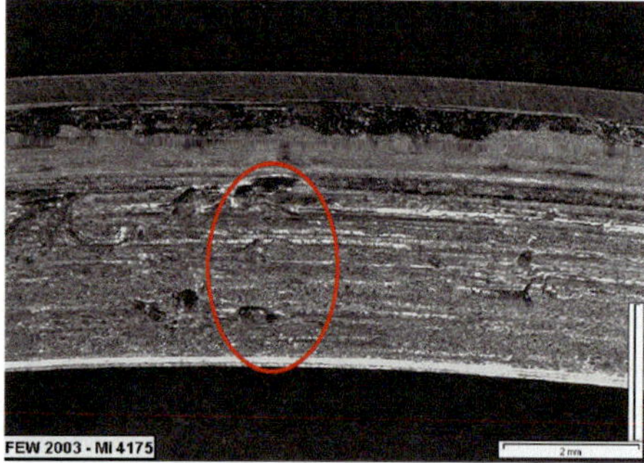
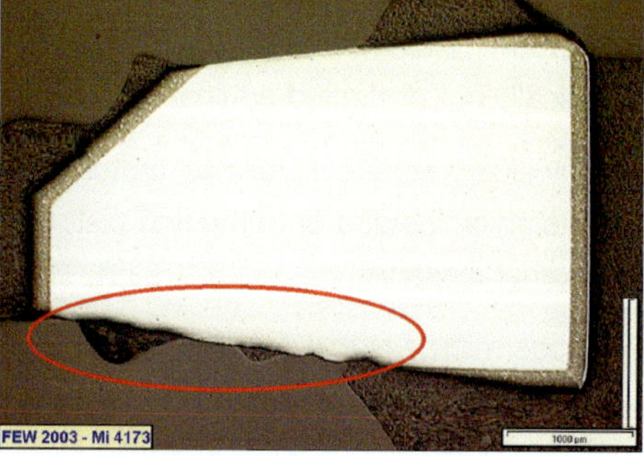

Figures 7.9 and **7.10:** Fretting damage on the lower flank of a 1st piston ring

of the effectiveness of the measure performed on the piston or the ring. Further tests are then performed as part of the standard endurance tests. Depending on the running time and assessment, measures such as full nitriding over the entire cross section of the ring, or chrome-plating of the lower flank of the ring may be considered.

7.1.3.4 Burning mark test

The burning mark test is used in the development of piston rings. Burning marks occur on the piston ring running surface under high thermal loads. This is a type of damage in which larger cracks and breaks can occur on the running surface; **Figure 7.11**.

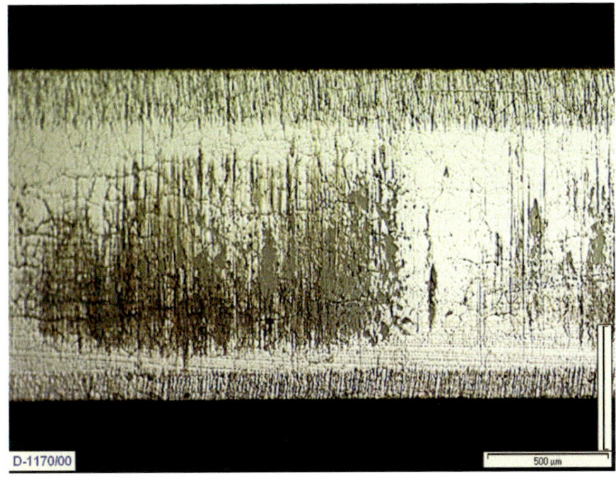

Figure 7.11:
Scuffing on the piston ring running surface

In advanced stages, the burning marks become more severe, such that the running surface of the piston ring starts to fall apart. This ultimately leads to damage to the cylinder running surface. In the end, this damage can cause seizing in the affected cylinder, resulting in total failure of the engine. In order to prevent such damage, it is necessary to develop piston ring materials and coatings that are highly resistant to burning marks.

One important step in this development is to be able to determine burning mark resistance directly in the engine under more difficult conditions. Tribometer tests cannot sufficiently simulate the entire complexity of the stresses in the actual engine. Therefore, a special test procedure, known as the burning mark test, had to be developed for this purpose. Although piston rings in diesel and gasoline engines have to fulfill similar tasks in the engine, there are differences with respect to geometry and material selection. It was therefore necessary to develop dedicated test procedures for both types of engine. The test conditions for a burning mark test are significantly more severe than in conventional test run program. They are listed in **Figure 7.12**, using the diesel engine as an example.

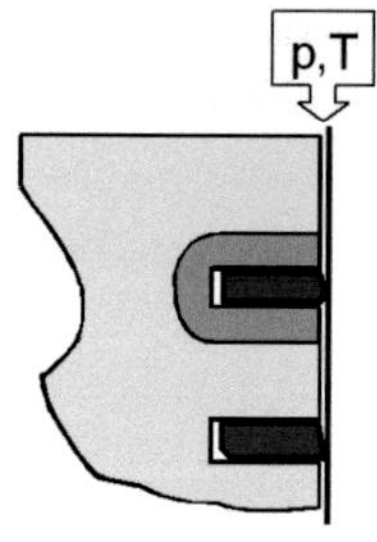
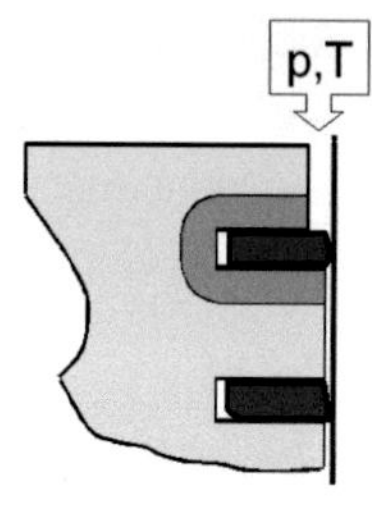

Type of test	Standard	Burning mark test
Piston type	Cooling gallery	Spray jet cooling
Cooling oil supply	Large	Small
Top land width	Large	Small
Top land clearance	Small	Large
Gap clearance, 1st ring	Small	Large
Axial clearance, groove 1	Small	Large
Axial clearance, groove 2	Small	Large
Cooling fluid	50% water / 50% antifreeze	100% antifreeze
Cooling fluid temperature	90°C	**100, 110, 120, 130°C**
Oil temperature	130°C	150°C
Charge air temperature	50°C	90°C
Test run program	Customer-defined	10 h at rated power

Figure 7.12: List of intensified test conditions for diesel engine burning mark test

What sets the burning mark test apart from all other laboratory test procedures, such as tribometer tests, is the ability to evaluate and classify ring surface materials and especially ring coatings with respect to burning mark resistance under actual operating conditions.

In order to determine the burning mark limit of a piston ring material, a new test setup (new engine block, new piston rings) is used to run each test with the cooling fluid temperature in the range of 100, 110, 120, and 130°C, increasing by 10°C each time. To evaluate burning mark resistance, four classification levels are applied for the severity of the burning marks that occur within the cooling fluid temperature level achieved in the burning mark test. These four classification levels are assigned the colors white, light gray, dark gray, or black. This allows simplified visual representation in tables and charts. The entire circumference of the ring is used for this analysis.

- "White" classification ⬜ – no burning marks
  In the best case, the ring running suraces exhibit no burning mark or other changes whatsoever. Individual scratches are permissible.

- "Light gray" classification ▨ – light burning marks
  This level allows isolated, light burning marks.

- "Dark gray" classification ▬▬▬▬ – severe burning marks

  The burning marks that occur extend over a wide area of the running surface and are significantly more pronounced than for the classification "light gray." Clearly visible stripes or even slight seizure marks typically occur on the associated cylinder running surface at the same time.

- "Black" classification ▬▬▬▬ – seizure marks over a wide area of the circumference

  This evaluation is applied if the burning marks affect nearly the entire circumference of the running surface and have already caused some disintegration of the surface. The associated cylinder typically shows very pronounced seizure marks over large areas of the circumference.

After a test of the running surface performed in this manner, the results can be represented simply in chart form. **Figure 7.13** shows an example of the burning mark limit reached for different running surface materials for a diesel engine application.

As a summary of all the tests, a final assessment can be made as to which running surface material is best suited to which engine type, on the basis of the specific power output, with respect to burning mark resistance. An example of this overview for diesel engines is shown in **Figure 7.14**.

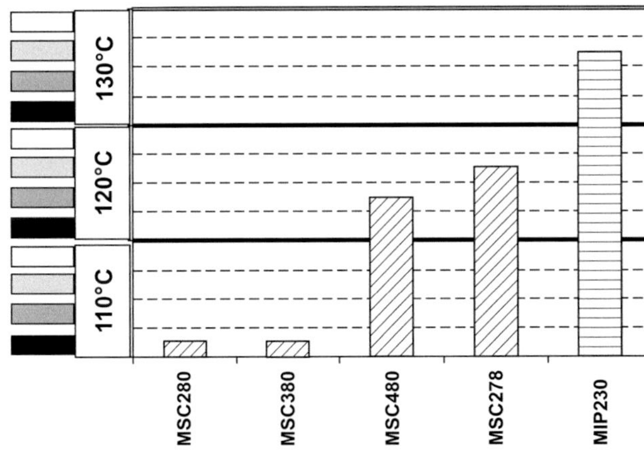

Figure 7.13:
Burning mark resistance of thermal spray coatings (MSC280, MSC380, MSC480, MSC278) in comparison with a PVD coating (MIP230)

Ring running surface material	Application range for diesel engines
PVD coatings: MIP230	Engines with very high specific power output
Galvanic chrome coatings: MCR236, MCR256 **Thermal spray coatings:** MSC480, MSC278	Engines with medium to high specific power output
Thermal spray coatings: MSC280, MSC380	Engines with low to medium specific power output

Figure 7.14: Burning mark resistance for various running surface materials

7.2 Applied measurement methods for determining the piston temperature

Engine combustion is inseparably linked to substantial rises in pressure and temperature in the combustion chamber, which puts severe mechanical and thermal stresses on the piston.

Mechanical stresses on the piston are caused primarily by the gas force acting on the piston crown, the movable part of the combustion chamber, and inertia and lateral force loads.

The thermal stress arises from the impingement of hot combustion gases on the combustion chamber side of the piston crown. This produces a heat flow from the combustion chamber, through the piston crown, and into the piston material. A large part of the heat flowing into the piston material is dissipated via the piston rings, particularly the first piston ring, into the cylinder wall, and then to the coolant that surrounds the cylinder. Another portion of the heat produced is dissipated to the engine oil, in engines with piston cooling.

These heat flows give rise to a specific temperature field, shown as an example for a passenger car diesel engine in **Figure 7.15**.

The permissible stresses for the aluminum alloys (AlSi) typically used as piston materials are greatly dependent on the temperature. At first, however, the stress field and the temperature field should be estimated using the engine data provided by the manufacturer and the safety factors should be calculated for highly stressed regions of the piston, on the basis of the required load spectrum.

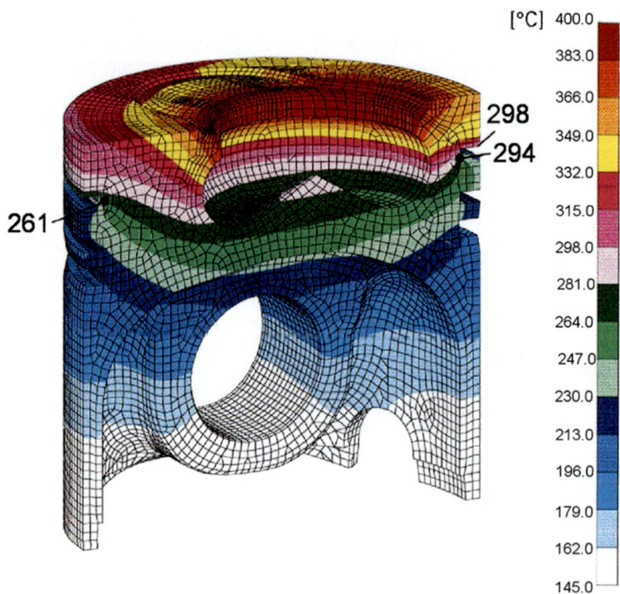

Figure 7.15:
Temperature field using the example of a passenger car ring carrier piston with eccentric combustion chamber bowl for a diesel engine

If these safety factors are not sufficient, despite additional optimization steps, then suitable measures can be suggested, as a result of piston temperature measurements, in order to obtain sufficient safety by reducing the component temperatures (e.g., by modifying the engine calibration).

Temperatures measured at various positions on the piston are needed as input variables for the FE analysis of the temperature field. Particularly when developing pistons for highly stressed passenger car diesel engines, measuring the piston temperature is an important and common step.

7.2.1 Methods for measuring the piston temperature

A distinction is made between thermomechanical (fusible plugs and templugs) and thermo-electric measurement methods (NTC and thermocouples).

7.2.1.1 Thermomechanical methods for measuring the piston temperature

7.2.1.1.1 Use of fusible plugs

Fusible plugs help to determine the temperature of the piston during steady-state engine operation. The process involves placing several–typically three–small metal pins made of suitable alloys with graduated melting points in a range of 10 to 15°C at the appropriate measurement points on the piston. The melting points of the plugs are selected so that they cover the expected temperature range.

After the measurement run at a steady-state operating point, the fusible plugs are evaluated to see if they have started to melt. This allows the prevailing temperature at the measurement point to be estimated, as it must lie within the interval between the melting points of the melted and unmelted fusible plugs.

Advantages:
- Little outlay for equipment and measuring instruments
- Short preparation and measurement time

Disadvantages:
- Expected temperature must be estimated beforehand.
- Low measurement precision due to the temperature intervals between melting points
- Measurement of only one operating point for each run

The fusible plug method has been replaced by a more suitable process using templugs.

7.2.1.1.2 Use of templugs

Templugs are small, hardened pins made of a special metal alloy; **Figure 7.16**. After a measurement run at the desired operating point under constant conditions, the mean temperature during the measurement run is determined on the basis of the decrease in hardness of the templugs after removal.

A single type of alloy is sufficient for the required temperature range for pistons. Templugs are available in various sizes.

The temperatures to which the templugs were exposed in the measuring piston are determined–after they have been removed from the piston–as a function of the decrease in hardness and the measurement running time.

Advantages:

- Little outlay for measuring instruments, preparation, and equipment
- Relatively high measurement accuracy
- On account of their small dimensions, a large number of measurement points can be implemented for each piston.

Disadvantages:

- Loss of time due to external analysis
- Measurement of only one operating point for each run
- A measurement run typically lasts ten hours.

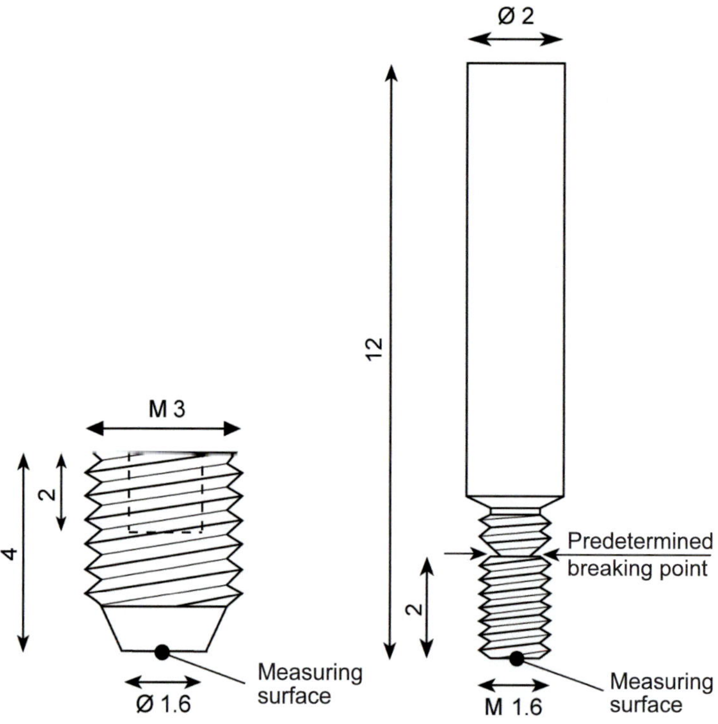

Figure 7.16: Dimensions of templugs, left: standard templug with Allen socket, right: templug with predetermined breaking point

7.2.1.2 Thermoelectrical methods for measuring piston temperature

7.2.1.2.1 Use of NTC resistors

The transducers are NTC resistors (thermistors, NTC = Negative Temperature Coefficient), which vary in electrical resistance as a function of the temperature; **Figure 7.17**.

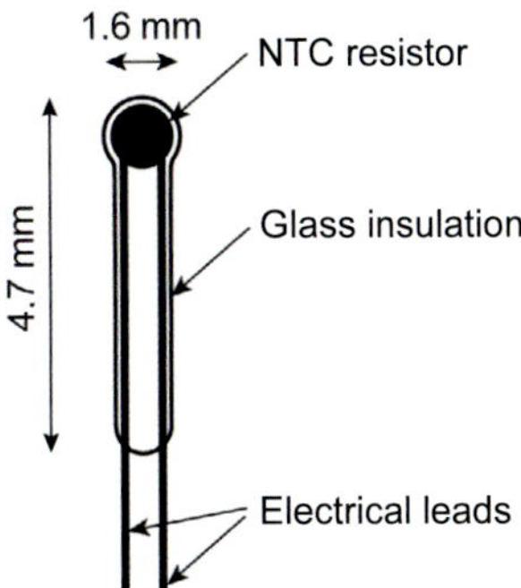

Figure 7.17:
Example of dimensions of an
NTC resistor with glass insulation

The measurement value is transferred without contact by means of inductive coupling of two oscillating circuits at the bottom dead center of the piston. Ring coils are used on the piston and pin coils are used on the engine block. **Figure 7.18** shows an example of a measuring piston equipped with NTC resistors.

Advantages:
- Proven standard measurement method
- Well suited to high speeds

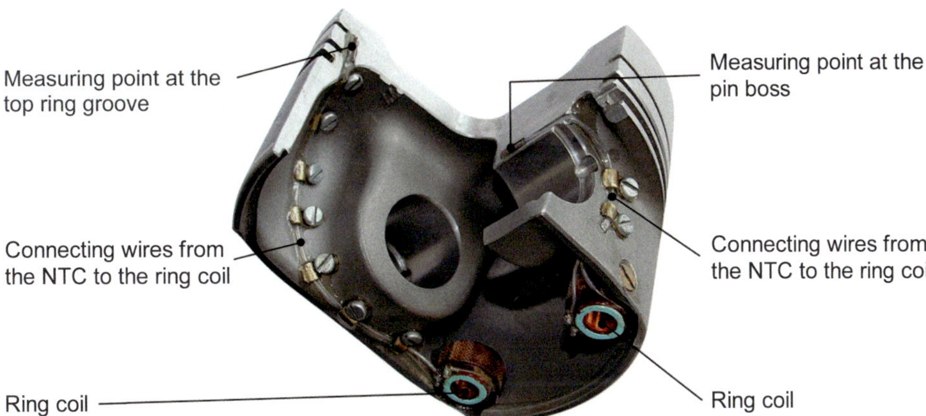

Figure 7.18: Measurement piston fitted with NTC resistors

Disadvantages:

- Limited physical and mechanical stability of the sensor
- Measurement range limited to 400°C, because of adhesives used in the application
- Sensitive to interference signals and impact on inductive coupling from metal near the transmitters, because of the oscillating circuit principle of measurement using amplitude modulation
- Depending on spatial conditions (piston diameter), a maximum of three sensors can be used.
- It may be possible in some circumstances to install only one ring coil in the piston, if the piston has a very small diameter.
- Different types of NTCs are needed, depending on the expected temperature loads at the measurement point.

7.2.1.2.2 Use of NiCr-Ni thermocouples

In contrast to NTC resistors, NiCr-Ni thermocouples can be used universally, i.e., throughout the entire temperature range that occurs in the combustion engine. **Figure 7.19** shows an application example of such a thermocouple.

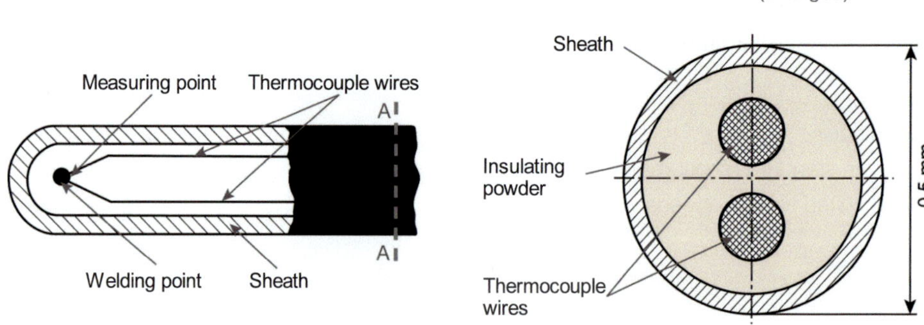

Figure 7.19: Schematic construction of a NiCr-Ni sheath thermocouple

Advantages:

- Sensor positions can be selected almost without restriction, thanks to the small diameter of the thermocouples.
- The measurement range is typically between −200°C and 1,150°C. However, calibration for measuring the piston temperature is usually performed only up to 650°C.
- Very well suited to measuring large temperature amplitudes in transient test run programs

7.2.1.3 Transferring the readings from thermocouples

7.2.1.3.1 Transferring the readings from thermocouples with measurement leads supported by linkage systems

The readings are transmitted out from the thermocouple in the piston through measurement leads supported by a linkage construction; **Figure 7.20**. One side of the linkage system is movably attached to the piston pin or the bottom side of the piston pin boss, and the other side to the crankcase.

The number and type of NiCr-Ni thermocouples can be selected freely.

Advantages:
- The number of measurement points is limited only by the geometric conditions.
- Rapid changes in temperature can be captured (resolution based on degree of crank angle is also possible).
- Good for steady-state and transient measurements
- The required power supply can be provided using leads supported by the linkage system, so no battery is required.

Disadvantages:
- High level of design effort and mechanical modifications to the engine
- Installation of a linkage system in individual cylinders is limited by the engine design.
- Typically installation is possible at only one cylinder.
- Maximum measurement speed is limited by the mechanical design of the linkage system.
- Measurement duration is typically limited to a few hours because of the extreme stress on the measurement cables.

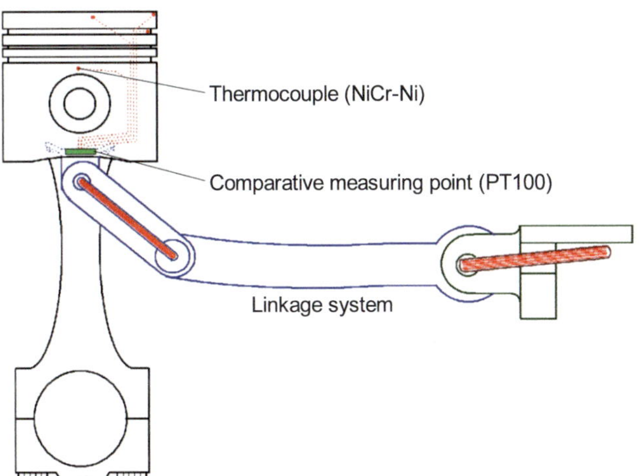

Figure 7.20:
Principle of signal transmission with measurement leads supported by linkage systems

7.2.1.3.2 Transferring the readings from thermocouples using telemetry

Readings are transmitted from the thermocouple in the piston via an RTM (Real-Time Telemetry Temperature Measurement) measurement system newly developed by MAHLE. The integrated sensor signal amplifier, **Figure 7.21**, receives the power it needs through inductive coupling.

The readings are transmitted with each revolution in the range of the bottom dead center. The minimum coupling time per revolution is 0.8 ms.

Advantages:

- Technical effort is lower for the same quality of results as for transmission of readings via measurement leads supported by a linkage system.
- No limit on speed for series production applications
- Temperature changes can be captured rapidly (data transmission with each crankshaft revolution).
- Only slight changes to the engine block are required.
- Up to seven temperatures can be captured on each piston. If several sensor signal amplifiers are used in one piston, the number increases accordingly.
- The sensor signal amplifiers used are temperature-stable up to 175°C.
- In comparison with measurements using a linkage system and measurements using NTC resistors, a much longer service life is possible.
- The temperature-compensated sensor signals are digitized and thus insensitive to interference when transmitted to the analysis unit.
- With real-time visualization, the RTM system is well suited, particularly for optimizing combustion parameters.

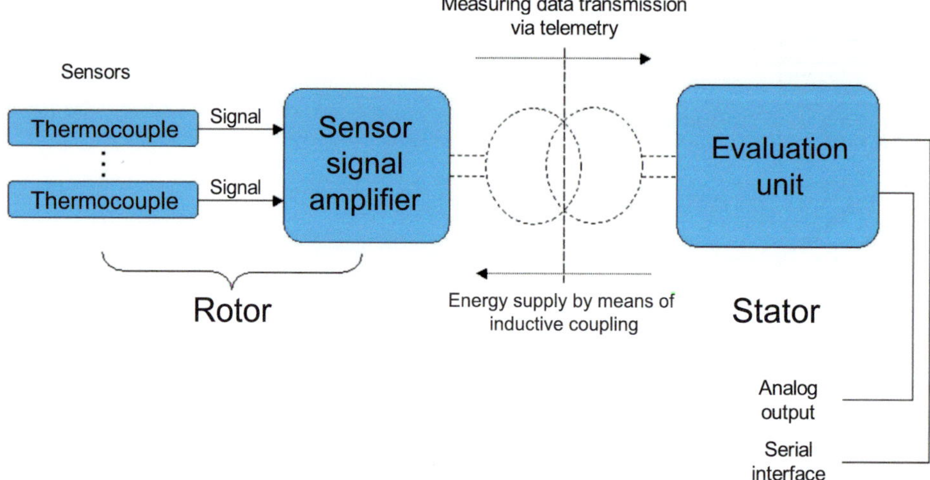

Figure 7.21: Schematic diagram of the principle of measuring and transmission of readings using telemetry (RTM system)

Disadvantages:

- Relatively high effort in applying the measuring equipment to the piston
- High financial cost for electronic components (sensor signal amplifiers)

7.2.1.4 Evaluation of the methods used at MAHLE for measuring piston temperatures

The selection of the right method depends on the requirements profile. If information is needed only for one defined operating point (e.g., the rated power point), then templugs can be used. If, however, a large number of measurements in the operating map or real-time visualization is needed, then the use of sheath thermocouples and telemetry is recommended, i.e., the RTM system described above. **Table 7.1** shows corresponding options, considering the preferred parameters in each case.

Table 7.1: Methods used at MAHLE for measuring piston temperatures

Measurement method		Templug	NTC	NiCr-Ni	NiCr-Ni
Transmission method			Inductive coupling	Linkage system	Telemetry
Process comparison		Residual hardness measurement			
Classification of methods		Thermo mechanical		Thermoelectric	
		Steady-state		Transient	
Operating points per measurement run		1		No limit	
Typical number of measurement points per piston	Passenger cars	8	2–3	10–20	7
	Commercial vehicle	15	4		
Robustness	of the sensor	High	Low	High	High
	of the reading transmisssion	–	Medium	Low	High
Suitability for high engine speeds		High	High	Low	medium
Precision		Medium	High	High	High
Requirements for the workshop equipment		Low	Medium	High	High
Project lead time		50%	70%	100–200%	100%

7.2.2 Piston temperatures in gasoline and diesel engines

When measuring the piston temperature, very specific positions are preferred for the sensor locations. They are shown in **Figure 7.22**, using the example of a passenger car diesel piston.

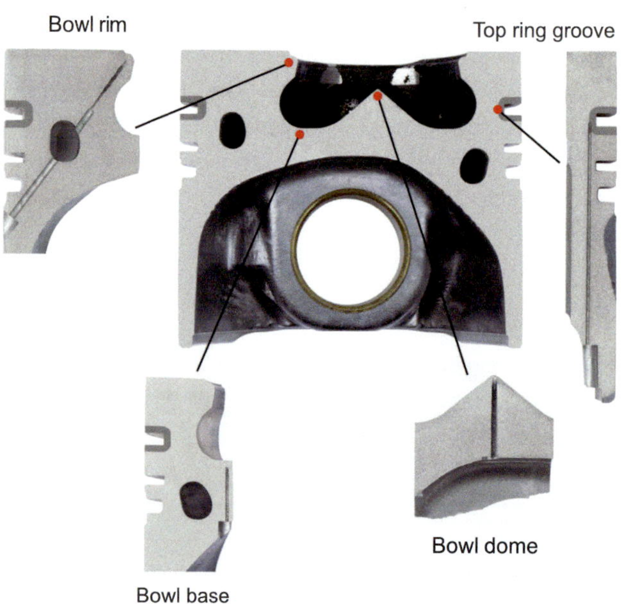

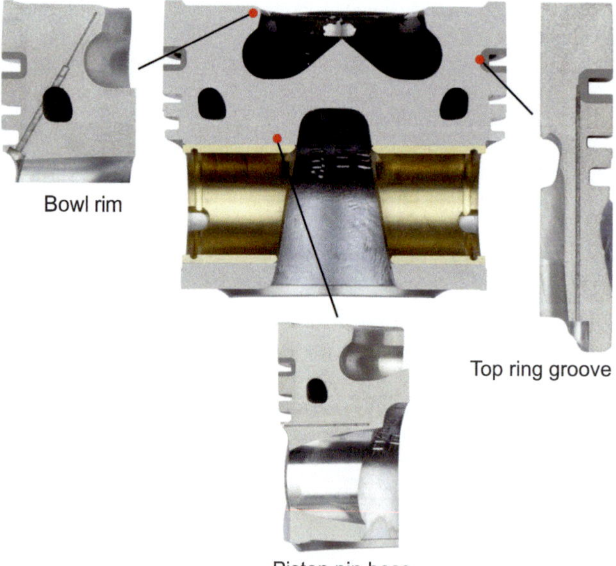

Figure 7.22:
Examples for equipping a passenger car diesel piston with sensors for measuring piston temperature

The criteria for selecting these measurement points are as follows:

- The bowl rim and base are subjected to heavy thermal and mechanical stresses, so cracks can occur if they are sufficiently overloaded.
- The temperature in the top ring groove is essential to the function of the piston rings (resistance to burning marks on the ring running surface). Buildup of residues in the top ring groove is also affected, and therefore the susceptibility to ring sticking.
- The piston pin boss transmits the entire gas force to the piston pin and is highly mechanically stressed. This can lead to the occurrence of pin boss cracks.

7.2.2.1 Typical temperature maxima on the piston

It is not possible to set an absolute limit for maximum permissible temperatures at various positions on the piston, because the load spectrum specified by the engine manufacturer or expected can vary greatly. The full-load portion in the range of the nominal speed over the engine service life will be substantially lower for a high-performance sports car than for a minivan with an entry-level engine.

Furthermore, owing to design boundary conditions, such as the location of the combustion bowl and the valve pockets, conrod design, etc., and the maximum gas pressure, different stresses can arise. The safety factors can therefore vary at the same temperature. The maximum permissible temperatures are therefore, of necessity, dependent on the application.

The values shown in **Table 7.2** are not a recommendation for threshold values, but are simply orientation points for maximum measured values in current, highly stressed engines.

Table 7.2: Typical values for maximum temperatures on the piston for various applications

Application	Commercial vehicle		Passenger car		
			Diesel engine		Gasoline engine
Material	Al	Steel	Al	Steel	Al
Bowl rim	340 °C	470 °C	380 °C	500 °C	290 °C*
Top ring groove	260 °C	260 °C	300 °C	280 °C	270 °C
Piston pin boss	190 °C	180 °C	235 °C	310 °C	240 °C

* = on piston crown

7.2.2.2 Influence of various operating parameters on piston temperature

When measuring temperatures with templugs, because of the limitation to one operating point that is inherent to the principle, the rated power point under standard operating conditions is typically selected. The highest temperatures are also expected here in nearly all cases.

For methods that allow several operating points to be measured, the full-load curve under standard operating conditions is first measured as a baseline. The result provides the maximum temperatures at the various measurement points, which also flow into the numerical calculation as input variables. The temperature differentials between individual cylinders and the engine speed at which the maximum temperature occurs can also be determined.

If needed, the influence of various engine operating conditions of interest on the piston temperature, while keeping the other parameters constant, can be determined as part of a measurement program.

Table 7.3 shows the average influence of the most important operating parameters on the temperature in the top ring groove of a diesel engine piston.

The influence on piston temperatures of the knock control characteristic in gasoline engines and that of pre-injection in diesel engines is also of interest, for example. Such a typical measurement program is often supplemented with a measurement under "worst case" conditions. In this case the coolant, engine oil, intake and charge air temperatures are all set to maximum values, such as would occur in a hot country test, for example.

Table 7.3: Influence of various operating parameters on the temperature in the top ring groove of a diesel engine piston

Operating parameters	Change in engine conditions	Change in temperature in the top ring groove
Coolant temperature	10°C	4–6°C
Water cooling	50% antifreeze	5°C
Lubricating oil temperature (without piston cooling)	10°C	1–2°C
Charge air temperature	10°C	1.5–3°C
Piston cooling with oil	Spray nozzle at conrod big end	−8°C to −15°C on one side
	Fixed nozzle	−10°C to −30°C
	Salt core cooling gallery	−25°C to −50°C
	Cooled ring carrier	−50°C (Additional reductions in temperature, relative to the salt core cooling gallery)
Cooling oil temperature	10°C	4–7°C
Mean effective pressure MEP (n = constant)	1 bar	4–8°C
Engine speed (MEP = constant)	100 rpm	2–4°C
Ignition point, port closing	1°CA	1.5–3°C
Air ratio λ	Range of variation $\lambda = 0.8-1.0$	< 10°C

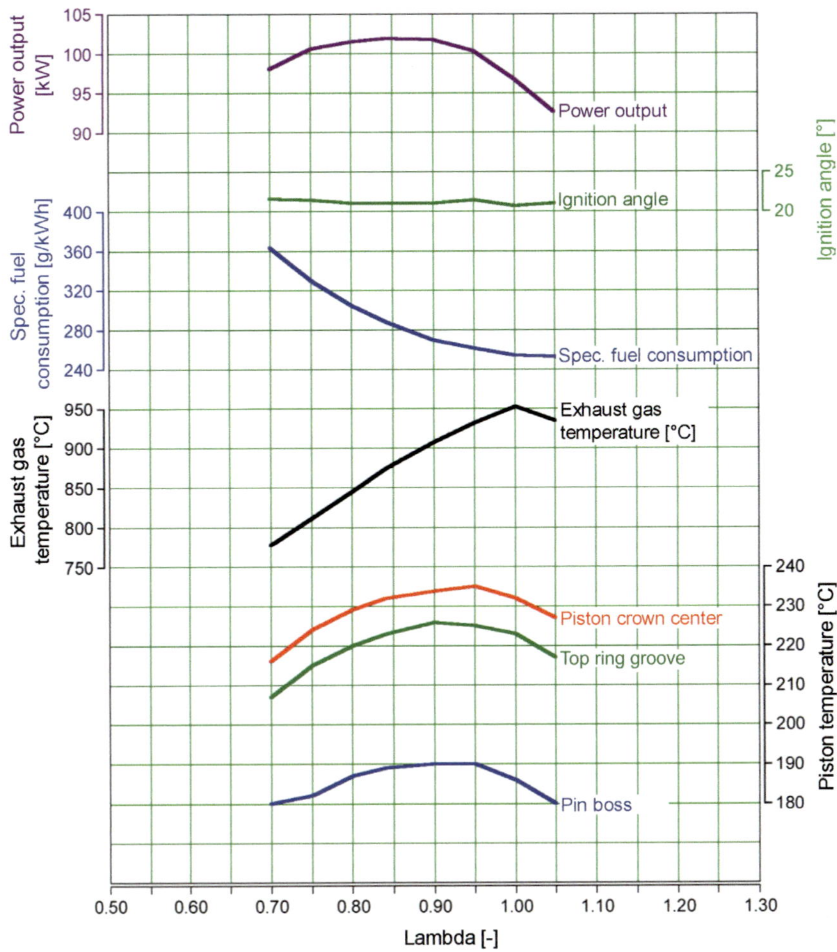

Figure 7.23: Relationship between engine operating values for the air ratio λ using a gasoline engine

The result of a variation of the air ratio λ for a passenger car gasoline engine is shown as an example in **Figure 7.23**. It is evident that the maximum piston temperatures occur at approximately the same air ratio λ as the maximum power output.

The lowest specific fuel consumption is seen at approximately λ = 1. However, the highest exhaust gas temperatures occur at this point. Owing to the maximum permissible exhaust gas temperatures at the inlet of the exhaust gas aftertreatment system (such as a catalytic converter), or at the turbocharger inlet in a turbocharged engine, this "lean" range may be limited.

The lowest piston temperatures, and simultaneously the lowest exhaust gas temperatures, occur in the "rich" range, such as at λ = 0.8 or less. The specific fuel consumption, however, increases drastically at this point.

For engine calibration, the best possible compromise should be made while considering the various boundary conditions. Varying other parameters, such as the ignition angle or start of injection, can also lead to such conflicting goals as a result of the different requirements for the operating values.

7.2.2.3 Influence of cooling oil quantity on the piston temperature

For engines equipped with piston cooling, the optimal amount of cooling oil is often a topic of discussion. The engine manufacturers always tend toward the lower values, with the objective of reducing the required oil pump power and oil foaming. From the piston manufacturer's point of view, on the other hand, a large amount of cooling oil is desirable for optimal piston cooling. This must be ensured even for engines with a long service life and low oil pressure associated with advanced bearing wear. As is shown as an example in **Figure 7.24**, the specific cooling oil quantity should not fall below 3 kg/kWh.

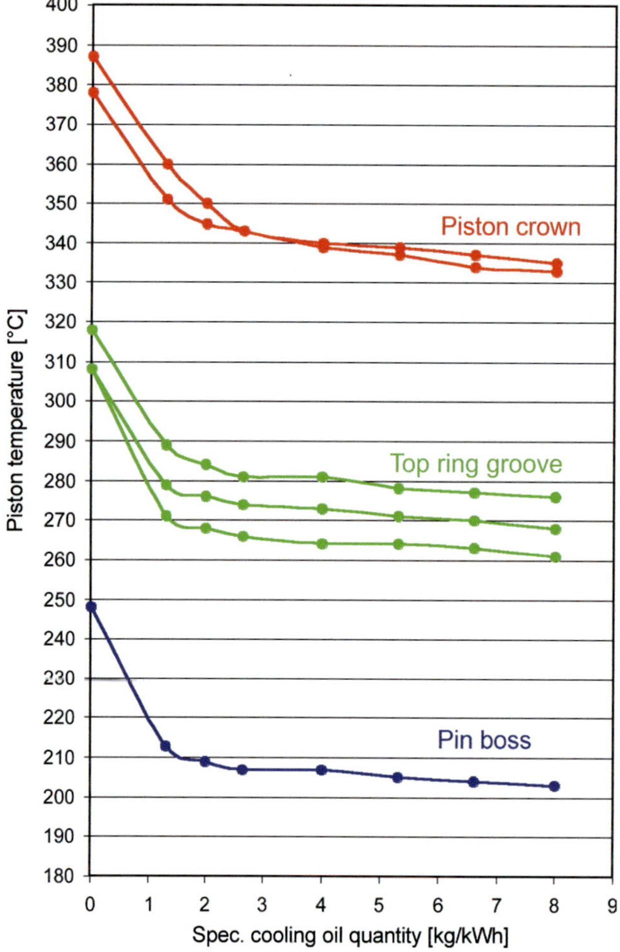

Figure 7.24:
Influence of specific cooling oil quantity on piston temperature (six measurement points) in a passenger car piston with cooling gallery

Table 7.4: Recommended specific cooling oil quantities

	Spray jet cooling (passenger cars)	Cooling gallery piston (aluminum, passenger cars and commercial vehicles)	Cooling gallery piston (steel, commercial vehicles)
Recommended specific cooling oil quantity	3 kg/kWh	5 kg/kWh	5–7 kg/kWh

The optimum point can be determined using appropriate piston temperature measurements and variable quantities of cooling oil.

The ratio of the cooling oil mass flow to the effective power output is known as the specific cooling oil quantity. The recommendation from **Table 7.4** for specific cooling oil quantities is based on the operating oil pressure at nominal speed and the engine oil temperature specified for the rated power point.

In addition, the cooling oil nozzles must meet specific requirements with regard to jet formation and the function of the pressure retention valve. It is therefore important to calibrate the cooling oil nozzles, including the pressure retention valves, on a suitable test bench prior to measurement of piston temperatures.

The opening and closing pressures of the cooling oil nozzles are controlled by pressure retention valves that are either central (in the main oil gallery) or inside the nozzle, and must be tuned to the boundary conditions of the engine. Using this type of valve, the minimum oil pressure needed for the bearings is ensured at low rpm, such as at idle speed.

Extensive piston temperature measurements are needed to optimize systems with demand-controlled piston cooling. The cooling oil nozzles are then supplied with cooling oil only in those engine operating map areas in which there is a verifiable need for a reduction in piston temperature. Reducing the oil pump power, and therefore fuel consumption, is the advantageous associated effect.

7.2.2.4 Piston temperature measurement in transient running programs

Figure 7.25 shows a cross section of a commercial vehicle aluminium diesel piston. Pistons of this type were used in an engine that was the subject of a so-called thermoshock run.

The engine is first run under overload conditions–increased injection quantity, higher peak cylinder pressure, increased coolant temperature. In a thermoshock run, a sudden change is then made from full load to no load, while switching from a hot coolant circuit to a cold coolant circuit.

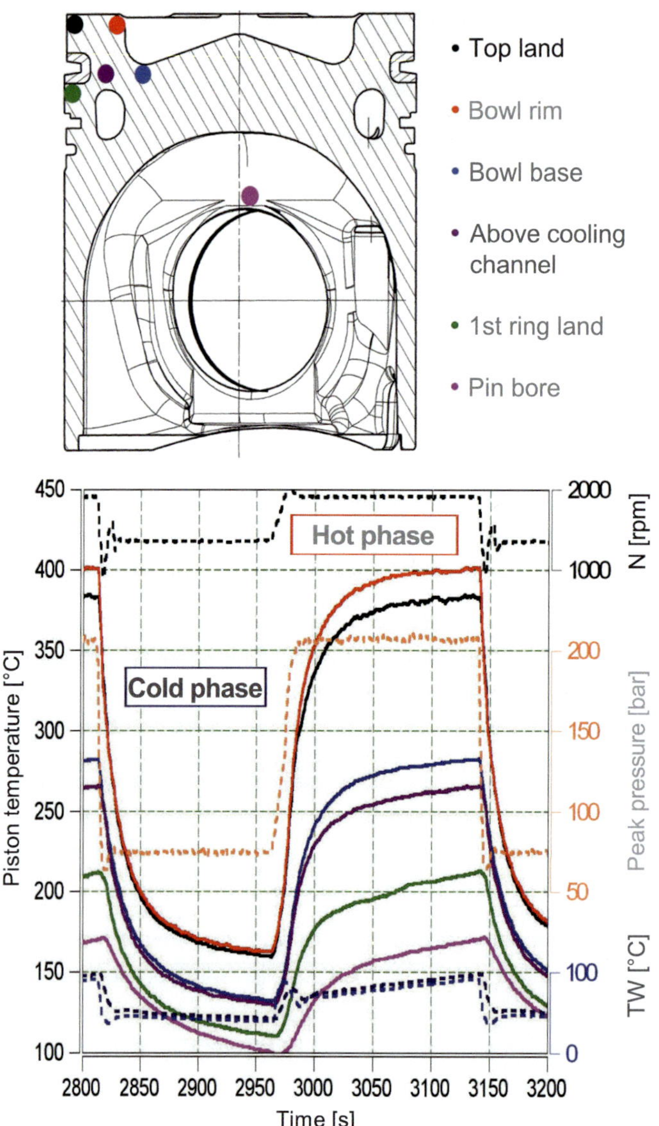

Figure 7.25: Section view of an aluminum commercial vehicle diesel piston with six temperature measurement points, and associated results of measurements made during a transient thermoshock cycle. Piston temperatures were measured using the RTM system.

The diagram shows the time trace of the temperatures at prescribed measurement points during the thermoshock cycle, using the piston mentioned above as an example. One of the results that can be seen here is that the bowl rim sees the highest temperature cycle stress, with a temperature amplitude of about 240°C. Together with the temperature gradients that occur, stresses on the piston can be calculated and used to determine service life.

7.3   Measurement of friction power losses on a fired engine

Continuously rising crude oil prices and the declared goal of reducing environmental impact by minimizing emissions of toxic exhaust gas components as well as carbon dioxide have led to the demand for drastically reducing fuel consumption in combustion engines.

In addition to downsizing with turbocharging, the development of new combustion processes, and dethrottling of the intake system, the reduction of mechanical losses is an important approach for reducing fuel consumption. It is important that the association between mechanical losses and various components is understood. Loss distributions have long been discussed in literature, but are mostly based on motored engines.

Figure 7.26 shows an example of the distribution of mechanical losses in gasoline engines, as determined back in 1984 by motored engine testing.

The proportions of mechanical losses shown on the right in **Figure 7.26**, obtained from motored engine tests, are small relative to the thermal losses in a fired engine, on the left in the figure. It can be assumed, however, that the actual distribution of mechanical losses changes significantly in the fired engine. The piston group, in particular, takes on a greater share of the mechanical losses with increasing engine load. It therefore makes complete sense to determine and optimize the friction power loss under load of this assembly, known as the power cell unit (PCU). The PCU includes the following components:

- Piston
- Piston rings
- Piston pin
- Piston pin circlips
- Connecting rod with bearings
- Cylinder liner

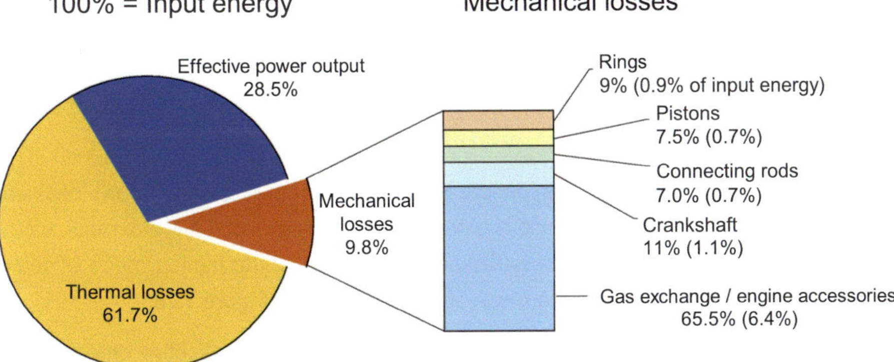

Figure 7.26: Loss distribution (averages of three different 2.0-liter gasoline engines at 5,000 rpm, motored operation)

The prerequisites for sensible optimization are very precise and reproducible measurements of the friction losses under real engine operation, using a suitable method of measurement and analysis.

7.3.1 Measurement methods for determining friction losses

A few measurement methods are described below that serve to determine the friction losses, but which vary greatly in precision and in the technical effort required. **Figure 7.27** shows a series of methods for determining friction in individual frictional pairings and in complete engines. These range from laboratory testing (e.g., tribometer testing) for individual engine components, to processes that use a motored engine, to several processes based on different principles of measuring on a fired engine. Depending on the parameter being investigated, the most suitable measurement method must be selected.

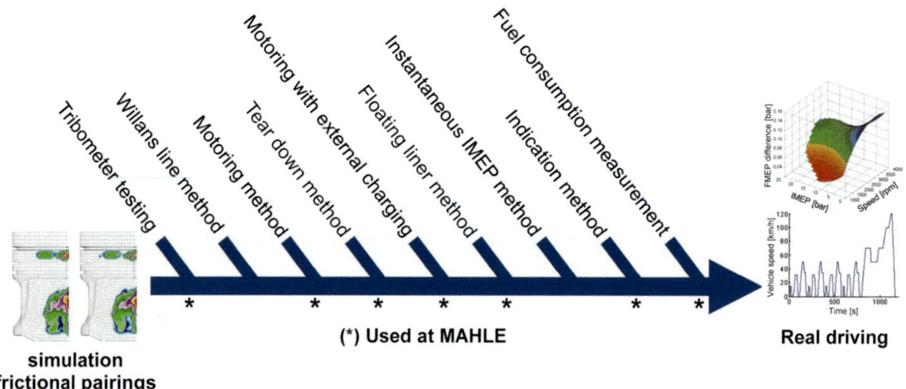

Figure 7.27: Measurement methods for determining friction in individual frictional pairings and complete engines

7.3.1.1 Willans line method

This method is based on a measurement of fuel consumption per operating cycle, at a constant engine speed, with varying brake mean effective pressure.

In order to determine the friction mean effective pressure, a tangent is drawn from the linear portion of the consumption curve, which is extrapolated to the axis of the brake mean effective pressure (x-axis). The intersection point yields the value of the friction mean effective pressure of the engine (negative brake mean effective pressure); **Figure 7.28**.

This graphical method is a simple way to estimate the friction mean effective pressure of a complete engine. On account of the extrapolation, however, only an approximate trend forecast can be made for comparing different engine types.

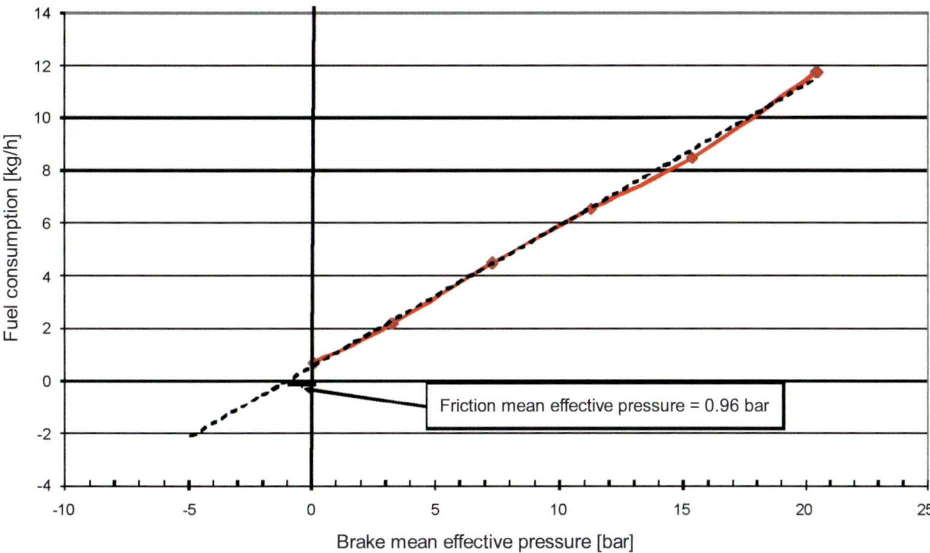

Figure 7.28: Willans line–4-cylinder diesel engine at 1,500 rpm

7.3.1.2 Motoring and tear down methods

When motored, the combustion engine is driven by an electric motor and the ignition and fuel supply are switched off to prevent combustion. The temperatures of coolant and oil are maintained at operating temperature by means of external conditioning units. The friction mean effective pressure of the engine, including gas exchange losses, can be derived from the driving power consumed by the electric motor.

One special form of motoring is known as the tear down method. The engine is motored and disassembled, or stripped, step by step. In this manner, friction losses can be determined for each individual engine component.

General disadvantages of motoring, compared with measurements on a fired engine, include the following:

- The loads on the affected components are very low because of the absence of gas force.
- The operating and component temperatures are lower, and the operating clearance conditions are significantly different.
- Gas exchange losses are assigned to friction.
- The high-pressure loop of the four-stroke engine does not do any work. Because of leakage and wall heat transfers, the indicated mean effective pressure has a negative sign, which is interpreted as friction.

7.3.1.3 Cylinder deactivation

In order to determine the friction mean effective pressure, the power output from the engine is measured, and then one cylinder of a multicylinder engine is switched off by interrupting the fuel supply. The power that is then output by the engine is measured as well. By comparing the two power output levels, the friction mean effective pressure of the engine can be determined.

The disadvantages described above for motoring also apply to cylinder deactivation.

7.3.1.4 Coast down test

In order to calculate the friction mean effective pressure using this method, the decrease in speed level (change in speed as a function of time) is measured, which yields the internal friction in an unbraked, coasting engine. If the mass moment of inertia of the engine is known, then the drop in speed level can be used to calculate the loss of engine torque, from which the friction mean effective pressure can be determined.

7.3.1.5 Floating liner method

In the floating liner method, the friction forces caused by the piston group are measured directly. The level of design and technical effort required is very high, which is why this method is used only on single-cylinder units. The cylinder liner is supported such that it floats with very low friction, and the lower end is supported on load sensors. This makes it possible to represent the friction force curve in the stroke direction over one operating cycle. **Figure 7.29** describes the design principle of a floating liner system.

The results, tracked against the crank angle, are very helpful for validating numerical calculations and for interpreting mechanisms of action. The friction mean effective pressure can then be derived from the measured friction force curve.

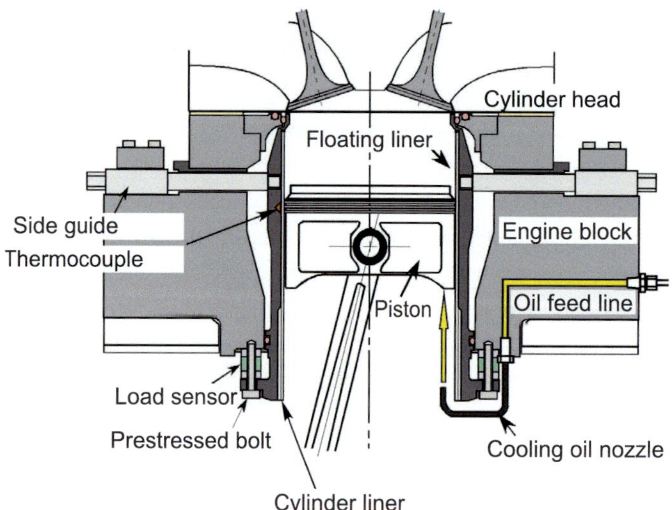

Figure 7.29:
Floating liner system [10]

A disadvantage of this measurement method is that the maximum achievable engine speeds and loads are severely constrained by the design. In addition, owing to the floating support of the cylinder liner, the cylinder distortions that arise are not representative of those for actual complete engines.

7.3.1.6 Indication method

The indicated mean effective pressure (IMEP) in the combustion chamber is determined using pressure sensors. Measurement of the output torque at the flywheel can be used to derive the brake mean effective pressure (BMEP). The friction mean effective pressure (FMEP) is the difference between these two values:

$$FMEP = IMEP - BMEP$$

The principle behind the derivation of the friction mean effective pressure is seemingly quite simple. The greatest challenge, however, is the very high measurement accuracy and reproducibility required for IMEP and BMEP, because they are very close in value and are subtracted to get the result.

Unlike the other measurement methods, this principle enables friction mean effective pressure to be measured during real engine operation. The ranges of speed, load, and temperature are not limited. Operating temperatures, and thus the local operating clearances and distortions, correspond to the actual conditions.

7.3.2 Friction mapping using the indication method

7.3.2.1 Profile of requirements

The friction power test bench is a development tool that can be used throughout the entire engine operating map to examine parameters for the following engine components and frictional pairings:

- Crank mechanism and bearing
  - Piston and cylinder bore
  - Rings and cylinder bore
  - Rings and piston grooves
  - Piston pin bearing in piston and connecting rod
  - Big end bore and crankshaft
  - Main crankshaft bearing
- Valve train
  - Camshaft and bearing
  - Cams and rocker arms

- Engine auxiliaries
  - Coolant pump
  - Oil pump
  - Generator
  - Vacuum pump
- Timing chains (belts) and drive chains (belts)
  - Crankshaft/camshaft connection
  - Connection to engine auxiliaries
- Influence of operating media and temperatures

The results of measurements of the power cell unit based on the measurement method described here are discussed as an example in Chapter 7.3.3. The main focus of attention is the crank mechanism and its bearing, and here in particular the piston friction, piston ring friction, and piston pin friction.

Parameters such as installation clearance, piston pin offset, piston profile, as well as skirt roughness and coating are of particular interest for piston friction.

In relation to the ring pack, the parameters of ring width and tangential force, as well as coating and geometry of the running surface of the rings, are of great interest.

For friction in the piston pin bearing, parameters such as coating and installation clearance are critical.

Parameter investigations should be feasible on all engine types and concepts.

The engines should be able to be measured under the following conditions, whether motored or fired:
- Fired operation is possible up to a maximum rated power of 200 kW.
- Motored operation should also be possible with external loading (max. combustion chamber pressure can be set continuously, up to over 200 bar).
- Basic prerequisites for obtaining reproducible results are precise conditioning and adjustability of the engine operating media.

The operating values for the engine, including the cylinder pressure curves, must be extensively documented in order to have sufficient input data for the numerical calculation.

7.3.2.2 Friction power test bench for passenger car engines

Figure 7.30 shows the schematic construction of the friction power test bench. It shows the engine being operated without auxiliaries. A detailed description of the test bench can be found in [11].

The oil and coolant pumps are replaced by external conditioning systems. An externally driven high-pressure fuel pump is used to generate pressure in the fuel rail.

The use of external systems ensures that reproducible operating conditions can be set up. Using the indication method, the friction losses in the engine auxiliaries would also be

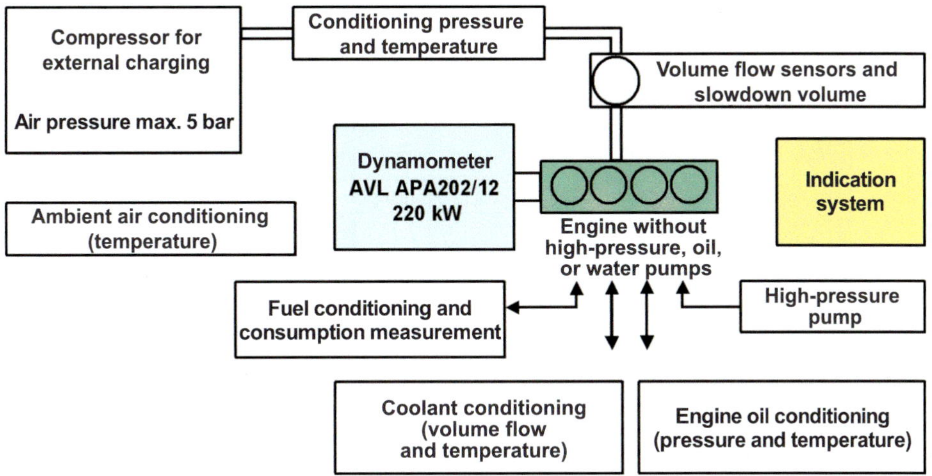

Figure 7.30: Schematic construction of the friction power test bench

included in the friction mean effective pressure calculation for the complete engine. Any fluctuations within these engine auxiliaries would therefore cause significant errors in the derivation of the friction mean effective pressure. The use of external conditioning systems and drives can minimize this error potential.

An indication system with water-cooled pressure sensors is used to determine IMEP with a high level of precision. Precise association of the cylinder pressure to the crank angle over time is critically important. The crank angle sensor used must therefore have the highest angular accuracy possible. The top dead center is determined dynamically with a capacitative sensor in the cylinder head.

A high-precision dynamometer is used to determine BMEP, which is calculated from the torque. Precise observance and reproducibility of the boundary conditions are indispensable for accurately determining the friction losses. In addition to the engine operating media of coolant, oil, and fuel, the ambient air temperature must also be precisely and reproducibly controlled. One external conditioning unit each is available for conditioning the engine oil and coolant. They replace the pumps and controllers on the engine; **Figure 7.31**. The fuel is also externally conditioned.

In motored operation, the engine can be run by means of an external charging system with a defined cylinder pressure (load). The pressure of the intake air in the intake manifold can be increased accordingly. The greater charge density in the combustion chamber that results after "close intake" can reach peak cylinder pressures of over 200 bar because of compression in so-called "motored externally charged" mode.

Figure 7.32 shows the setup of the test bench, with test engine and dynamometer.

The key data for the friction power test bench are compiled in **Table 7.5**.

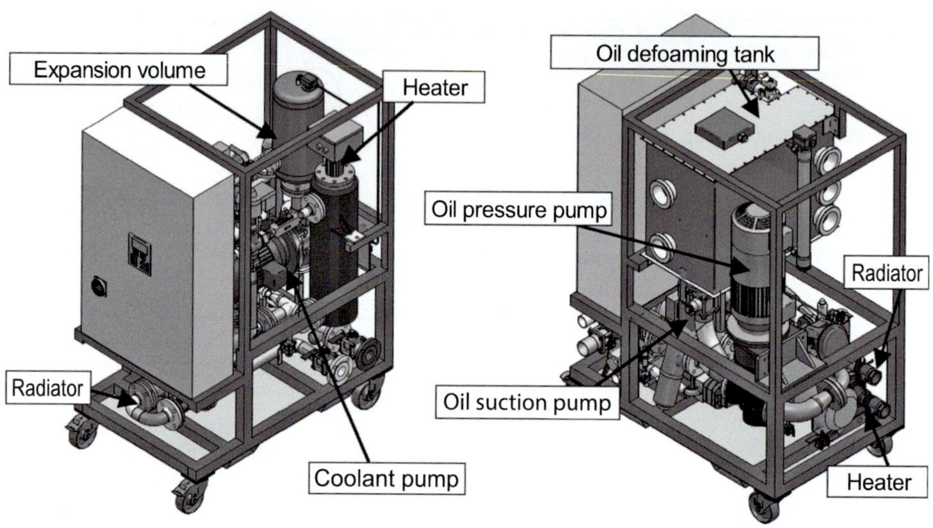

Figure 7.31: Conditioning systems for coolant circuit (left) and oil circuit (right)

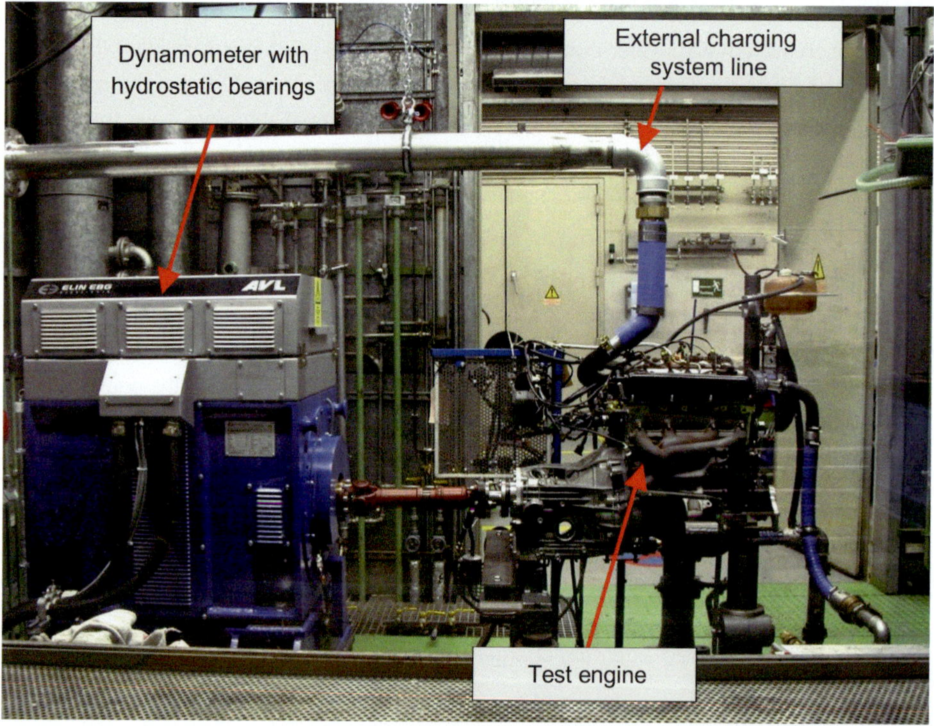

Figure 7.32: Test engine with external charging system connected

Table 7.5: Technical data of the friction power test bench

Dynamometer	
Engine speed	Max. 12,000 rpm
Torque	Max. 525 Nm ± 0.3 Nm
Coolant conditioning system	
Temperature setting	40 ... 120°C ± 1 K
Flow rate	50–200 l/min ± 5 L/min
Engine oil conditioning system	
Temperature setting	40 ... 120°C ± 1 K
Oil pressure	1 ... 10 bar ± 0.05 bar
Allows engine to be operated with dry sump lubrication	
External charging system	
Air flow rate	Max. 25 m^3/min
Air pressure	Max. 5 bar
Air temperature	25°C ± 1 K
Fuel conditioning and consumption measurement	
Temperature stability	± 0.1 K
Flow rate	Max. 125 kg/h ± 0.15 kg/h
Ambient air temperature controller	
Air temperature	25°C ± 2 K

7.3.2.3 Measurement and analysis method

In order to determine the influence of PCU design parameters on the friction mean effective pressure, different variants are analyzed. The measurement program includes measurement of the friction mean effective pressure during the run-in phase, over full-load curves, and for the engine operating map for each variant. Engine operating map measurements are made both in fired mode and in motored externally charged mode.

The operating maps, for which the influence of the parameters of load, speed, and engine temperature on the friction mean effective pressure of a combustion engine is to be determined, are created using a statistical test planning method, "design of experiments" (DoE). The core of this approach is the definition of a multidimensional experiment space and the optimal distribution of the measurement points in this space.

The advantages of DoE are that the results can be displayed as a function of all the selected parameters, and the test time required can be significantly reduced.

Figure 7.33 shows the variation parameters and desired response variables selected for the DoE analysis, as well as the set boundary conditions and the dependencies between the individual parameters.

Response parameter: **Friction mean effective pressure (FMEP)**

Variation parameters:

Quadratic factor

1. **Load**
   Indicated mean effective pressure IMEP / peak cylinder pressure
   No load–full load/40–180 bar

   Interaction included

2. **Engine speed**
   1,000–4,000 rpm

   Interaction not included

3. **Engine temperature**
   Oil and coolant temperature (50–100°C)

Interaction included

Figure 7.33: Variation parameters and response variables for compiling a DoE test plan

Figure 7.34 shows two friction mean effective pressure maps at different engine temperatures, 50°C and 100°C, which were compiled using a mathematical approximation from the DoE program. The friction mean effective pressure is shown as a function of the engine speed and load (indicated mean effective pressure). In this case, engine temperature means the same value for coolant temperature and oil temperature.

A comparison of the two operating maps shows that engine friction depends on the engine temperature. The higher the operating fluid temperatures, the lower the friction mean effective pressure. It can also be seen that the friction mean effective pressure increases with rising rpm, and that the friction mean effective pressure is significantly dependent on the load.

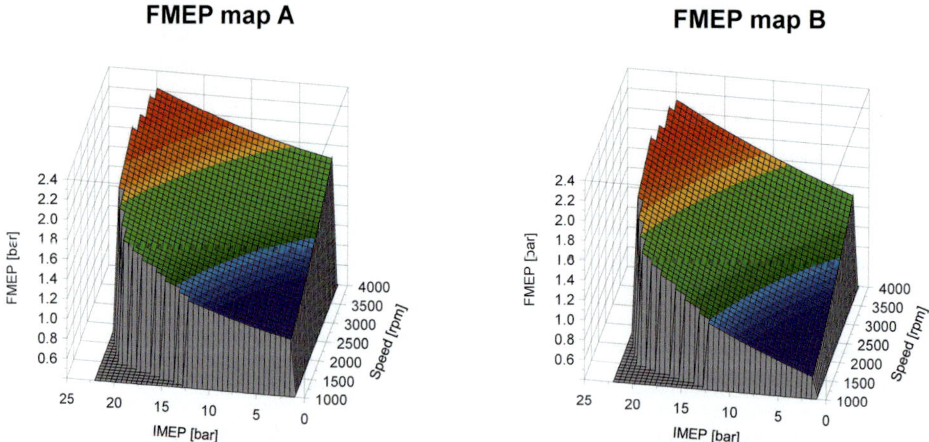

Figure 7.34: FMEP maps at different engine temperatures; Operating map A: water and oil temperature 50°C, operating map B: water and oil temperature 100°C

FMEP difference map A (50 °C) – B (100 °C)

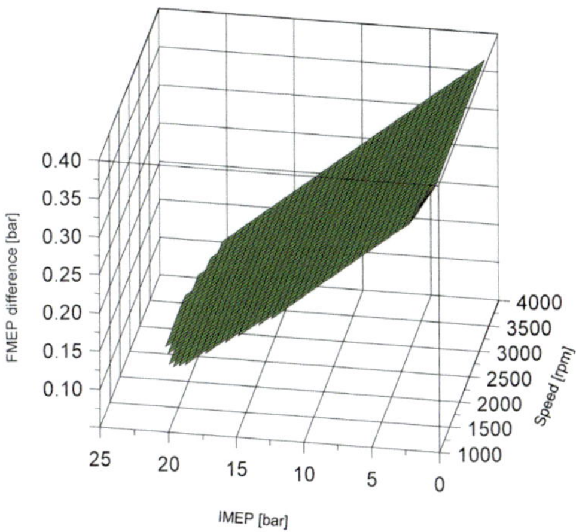

Figure 7.35:
Friction mean effective pressure difference map, calculated from the friction mean effective pressure operating maps of operating maps A and B. The differential can be attributed to a difference in engine temperature of 50°C.

A detailed interpretation by visually comparing the operating maps is difficult, because the differences are very slight. If the difference between the two is considered, however, even small differences can be displayed clearly. A positive differential represents the savings potential due to the higher temperature; **Figure 7.35**.

A comparison of the different engine temperatures with the aid of the friction mean effective pressure difference map, **Figure 7.35**, shows no significant dependence on speed, but a strong dependence on load.

The interpretation of such friction mean effective pressure difference maps allows fundamental statements about the influences of the parameter changes made on the characteristic curve of the friction mean effective pressure in the operating map.

Further information on the measurement and analysis method used, including error analysis, can be found in [11].

7.3.3 Selected results from tests on a passenger car diesel engine

The following results were obtained from a turbocharged 4-cylinder 2.0 L passenger car diesel engine with a gray cast iron block. The results presented were supplemented and expanded on with [11, 13].

The friction mean effective pressure values obtained are shown for an engine operating temperature of 100°C. The friction mean effective pressure difference map for the motored externally charged mode is compared with that of the fired mode in each case.

7.3.3.1 Piston installation clearance

If two extremely different installation clearances are run under identical external boundary conditions, then the result is the friction mean effective pressure difference maps shown below; **Figure 7.36**.

The variant with the enlarged installation clearance, as expected, shows a lower friction mean effective pressure in both operating modes. This is evident in the positive friction mean effective pressure difference.

A significant dependence on speed can be seen in the difference map for the motored externally charged mode, but only a very slight dependence on load is detected.

By contrast, a very significant dependence on load is present in the fired mode. With increasing load, i.e., increasing indicated mean effective pressure, increasingly great advantages are seen for the variant with enlarged installation clearance.

This difference in the characteristic, especially for increasing load, can be attributed to the fact that the motored externally charged mode has excluded the component temperature from the net influence of the gas force on friction, because the piston temperature increases only slightly at peak pressure; **Figure 7.37**. In fired mode, in contrast, this effect due to the gas force also has a thermal effect superimposed on it. This is evident in **Figure 7.37** in the significantly greater increase in the piston temperature with increasing load. Therefore, the pistons have different operating clearances at the same peak cylinder pressure in the two different engine operating modes. For the piston with the smaller installation clearance, under increasing load in fired operation, the operating clearance is used up sooner than that of the piston with the enlarged installation clearance. This leads to the severe increase in friction mean effective pressure difference in fired operation; **Figure 7.36, right**. In the motored externally charged operating mode, this increase does not occur, owing to the lesser increase in component temperature; **Figure 7.36, left**.

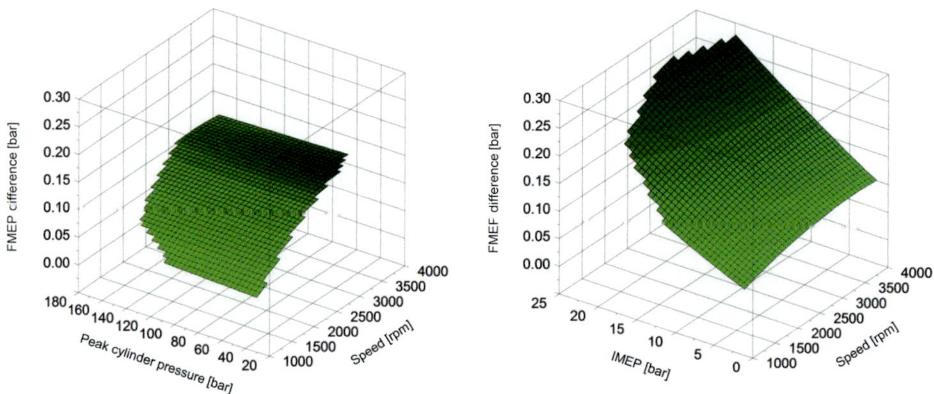

Figure 7.36: Friction mean effective pressure difference maps, different piston installation clearance (17 μm compared with 101 μm), motored externally charged (left) and fired mode (right), engine temperature 100°C

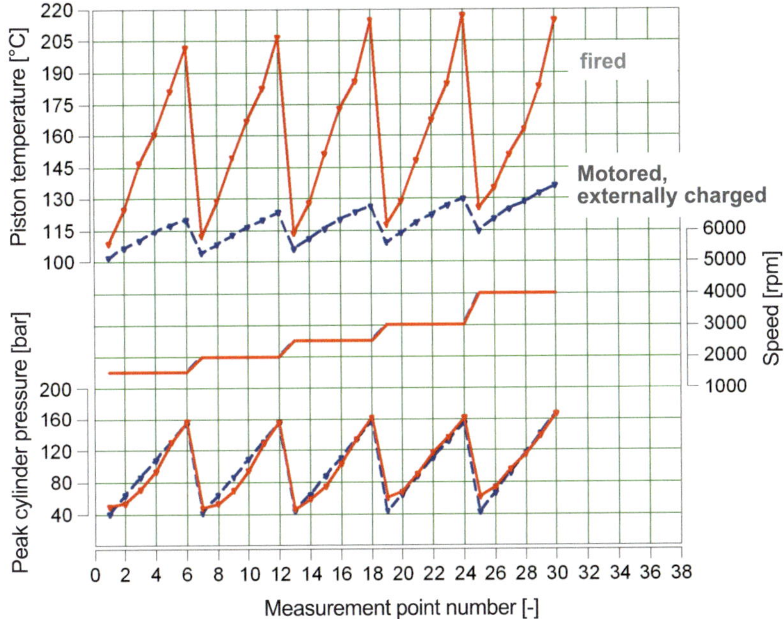

Figure 7.37: Measured piston temperatures (piston pin boss) for motored externally charged and fired engine operation, with equal peak cylinder pressures

For the reasons just listed, a motored externally charged friction mean effective pressure difference map is of only limited use in engine optimization for actual driving operation. However, because it can be compared with fired operation, it contributes significantly to the understanding of friction mechanisms, especially for detecting thermal influences. Difference maps for fired operation can be used unconditionally for determining the potential of a change to a parameter in real driving operation.

7.3.3.2 Surface roughness of the piston skirt

In order to investigate the influence of surface roughness, pistons without skirt coatings were used. This allowed precise adjustment of defined surface parameters. For the piston with a smooth skirt surface, an average roughness of Ra = 0.2 µm was selected, and for the piston with a rough skirt surface, it was Ra = 2.1 µm. In order to exclude any potential run-in effects, the values indicated were measured after the run.

An increase in the friction mean effective pressure difference, i.e., an improvement in friction mean effective pressure, occurred under increasing load for the variant with the smooth skirt surface; **Figure 7.38**. The cause for this is that a hydrodynamic lubricating film can form more easily on the piston with the smooth skirt surface. This positive effect is more extensive in fired operation, owing to the tighter operating clearances caused by the different thermal load.

Roughness measurements on the skirt surfaces before and after the test show clear run-in effects of the rough piston. The roughness peaks are severely worn away, which give the

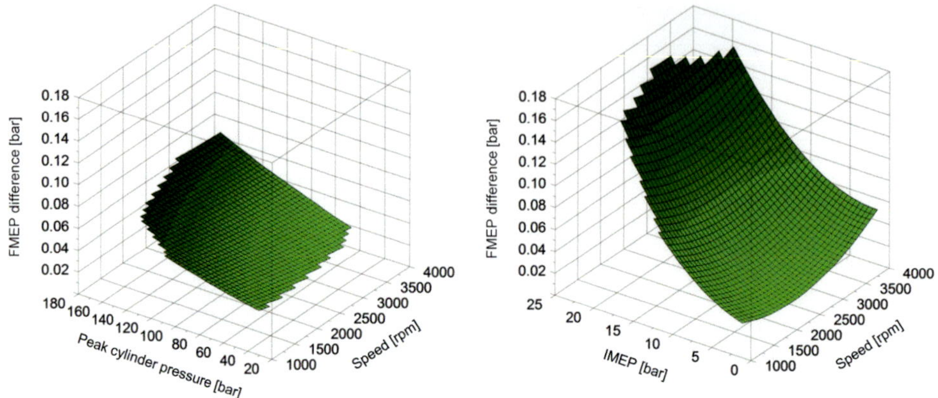

Figure 7.38: Friction mean effective pressure difference maps, different surface roughness of the piston skirt (Ra = 2.1 µm compared with Ra = 0.2 µm), motored externally charged (left) and fired (right), engine temperature 100°C

piston the opportunity to adapt its wear to the distortions in the individual cylinder. For the smooth piston, this effect is definitely lessened, or even undetectable.

For current production pistons with relatively tight installation clearance, the two described effects can be combined by using GRAFAL®. Coating a relatively rough piston skirt with GRAFAL® allows the piston profile to be adapted to the cylinder distortions by wearing off the remaining roughness. At the same time, the supporting area on the skirt smoothes out after just a short run-in period, leading to low friction power losses. In engines with greater clearance design, smoothing coating wear during run-in is undesired. For this reason, a more resistant coating, such as EvoGlide, is applied to the parts whose surface is less rough. This is the case in particular for friction power loss optimized engines with reduced cylinder distortions.

7.3.3.3 Piston pin offset

A piston pin offset can be used to affect the secondary motion of the piston, and therefore the formation of a lubricating film on the piston skirt. For this reason, a parameter study was performed with a stepwise change in the piston pin offset, from a value of 0.5 mm toward the antithrust side (ATS) to 0.5 mm toward the thrust side (TS).

The results of the comparison of piston pin offsets, from 0.5 mm toward the antithrust side and 0.5 mm toward the thrust side, are described below. The difference maps for fired and motored externally charged operation, **Figure 7.39**, yield very different curves. This leads to the conclusion that the piston pin offset is a parameter that is strongly dependent on the component temperature.

In motored externally charged operation, **Figure 7.39, left,** the friction mean effective pressure difference is highly dependent on the peak cylinder pressure at lower speeds. The dependence on load decreases as speed increases. The greatest apparent advantage of offsetting in the direction toward the thrust side is seen at low speed and high load.

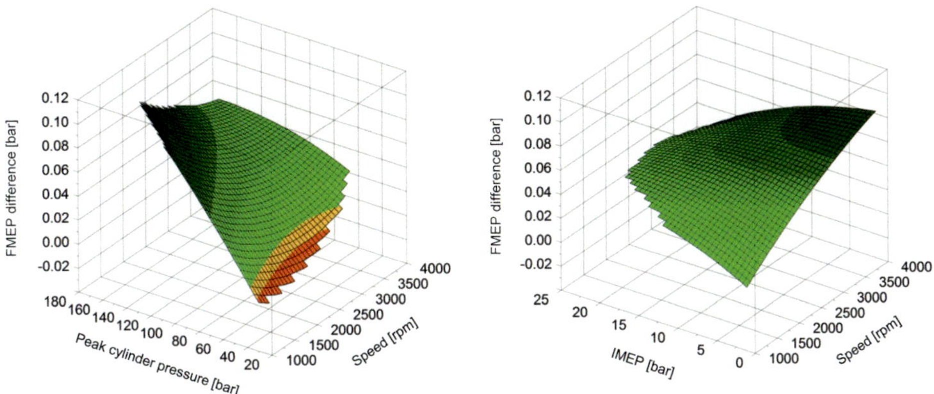

Figure 7.39: Friction mean effective pressure difference maps, different piston pin offset (0.5 mm toward the antithrust side compared with 0.5 mm toward the thrust side), motored externally charged (left) and fired mode (right), engine temperature 100°C

In actual fired operation, however, **Figure 7.39, right**, a clear dependence on speed can be seen for low loading. The dependence on speed drops off clearly as the load increases. The greatest potential of the thrust-side piston pin offset for reducing friction mean effective pressure can be detected at low load and high speed.

Figure 7.40 shows the friction power losses of the four piston pin offset variants that were tested, comparing different speeds. The variant with a piston pin offset of 0.5 mm toward the

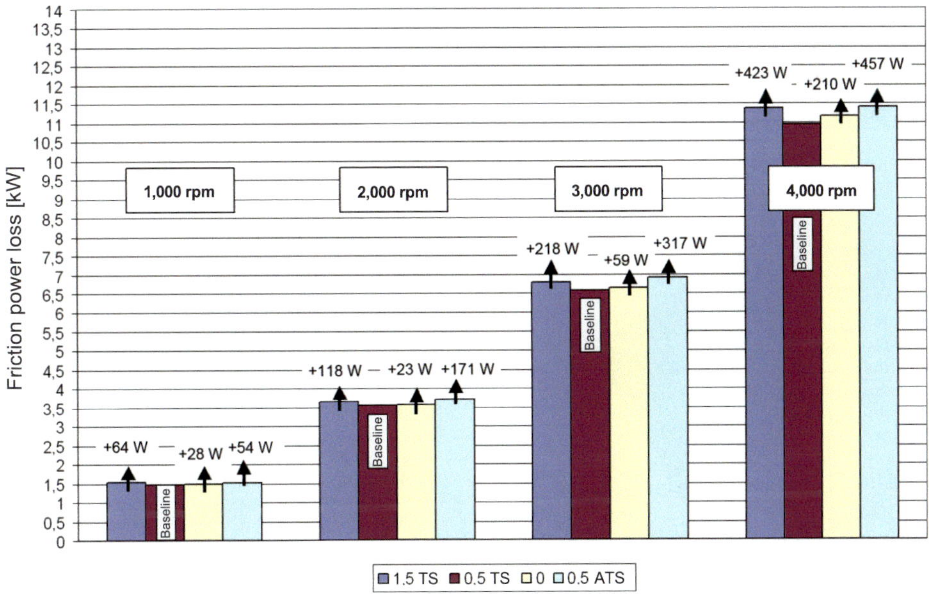

Figure 7.40: Friction power losses for four different piston pin offsets, at different speeds and IMEP = 10 bar, engine temperature 100°C

thrust side always exhibits the least friction power loss, regardless of the speed. A change in the piston pin offset in the direction of the thrust side, or in direction of the antithrust side, causes an increase in friction power loss.

7.3.3.4 Width of the piston ring in groove 1

In order to determine the influence of the width of the compression ring in groove 1 on the friction mean effective pressure, tests were run with piston rings having widths of 1.75 mm and 3.0 mm.

Because of design boundary conditions, the 3.0 mm wide ring also had the tangential force increased by 11 N per ring (44 N for the complete engine). This means that these two parameter changes are superimposed in the friction mean effective pressure difference map.

The friction mean effective pressure difference maps, **Figure 7.41**, show an advantage for the 1.75 mm wide compression ring, both for motored externally charged operation and for fired engine operation, across all speeds and loads.

The increase in savings potential with increasing speed, even at low loads or maximum pressures, can be attributed to the change in tangential force and was also detected in purely motored experiments.

The load-dependent portion of the friction mean effective pressure difference comes from the increased contact pressure in the contact surface between the ring and the cylinder wall. This increase for the 3.0 mm wide compression ring results from the greater inner surface area of the ring at the same gas force load behind the ring, compared with the 1.75 mm wide compression ring, and thus the increasing contact pressure.

Owing to this increase in contact pressure, the mixed lubrication component of the 3.0 mm wide compression ring increases under gas force load. This results in the friction mean effective pressure becoming worse with increasing load.

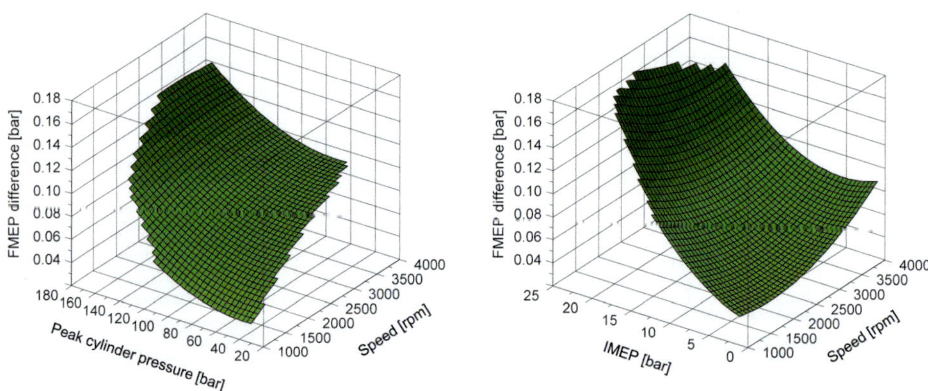

Figure 7.41: Friction mean effective pressure difference maps, different compression ring widths (3.0 mm compared with 1.75 mm), motored externally charged (left) and fired mode (right), engine temperature 100°C

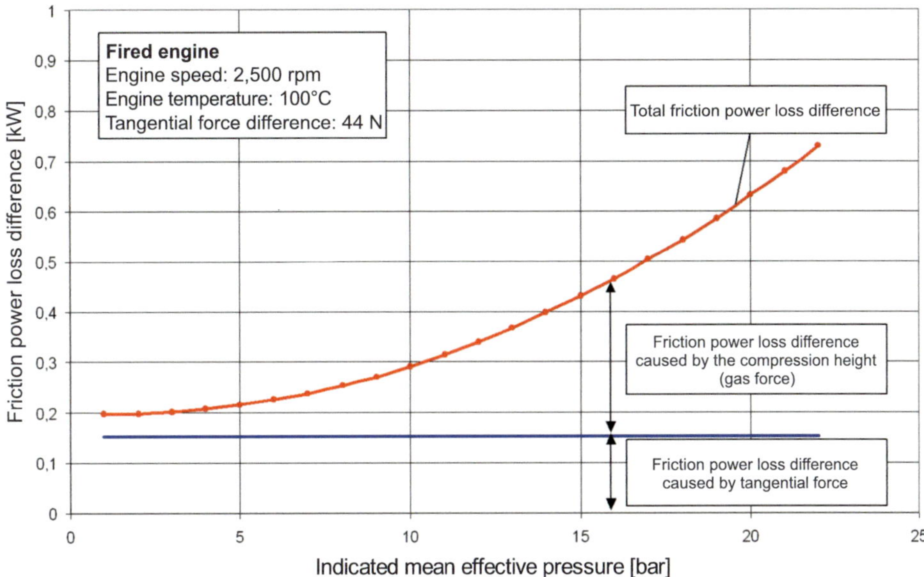

Figure 7.42: Division of the measured friction power loss difference into components for ring width and for tangential force for a compression ring, fired engine operation, engine temperature 100°C

It is notable that in this variant, the friction mean effective pressure difference maps for motored externally charged operation and for fired operation are very similar in terms of values and curves. This indicates a subordinate influence of the component temperatures–in contrast to the changes to the piston.

In **Figure 7.42**, the difference in friction power losses for fired operation at a speed of 2,500 rpm is divided into a portion for the compression ring width and a portion for the tangential force. Measurements with different tangential forces were performed in advance.

7.3.3.5 Tangential force of the oil control ring

Preliminary experiments helped to ensure that if the tangential force is reduced by 18 N per ring (72 N for the complete engine) with the identical ring type, the engine boundary conditions, such as oil consumption or blow-by, are not significantly affected.

An increase in the friction mean effective pressure difference with increasing speed can be observed for both the motored externally charged mode and the fired operation; **Figure 7.43**. A slight increase in the difference with increasing load is evident only for fired operation.

This characteristic leads to the conclusion that the mechanical load, i.e., the peak cylinder pressure, only slightly affects the savings potential in friction mean effective pressure obtained by reducing the tangential force on the oil control ring. Thermal effects, such as the reduction in operating clearance under increasing load, also exhibit only a subordinate influ-

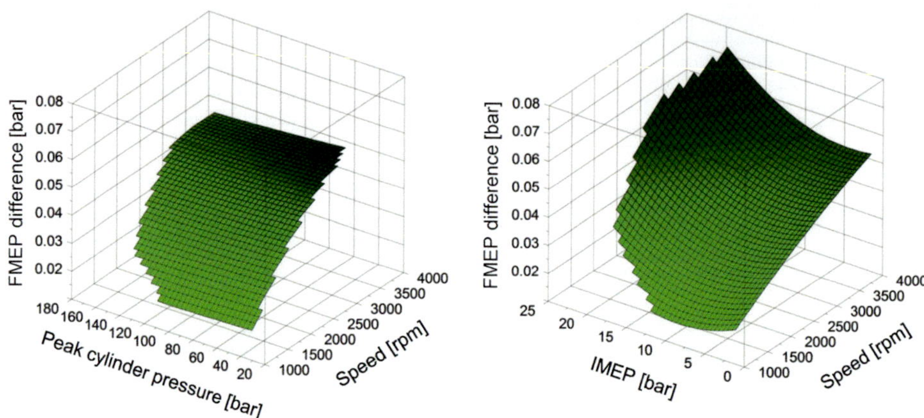

Figure 7.43: Friction mean effective pressure difference maps, reduction in tangential force of the oil control ring by 18 N, motored externally charged operation (left) and fired operation (right), engine temperature 100°C

ence. Results of additional variations of tangential force on the oil control ring, aimed at the conflicting goals of friction power loss and oil consumption, are presented in Chapter 7.9.5.

The comparison of the fired friction power loss difference map with the results of an approximation formula derived from motored measurements confirms both the value and the nature of this characteristic; **Figure 7.44**.

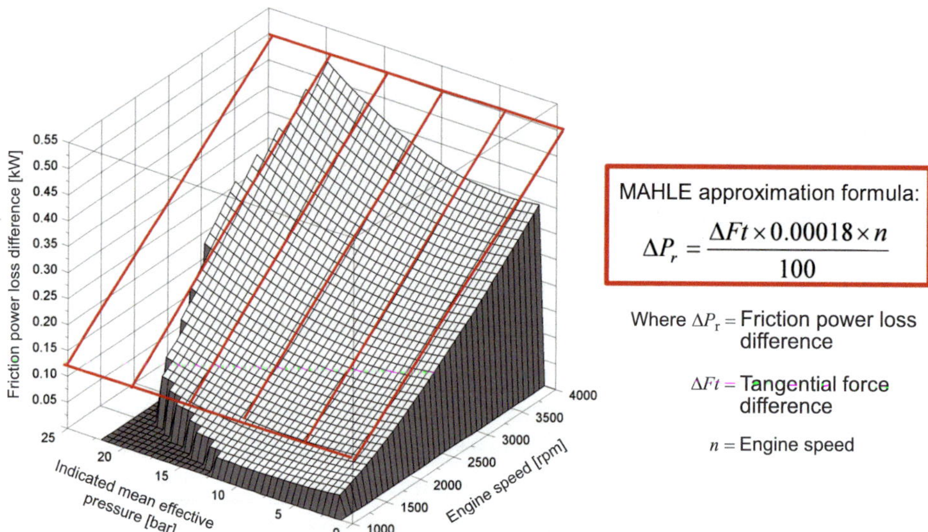

$$\Delta P_r = \frac{\Delta Ft \times 0.00018 \times n}{100}$$

Figure 7.44: Comparison of the friction power loss difference map measured in fired mode (black) to the results of the MAHLE approximation formula (red). The tested tangential force difference for the four oil control rings is 72 N for the complete engine, engine temperature 100°C.

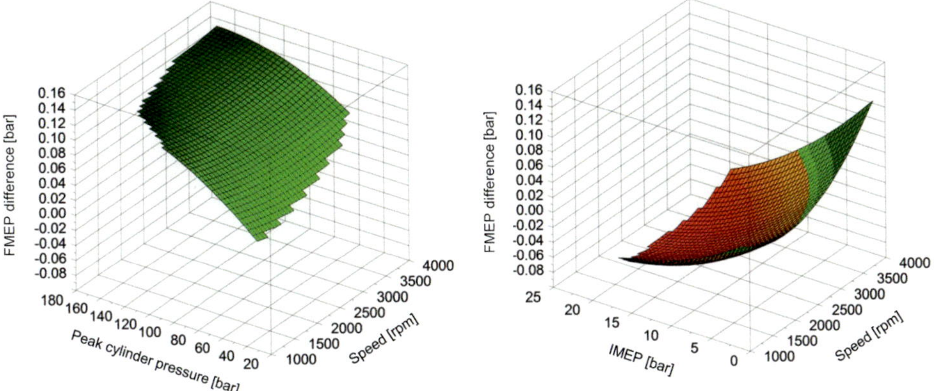

Figure 7.45: Friction mean effective pressure difference maps, various piston pin surfaces (steel compared with DLC coating), motored externally charged (left) and fired (right), engine temperature 100°C

7.3.3.6 Coating of the piston pin

In **Figure 7.45**, it is interesting to note that the response of the friction mean effective pressure differences relative to increased load is quite different between the motored externally charged mode and fired operation. While the largest potential is seen at low loads in fired operation, in motored externally charged operation it is found at high peak pressures, i.e., at high loads.

This characteristic indicates that the varying parameter is strongly dependent on the thermal load of the engine. As is known from the piston temperature measurements, **Figure 7.37**, component temperatures, particularly the piston temperature, increase only slightly with increasing peak pressure in motored externally charged operation. This means that the operating clearance in the piston pin bearing does not change significantly, despite the different thermal expansion coefficients of the piston and pin materials. In fired operation, however, the operating clearance in the boss increases with increasing load. This difference in the changes in operating clearance appears to lead to various mixed lubrication components, while the superior dry coefficient of friction for DLC can have a greater effect in motored externally charged operation under high loads.

7.3.3.7 Engine oil viscosity

In order to determine the influence of oil viscosity on friction losses, experiments were performed using typical commercial oils of the viscosity classes 10W60 and 5W30.

When interpreting the test results of experiments with different oils, care must be taken that the oil type is changed not only around the piston group, but throughout all the relevant friction sources in the engine, such as the main bearings and valve train. The changes in the friction mean effective pressure due to changing the oil type, therefore, must not be assigned to only one frictional pairing.

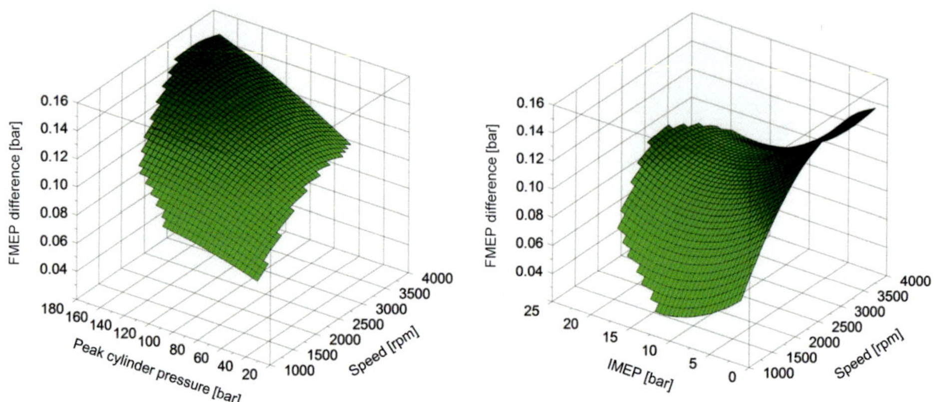

Figure 7.46: Friction mean effective pressure difference maps, different engine oil viscosities (SAE 10W60 compared with SAE 5W30), motored externally charged (left) and fired mode (right), engine temperature 100°C

Low-viscosity oil has a fundamental advantage with regard to friction power loss. This can be seen in that the friction mean effective pressure difference maps, **Figure 7.46**, have a similar frictional advantage for both motored externally charged operation and fired operation. However, this is expressed quite differently in the characteristic in the operating map. In motored externally charged operation, the greatest advantage of the low-viscosity oil is at high speed and high peak pressure, while in fired operation it is at very low loads. These difference maps may reflect the influence of the different component temperatures. In fired operation, the operating clearances between the piston and cylinder get smaller and smaller with increasing load because of the increasing component temperatures, and the load-bearing oil film thickness gets thinner and thinner. The advantage of low-viscosity, thinner oil thus becomes less significant as the load increases; **Figure 7.46**.

These results again show that only experiments under actual engine operating temperatures provide the information that is necessary for optimizing the friction power losses in actual driving operation.

7.3.3.8 Profile of the piston skirt

In the present case, the influence of the piston skirt contour on the friction mean effective pressure was examined. The piston profile design can affect the buildup and propagation of a hydrodynamic lubricating film. Systematically varying the piston profile leads to four basic piston skirt contours (**Figure 7.47**).

Piston profiles can be differentiated as "weak barrel-shaped", "top feed", "bottom feed", and "strong barrel-shaped." With the "top feed" piston profile, the upper end of the piston skirt is considerably reduced in the direction of the piston's central axis when compared with the weak barrel-shaped piston profile. With the "bottom feed" piston profile, the same applies to

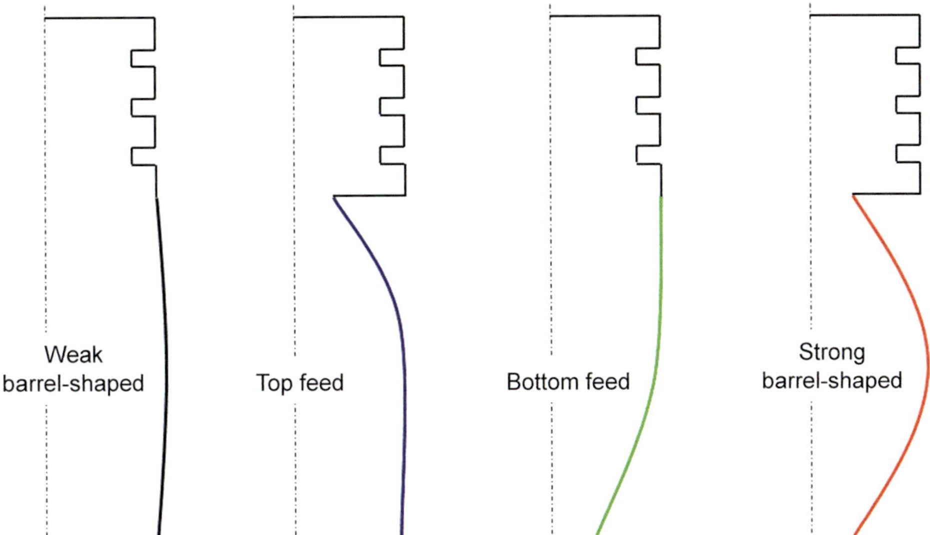

Figure 7.47: Schematic diagram of the examined piston skirt contours

the lower end of the piston skirt. The "strong barrel-shaped" piston skirt can be interpreted as a combination of the "top feed" and "bottom feed" piston profiles.

The "top feed" variant has the greatest friction mean effective pressure advantages across nearly the entire operating map for both types of engine operation. For the weak barrel-shaped piston skirt, however, there are disadvantages in the frictional behavior. **Figure 7.48** shows a direct comparison between the two piston profiles "weak barrel-shaped" and "top feed" with respect to their potential friction savings. The "top feed" variant is reduced by about 40 µm.

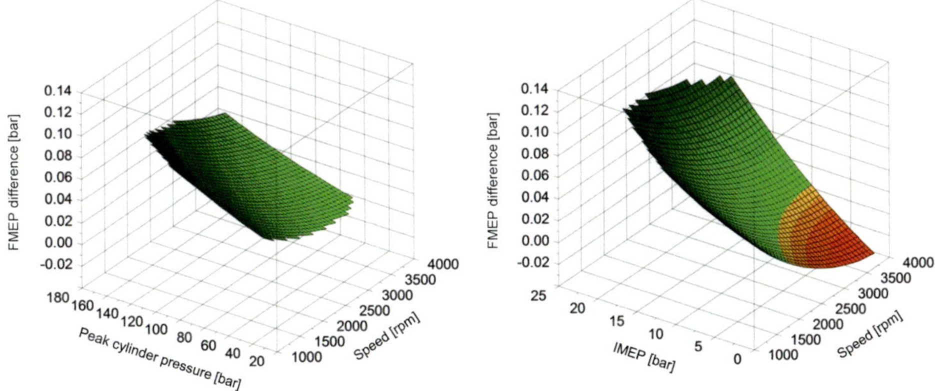

Figure 7.48: Friction mean effective pressure difference maps, different piston profiles ("weak barrel-shaped" in comparison with "top feed"), motored externally charged (left) and fired (right), engine temperature 100°C

In fired engine operation, slight disadvantages can be seen for the piston skirt with top feed, at high speed and low load only. Maximum savings are possible within the high load and low speed ranges.

Because of its proximity to the piston crown, which is directly exposed to the combustion, the upper piston skirt area is the zone of the piston skirt subjected to the highest thermal load. By reducing the upper skirt area ("top feed"), such interference can be clearly minimized, especially under high load. As a result, the formation of a hydrodynamic lubricating film occurs more easily, which consequently reduces friction.

The results from the motored externally charged engine operation confirm that the friction mean effective pressure advantages of the "top feed" variant can be attributed largely to the thermal expansion of the piston; **Figure 7.48, left**. The motored externally charged operating mode represents the savings that can be attributed solely to the influence of the piston profile on the buildup of the lubricating film, isolated from the thermal influence. The other variants of piston profile shown in **Figure 7.47** are described in detail in [11].

7.3.3.9 Coating of the piston skirt

The main purpose of a coating on the piston skirt is to prevent local welding between the piston and the cylinder, or piston seizing. Under extreme engine operating conditions, the risk of a piston seizing is particularly high because of the expected interference (operating clearance less than zero).

Figure 7.49 shows a comparison between an uncoated piston skirt (aluminum) and a piston skirt coated with GRAFAL® in fired engine operation. The installation clearance of the piston with no coating on aluminum and coated with GRAFAL® (measured after the run) was about 100 µm.

There is a clear dependence on load, whereas the dependence on speed is comparatively low. The greatest frictional advantage is shown by the coated piston under high load at high

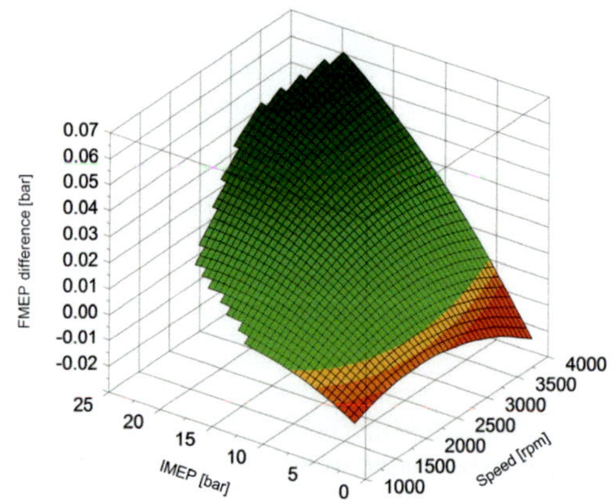

Figure 7.49:
Friction mean effective pressure difference map, different skirt coatings (uncoated in comparison with GRAFAL® coating), fired, engine temperature 100°C

speed. At high load, the heat input to the piston is greatest; therefore, the greatest thermal expansion of the components occurs here as well. If the piston encounters interference, then the mixed lubrication component on the piston skirt presumably increases. In this case, the lower friction coefficient of the coating in comparison with aluminum has a positive effect. The appropriate tests in motored externally charged operation–that is, without heat input–show no advantage after the run for the coated piston with the same roughness, as expected.

The friction differences with and without coating are based on the different series conditions with either rough machined aluminum, or the relatively smooth GRAFAL® coating, which also have clear differences in surface roughness in the run-in condition. If the parameter of skirt roughness is investigated in isolation, then it can be seen that a smooth piston skirt surface leads to substantial frictional advantages under high loads (see Chapter 7.3.3.2). A direct comparison of coated and uncoated piston skirts, with comparably smooth surfaces, indicates only a slight potential for improvement with skirt coating. The appreciable advantage of a coating can mainly be attributed to the change in surface roughness shown in **Figure 7.49**. This result leads to the conclusion that the friction on the piston skirt can be chiefly improved by increasing the hydrodynamic proportion, less so by the coating material and its friction coefficient. Particularly at low piston speeds in the region around the dead centers, however, a coating on the piston skirt can significantly improve wear resistance.

The potential for a skirt coating to reduce piston friction is relatively low, considering the other parameters that have been investigated. Nevertheless, a coating for the piston skirt is indispensable in order to ensure operating safety under extreme conditions [14].

7.3.3.10 Stiffness of the piston skirt

In the case presented, the variation in skirt stiffness is the result of the finish machining of a base piston. The areas highlighted in blue in **Figure 7.50** were milled to reduce the base piston's skirt stiffness. In addition, the cooling oil gallery was slotted (red). The fatigue resistance was not taken into consideration in this context.

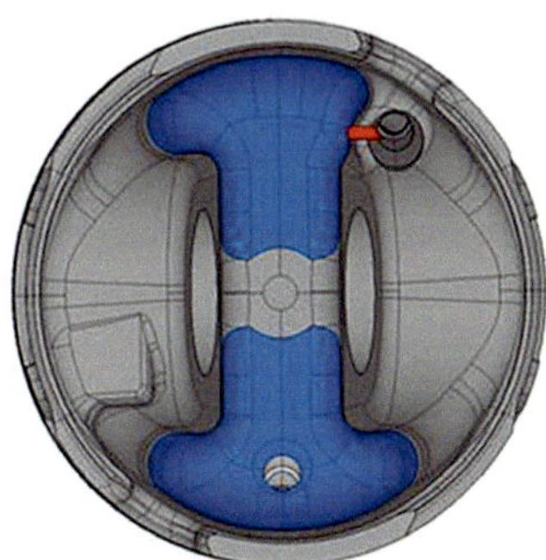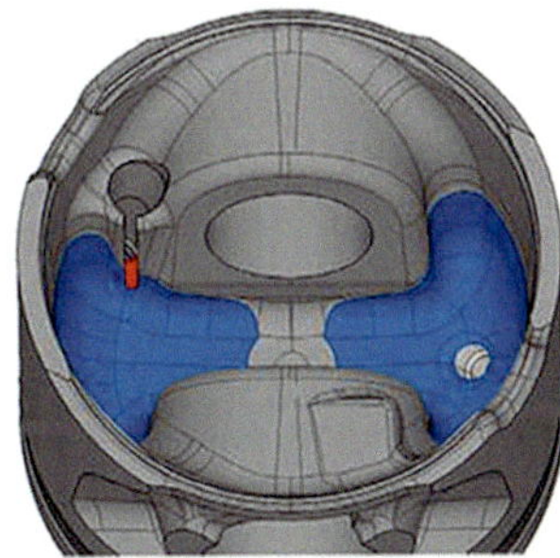

Figure 7.50: Reduction of the skirt stiffness by milling the base piston (blue) and slotting the cooling oil gallery (red)

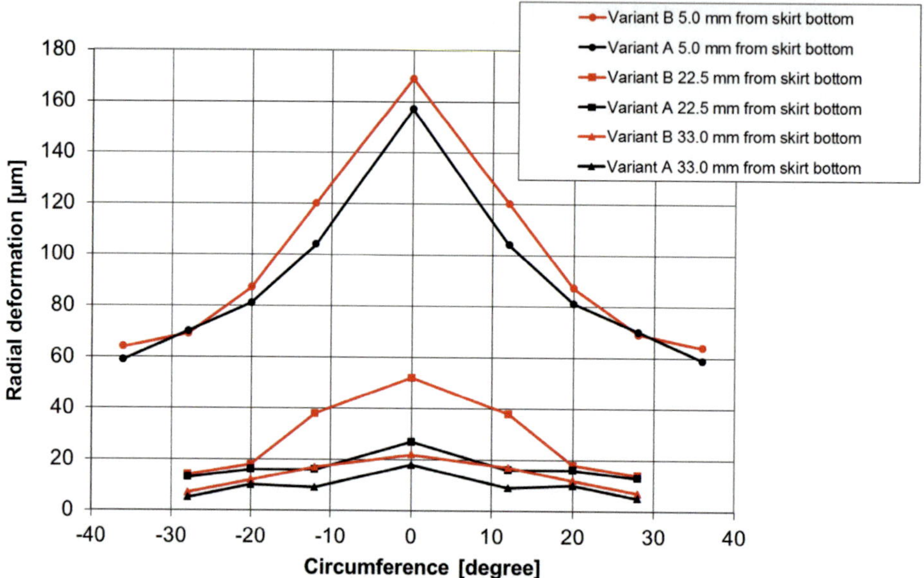

Figure 7.51: Measurements of radial deformation on the thrust side; test load 2,000 N; variant A = base piston (black), variant B = piston with reduced skirt stiffness (red)

The reduction in skirt stiffness achieved by means of these measures can be determined by measuring the radial deformation at several points across the skirt contour and skirt height. The radial deformation of the piston skirt under a test load of 2,000 N around the skirt contour is shown in **Figure 7.51**. The radial deformations of the finish machined piston are considerably greater than those of the base piston. The skirt stiffness has been substantially reduced, particularly on the thrust side.

The variant with reduced skirt stiffness shows friction mean effective pressure advantages across the entire engine operating map in fired engine operation compared with the base variant, **Figure 7.52, right**. There is a clear dependence on load, whereas the dependence on speed is comparatively low. Maximum reduction can be achieved at high load and high speed.

Under high load, the thermal expansion of the piston is particularly large, so it can be assumed that the piston encounters interference. The increased mixed lubrication component of the piston skirt friction leads to increased friction losses. However, a soft piston skirt can counteract the thermal expansion and thus reduce the contact pressures. This makes it easier to maintain a largely hydrodynamic lubrication. The friction losses resulting from piston skirt friction decrease accordingly. The reduction in skirt stiffness thus allows a local increase in operating clearance for the piston skirt. Particularly at the upper end of the skirt, where the heat input is greatest as a result of combustion, a reduction in stiffness has a positive effect on friction losses.

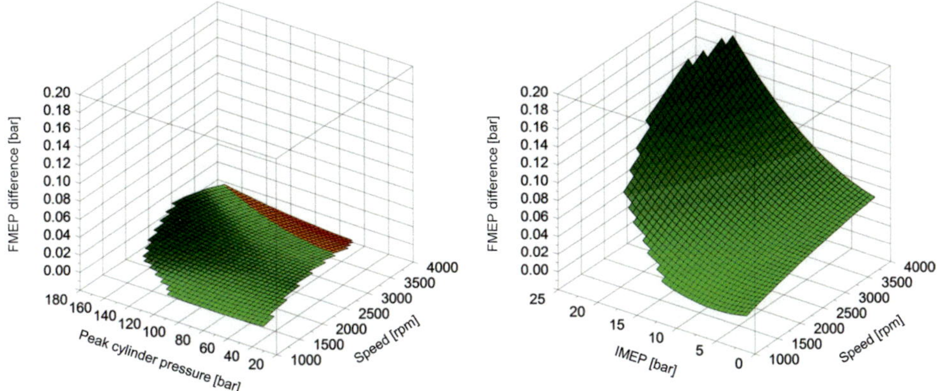

Figure 7.52: Friction mean effective pressure difference maps, varying skirt stiffness (skirt stiffness of the production condition in comparison with reduced skirt stiffness), motored externally charged (left) and fired (right), engine temperature 100°C

During motored externally charged engine operation, **Figure 7.52, left**, there is no substantial influence of the skirt stiffness on the friction mean effective pressure. Triggered by the lacking heat input, the piston only heats up minimally, which in turn only leads to limited thermal expansion. The operating clearance differs significantly from the clearance in fired operation. The mechanisms described above for fired operation therefore do not apply to motored externally charged operation.

7.3.3.11 Area of the piston skirt

In order to determine the influence of skirt area on friction mean effective pressure, the support area of the skirt was significantly reduced. The location of the area was also varied while maintaining almost the same surface area. A distinction is made between a horizontal and a vertical orientation of the supporting skirt area; **Figure 7.53**.

According to Newton's law of friction, the area's orientation has no influence on the friction losses. It is assumed, however, that the location of the support area influences the buildup of the lubricating film on the piston skirt. This makes it possible to change the shear rate and therefore influence the friction force. In addition, effects on the displacement and oil transport mechanisms can be expected.

In motored externally charged engine operation, no significant differences are evident for the horizontally oriented, reduced skirt area in comparison with the base; **Figure 7.54**.

In fired operation, slight frictional advantages can be seen for the reduced, horizontally oriented skirt area. As the load increases, the frictional advantages increase, and the maximum is under high load. The engine speed appears to have no significant influence.

In motored externally charged operation, again no significant friction mean effective pressure difference is evident for the reduced skirt area. In fired operation, the reduced, vertically

Figure 7.53: Horizontal orientation (left, green) and vertical orientation (right, blue) of the reduced bearing skirt areas (height of the supporting profile 100–200 µm)

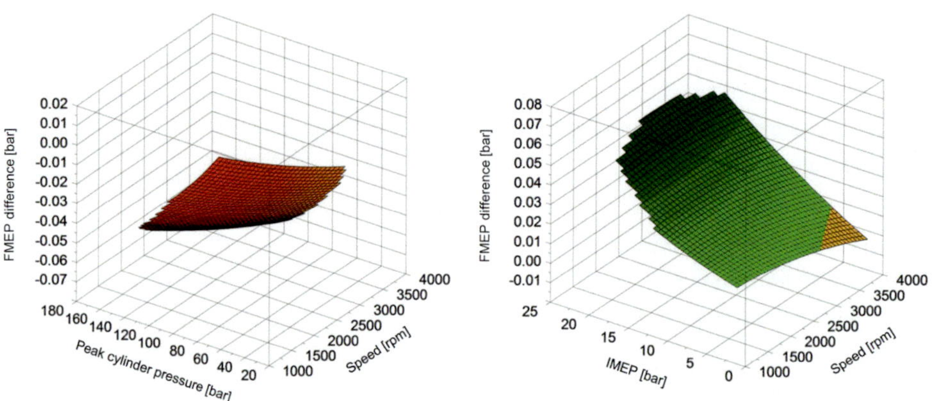

Figure 7.54: Friction mean effective pressure difference maps, different skirt area (base skirt area in comparison with reduced skirt area, horizontal orientation), motored externally charged (left) and fired (right), engine temperature 100°C

oriented skirt area also experiences slight frictional advantages, which become greater as the load increases and reach their maximum undor high load at high speed; **Figure 7.55**. The variants with reduced skirt area thus show very similar frictional behavior, regardless of the orientation. It can be assumed that a reduction in bearing skirt area leads, in principle, to a reduction in friction. When optimizing the skirt area, however, care must be taken to ensure that the reduction in skirt area does not cause an excessive increase in the contact pressure on the skirt, because otherwise the mixed lubrication component at the piston skirt increases disproportionately, which can result in a frictional disadvantage. As the wear pattern becomes smaller, piston profile optimization is gaining in importance.

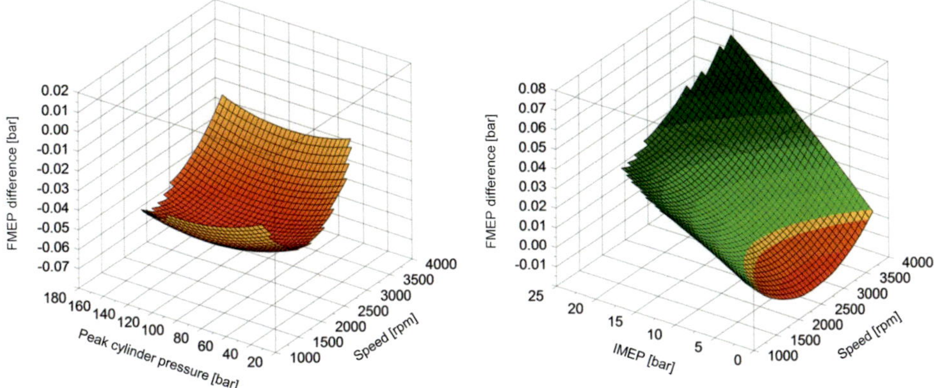

Figure 7.55: Friction mean effective pressure difference maps, different skirt area (base skirt area in comparison with reduced skirt area, vertical orientation), motored externally charged (left) and fired (right), engine temperature 100°C

7.3.4 Simulation of fuel consumption and CO_2 emissions values in the cycle

Evaluation of friction power loss potential with respect to the fuel and CO_2 savings that can be achieved is only possible with the use of driving cycle calculation programs. A detailed description for determining CO_2 emissions in the driving cycle and for validating the model can be found in [12].

In the present case, the New European Driving Cycle (NEDC) is used as the driving cycle. A kinematic calculation method is used, where a vehicle speed profile is provided to the model, as well as the gear ratio, depending on the driving cycle being investigated. The angular velocities, efficiencies, and torques of the various vehicle components are then calculated backward along the powertrain, until the corresponding engine operating map points have been determined. In addition to the friction mean effective pressure maps derived from engine testing, geometric vehicle data and experimentally derived driving resistances are required. Using this approach, the fuel consumption for every investigated parameter can be calculated in the cycle under analysis. The resulting CO_2 emissions can be derived from the cumulative fuel consumption over the cycle. The difference between both variants yields the corresponding CO_2 savings in the driving cycle being investigated. This method is depicted schematically in **Figure 7.56**.

Using the existing vehicle model, additional driving cycles can be investigated, such as the ARTEMIS cycle, or customer-specific running profiles. In addition, it is possible to demonstrate the influence of a stop-start system on fuel consumption and CO_2 emissions in a driving cycle.

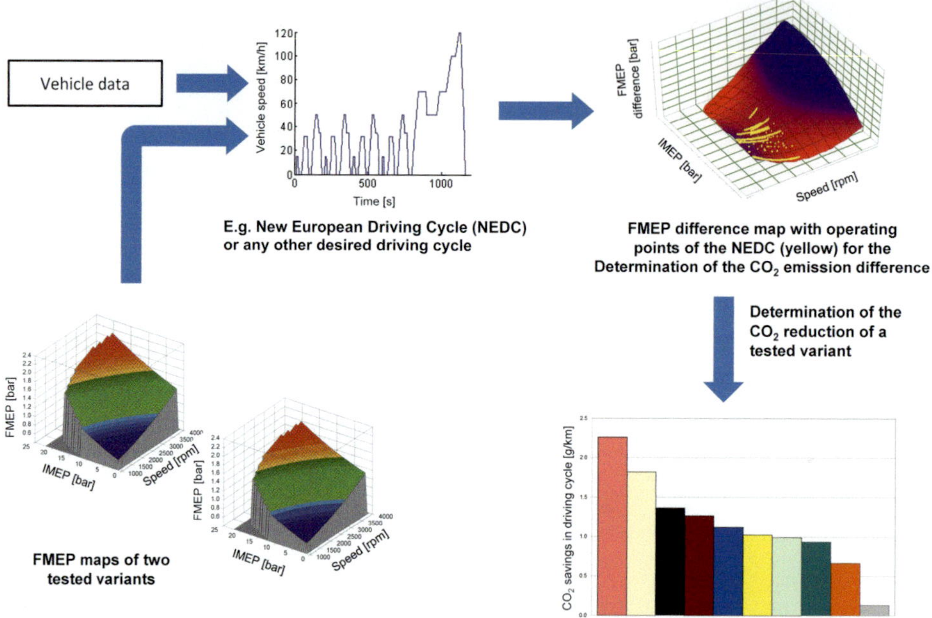

Figure 7.56: Driving cycle-relevant CO_2 savings using friction mean effective pressure maps

7.3.5 Comparison of results

The maximum values of the measured friction mean effective pressure differences indicate the potential for the parameter change being investigated. A summary of these maximum potentials that can be achieved in the operating map is shown in **Figure 7.57, top**, for some of the investigated parameters. It is evident that the parameters installation clearance, skirt roughness, piston profile, and piston pin offset have various degrees of potential for improving the friction behavior of the piston. In the ring pack, the tangential force of the oil control ring and the ring width, especially of the top ring, can be considered significant influence variables.

Converting the friction mean effective pressure differences into friction power loss differences yields the ranking shown in **Figure 7.57, center**. Because there are potential interactions between different parameters, the superposition of individual measures is only somewhat accurate. In order to obtain an estimate of the entire friction power loss potential for the piston group, this must be taken into consideration. In the four-cylinder diesel engine investigated here, for a speed of 4,000 rpm and power output of about 100 kW, the findings indicate that there is a potential for improving the piston group to save about 3.5 kW.

Figure 7.57, bottom, shows the ranking of CO_2 savings calculated in the NEDC for the selected variants. A comparison of the listed rankings shows that there are shifts with respect

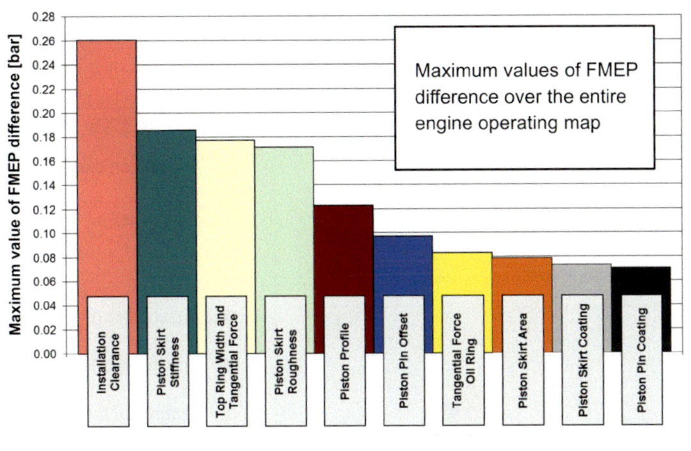

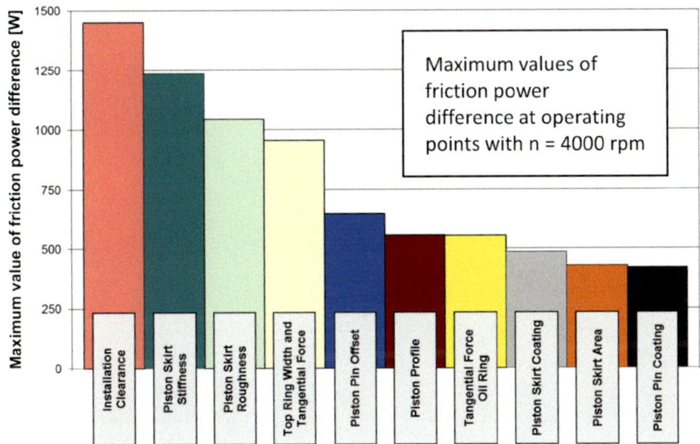

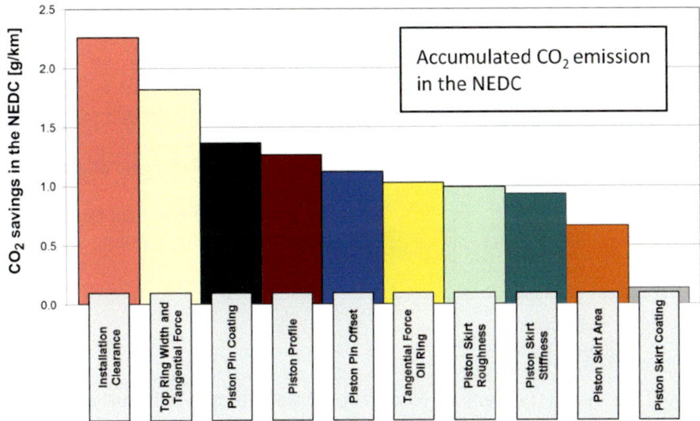

Figure 7.57: Savings potential of various design parameters for the piston group in a 2.0-liter diesel engine. Top: maximum values from friction mean effective pressure difference maps; center: maximum values from friction power loss difference operating maps at 4,000 rpm; bottom: accumulated CO_2 savings in NEDC

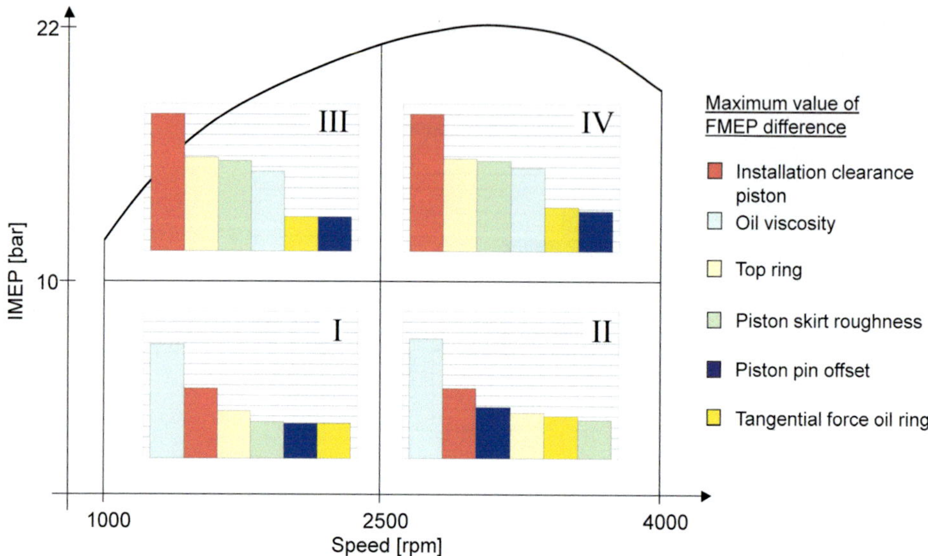

Figure 7.58: Quadrant ranking of the friction power loss results obtained (fired), where
Diagram I: relevant to the cycle;
Diagram II: representative for gasoline engines;
Diagram III: representative for commercial vehicle diesel engines;
Diagram IV: representative for passenger car diesel engines.

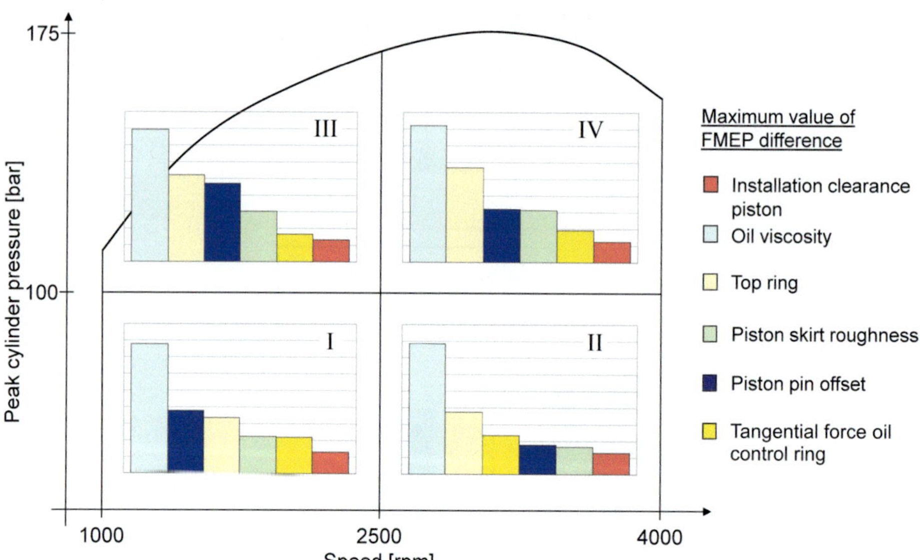

Figure 7.59: Quadrant ranking of the friction power loss results obtained (motored externally charged), where
Diagram I: relevant to the cycle;
Diagram II: representative for gasoline engines;
Diagram III: representative for commercial vehicle diesel engines;
Diagram IV: representative for passenger car diesel engines.

to the valuation of the individual parameters. The installation clearance is the parameter with the biggest influence in all three display types. The parameters of skirt roughness, skirt stiffness, and pin coating are particularly notable, and have distinct frictional advantages, particularly in the load spectrum of the NEDC.

Superposition of CO_2 savings in the NEDC is possible, at least with some conditions, because each parameter under test was based on the same operating points. Any possible interactions between the individual design parameters must, however, be noted. It is therefore recommended that the set of measures derived accordingly then be tested as an engine concept variant. The results obtained show that, depending on the customer requirements, appropriate friction optimization can be performed.

In order to allow the measures for various engine concepts and load spectrums to be optimally combined, the engine operating map is divided into four quadrants, each of which has typical operating conditions for various engine concepts. An internal ranking of the measures can be generated for each quadrant; **Figure 7.58**.

For low loads, the oil viscosity is also a dominant influence factor besides design parameters. In the high load range, the oil viscosity loses significance, while piston installation clearance can be seen as the most important design optimization measure in this range. Measures on the top ring and the skirt surface also increase in significance at higher loads. The conditions described apply only to fired operation. The valuation of the parameters is very different for motored externally charged operation. This change in valuation is largely due to the different temperature distribution in the crank mechanism and the associated changes in operating clearance and distortion; **Figure 7.59**.

The test bench and systematic investigation of numerous individual design parameters, as shown, allow an extensive array of measures for minimizing friction to be developed. Together with an evaluation of the various operating ranges, they allow the targeted optimization of friction losses.

7.4 **Wear testing of the piston group**

This article addresses wear in the context of the piston group. The affected components are
- pistons;
- piston rings;
- piston pins;
- circlips;
- cylinder surfaces.

Figure 7.60:
Piston with GRAFAL® coating, with
recesses for measuring the skirt profile

7.4.1 Piston skirt

7.4.1.1 Skirt collapse and coating wear

In order to be able to detect skirt collapse and wear of a piston skirt coating (such as GRAFAL®) that may occur during operation, the shape of the piston mantle curve is charted.

Figure 7.60 shows a passenger car diesel piston that has been run, with the surface of the skirt being coated with GRAFAL®.

Figure 7.61 shows the result of the measurement of the mantle curve of a piston as delivered, where the second ring land is set back so far that it does not appear in this view. The smaller diameter of the first ring land in the pin axis is due to the specified ovality of the piston profile.

At the level of the diameters D_N and D_1, the measured curve for the unused piston shows the recesses in the GRAFAL® coating on the skirt. The base piston material can be accessed at these locations in order to take the measurements. These windows can also be seen on the used piston in **Figure 7.60**. The thickness of the GRAFAL® coating is about 20 µm radially in the "as delivered" state.

Figure 7.62 shows the result of measuring the mantle curve of the used piston.

According to this example, the skirt collapse is approximately 10 µm diametrically at the level of the diameter D_1. This cannot be attributed to wear at this location, because the measurement point is within the GRAFAL® window and therefore is not exposed to wear. At the same time, the GRAFAL® coating at this location has been ablated by 20 µm diametrically, or 10 µm radially.

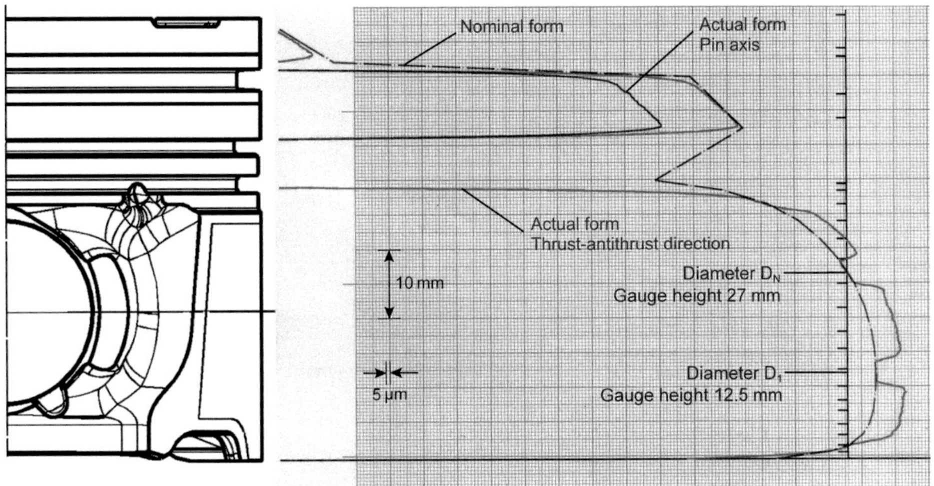

Figure 7.61: Results of measuring a piston in the "as delivered" state. Measurement results apply diametrically.

D_N: nominal piston diameter

D_1:   largest diameter of the piston

Green: nominal form of the mantle curve in the thrust-antithrust direction

Red: measured curve in the thrust-antithrust direction

Blue: measured curve in the pin axis

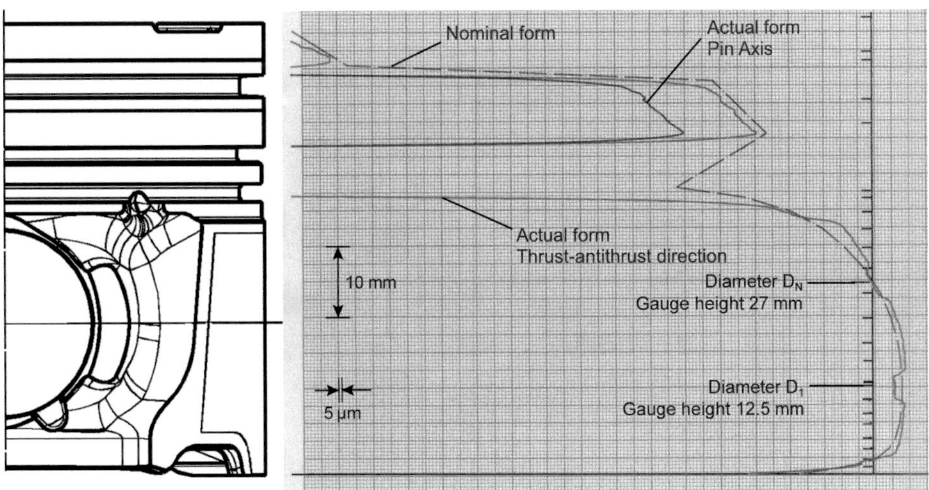

Figure 7.62: Results of measuring a used piston. Measurement results apply diametrically.

D_N: nominal piston diameter

D_1:   largest diameter of the piston

Green: nominal form of the mantle curve in the thrust-antithrust direction

Red: measured curve in the thrust-antithrust direction

Blue: measured curve in the pin axis

7.4.1.2 Ovality

Another way to show and evaluate potential deformation of the piston skirt and potential coating wear is to record the ovality at prescribed heights on the skirt, according to **Figures 7.63** and **7.64**. This allows the circumferential deviations of the real piston profile from the nominal form to be evaluated (ovality and profile distortion).

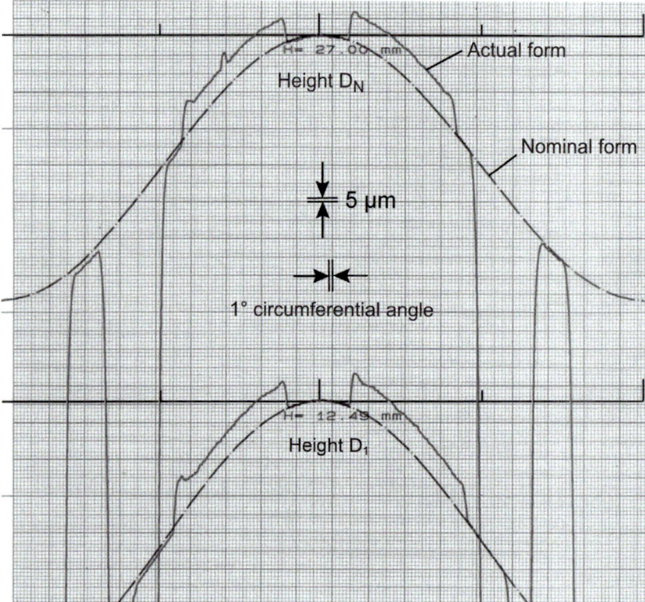

Figure 7.63:
Ovality record around the circumference of the piston as delivered

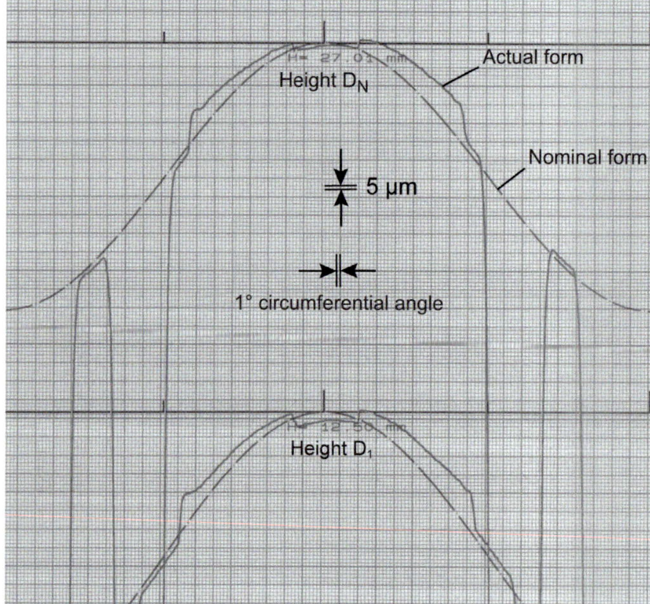

Figure 7.64:
Ovality record around the circumference of the piston in used condition in order to determine deformation and wear of coating

The values of diameters D_N and D_1 serve as reference dimensions for such ovality records, shown as a flat projection.

Similar to the representation of the mantle curves, here again the skirt collapse can be measured as 10 µm diametrically at the level of diameter D_1, and the thickness of the GRAFAL® coating as about 10 µm radially.

7.4.2 Piston ring and cylinder surface

7.4.2.1 Piston ring running surface

In order to fulfill its function, the piston ring exerts a defined radial force on the cylinder surface, acting perpendicular to it. To define it, a band encompassing the piston ring is contracted so far that the prescribed gap clearance is achieved. The necessary force to contract the band–called tangential force–is measured. The tangential force of the piston rings is directly related to friction and wear.

The gas pressure also acts on the back side of the compression rings and presses them against the cylinder surface. Owing to the high gas force load under load conditions, the first piston ring is most affected. This increase in radial force leads to severe loading of the running surface of the compression rings, particularly in the area of the TDC fired. Another difficulty is that the piston briefly stops moving at the dead center, causing the hydrodynamic pressure in the lubrication wedge between the two friction partners to drop and leading to metallic contact.

Figure 7.65 shows the profile of the piston ring in its "as delivered" and used condition. Wear can be clearly detected in the area of the greatest barrel shape. This wear can also be veri-

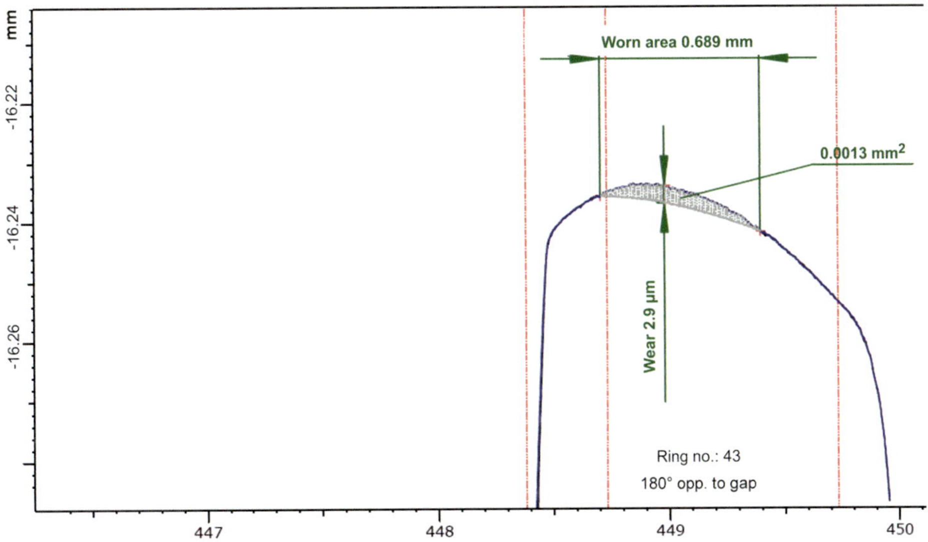

Figure 7.65: Profile measurement of 1st piston ring with an asymmetrical, barrel-shaped running surface

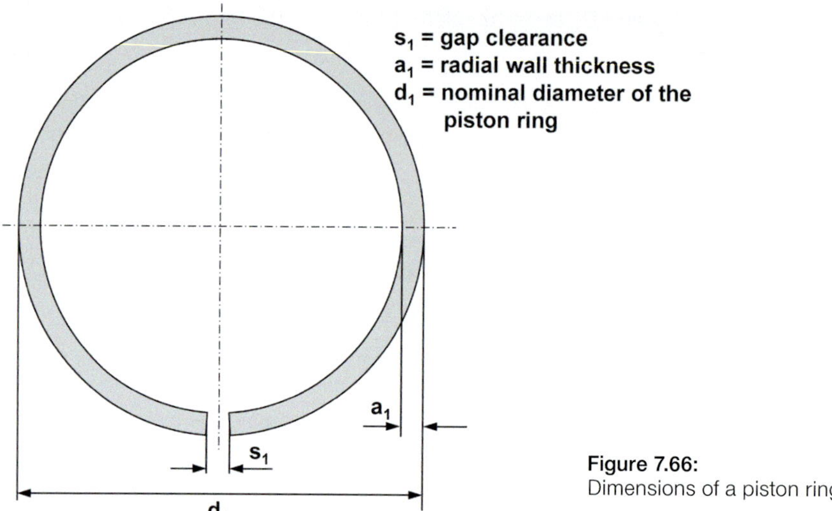

Figure 7.66:
Dimensions of a piston ring

fied by measuring the radial wall thickness of the piston ring and is typically in the range of a few microns, depending on the running time. The layout of the major dimensions sketched in **Figure 7.66** shows the direct association between the reduced radial wall thickness and increased gap clearance.

If the radial wall thickness (a_1) of the piston ring is reduced, it will result in the ring gap (s_1) being opened up. Because this causes the ring to relax, the tangential force also decreases. The type of wear described here can therefore also change the engine operating parameters, such as blow-by and oil consumption.

For compression rings that do not exhibit extensive barrel shape, chamfering, or other measures that define a contact point of the running surface on the cylinder, this can lead to so-called "double contact." This results in two different running levels at different heights on the ring running surface, due in part to the twisting of the ring and in part to the flat contact under peak cylinder pressure. In the interest of good sealing and oil control, however, a defined running level is generally aimed at.

7.4.2.2 Coil springs

Coil springs are used to support so-called two-piece oil control rings. After long running times, it is possible for the ends of the oil control ring to dig into the coil-spring; **Figure 7.67**.

7.4.2.3 Abnormal wear patterns

Figure 7.68 shows an example of a piston ring with extreme wear on the running surface, where the radial wall thickness has been partially severely reduced diametrically opposite the ring gap. The oil control rings shown in **Figure 7.69** also exhibit extreme wear on the running surface. The uppermost ring has no scraping edge left at all.

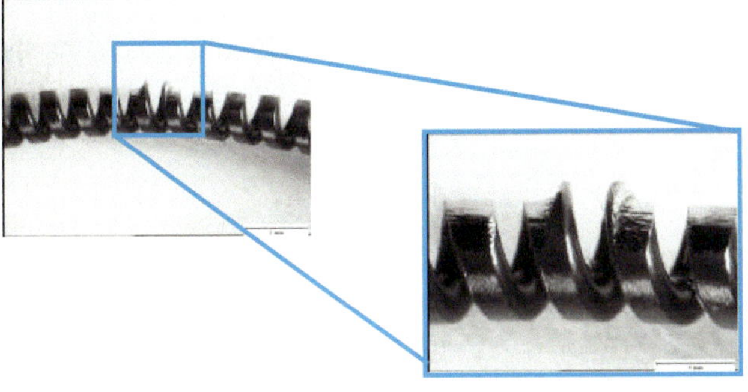

Figure 7.67:
So-called burial in the coil spring, caused by the ends of the oil control ring

Figure 7.68:
Example of extreme single-sided wear of 1st piston ring

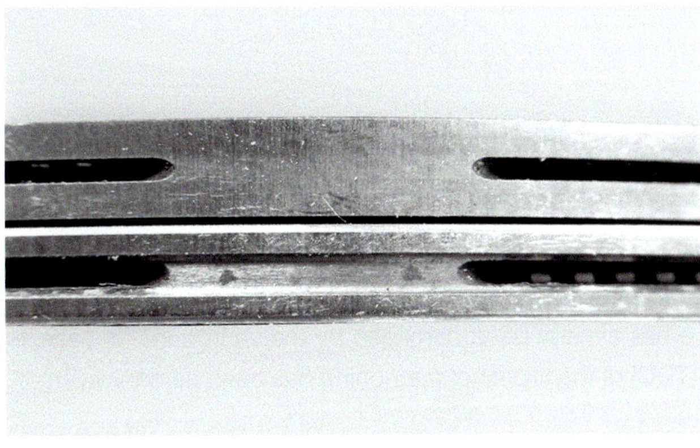

Figure 7.69:
Example of extreme wear of two oil control rings

Abnormal wear rates, as shown here, can typically be attributed to abrasive particles in the intake air, due to a missing air filter element, for example. The associated running surfaces of the cylinder liners have a matte gray appearance in such cases, typically with many fine grooves in the direction of the stroke.

7.4.2.4 Cylinder surface and cylinder polishing

The greatest wear on the cylinder surface caused by the piston ring occurs in the running area around the top dead center (TDC). This wear pattern, which is evident from the cylinder measurement as a locally limited area with an above-average increase in diameter at the dead center of the 1st piston ring, is known as top dead center wear. **Figure 7.70** shows the shape record of the cylinder opposite the pin axis, clearly showing the top dead center wear.

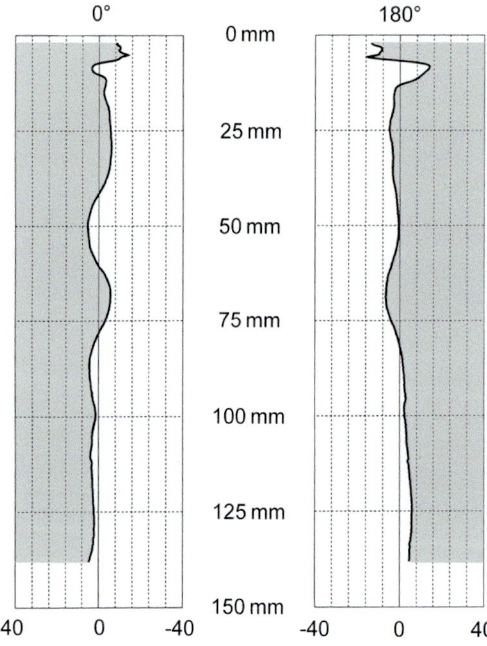

Figure 7.70:
Measurement record of vertical cylinder mantle curves in the pin axis. The shape of the cylinder surface is shown greatly exaggerated as a result of magnification.

Another type of wear arises from the occurrence of polished areas on the cylinder surface. This happens when hard and abrasive carbon builds up on the top land of the piston under unfavorable operating conditions. **Figure 7.71** shows the top land of a piston, with hard carbon in a locally limited area.

The influence of this local carbon buildup on the associated cylinder surface is shown in **Figure 7.72**. The cylinder surface has already been damaged by the carbon in this case. In the area of the top dead center (TDC) of the top land, the honing has been partially worn off.

Figure 7.71: Carbon buildup on the top land of a commercial vehicle piston

Figure 7.72: Cylinder area polished by the buildup of carbon on the top land

Directly below that, within the running range of the piston rings, another so-called polish site can be detected.

The function of the honing, which is to retain lubricating oil and to ensure lubrication between the piston ring and cylinder surface, particularly at TDC, can no longer be ensured at these locations. The consequences are local disturbances in the tribological system of the piston ring and cylinder surface. This causes a risk of burning marks on the piston ring running surfaces, and therefore also a risk of piston ring seizure.

With suitable design measures, the buildup of carbon on the top land, and thus the occurrence of such cylinder polishing, can be specifically prevented.

7.4.3 Piston ring side face and piston ring groove

Similar to the wear on the running surface, the relative motion between the piston ring and the piston can cause a loss of material, both at the side faces of the piston rings and at those of the top ring groove.

7.4.3.1 Side faces of the 1st piston ring

Wear occurs particularly at the lower flank of the piston ring. **Figure 7.73** shows the shape record of the upper and lower flanks of the first piston ring in the "as delivered" state.

Figure 7.74 shows the shape record of the first piston ring shown above, after its use in an engine. A clear wear zone has formed on the lower flank of the piston ring (right), on the side facing the running surface of the cylinder liner. The depth of this burial is typically in the range of a few microns.

7.4.3.2 Side faces of the top ring groove

The lower flank of the top ring groove is particularly highly stressed. The results of a test on a piston with keystone ring are shown here as an example.

Figure 7.75 shows the result of a measurement of the side faces of the keystone groove in "as delivered" state. In order to capture the measurement record, the groove flanks are tilted

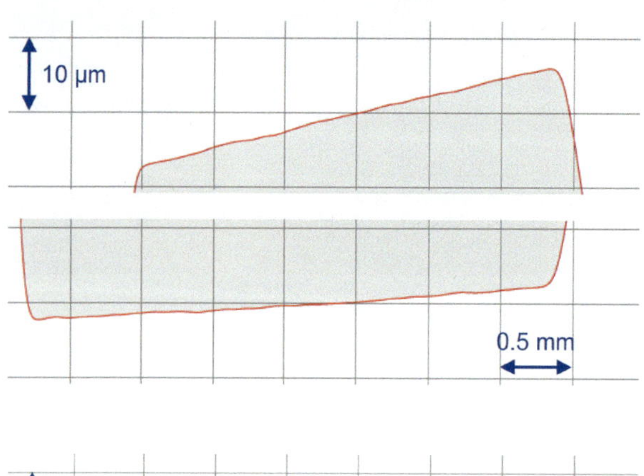

Figure 7.73:
Shape record of the upper and lower flanks of 1st piston ring in "as delivered" state (not under tension). The vertical spacing of the two records is arbitrary and does not represent the actual ring width.

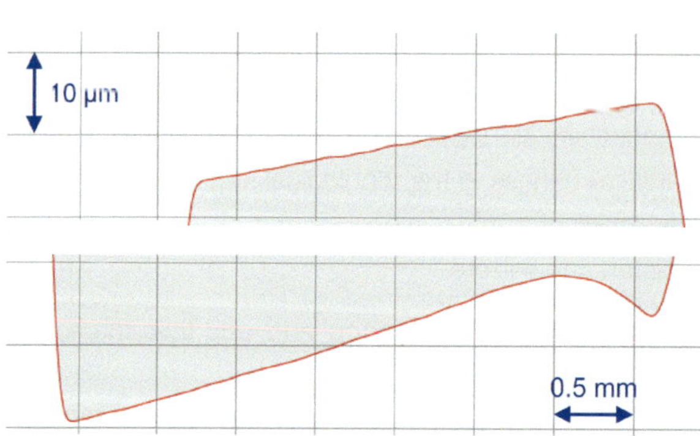

Figure 7.74:
Wear on the lower flank of used 1st piston ring (not under tension). The vertical spacing of the two records is arbitrary and does not represent the actual ring width.

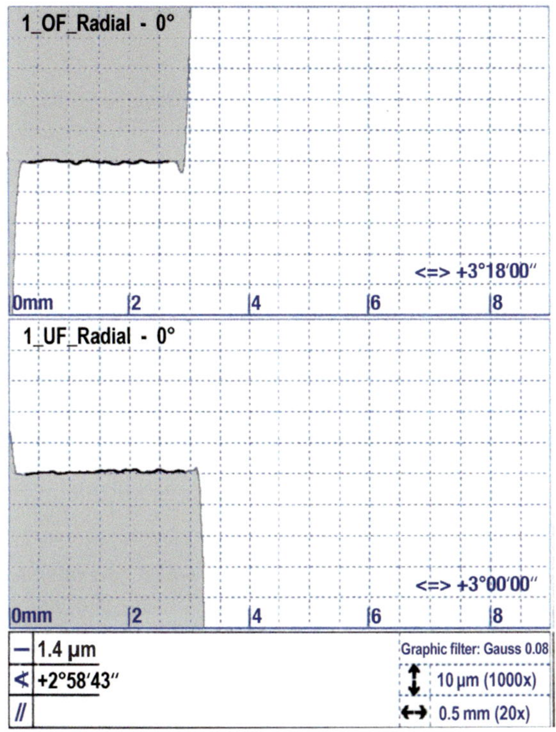

Figure 7.75:
Measurement record of the side faces of a keystone groove in "as delivered" state. The nominal angle that is preset for measurement probing is noted on the measurement sheet.

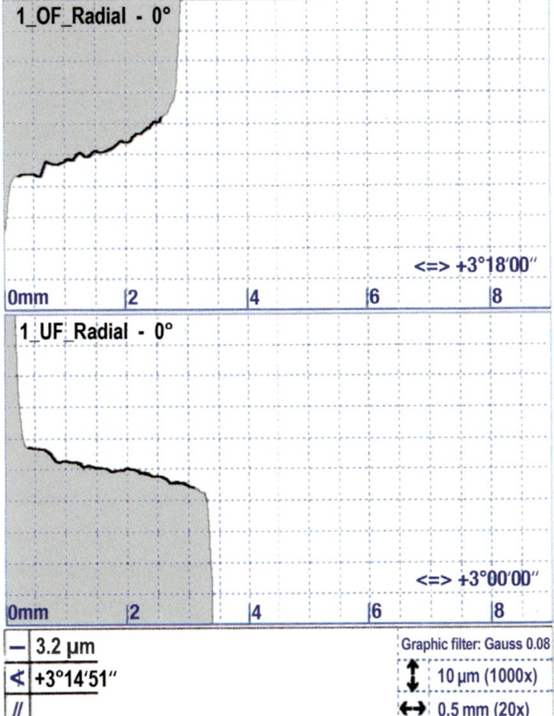

Figure 7.76:
Measurement record of the side faces of a keystone groove in used condition. The nominal angle that is preset for measurement probing is noted on the measurement sheet.

by the nominal angle, so that the shape record of the groove flanks does not deviate from the horizontal on the measurement sheet if the inclination is correct.

Figure 7.76 shows the profile of the groove flanks in used condition. The deviations due to wear are once again in the range of a few microns.

7.4.4 Piston pin and piston pin boss

7.4.4.1 Piston pin

The piston pin transmits the force due to the combustion chamber pressure from the piston to the connecting rod, and thus enables the necessary relative motion between piston and small end bore due to the movement of the connecting rod. According to the effective forces, the piston pin is deformed within its elastic range by bending and ovalization. Wear also occurs at the affected contact surfaces as a result of relative motion, though this wear is typically low.

A corresponding measurement result for the piston pin is shown in **Figure 7.77**. The orientation of the degree scale is arbitrary. There is no reference to the installation orientation of the pin in the boss, because the pin typically rotates during operation.

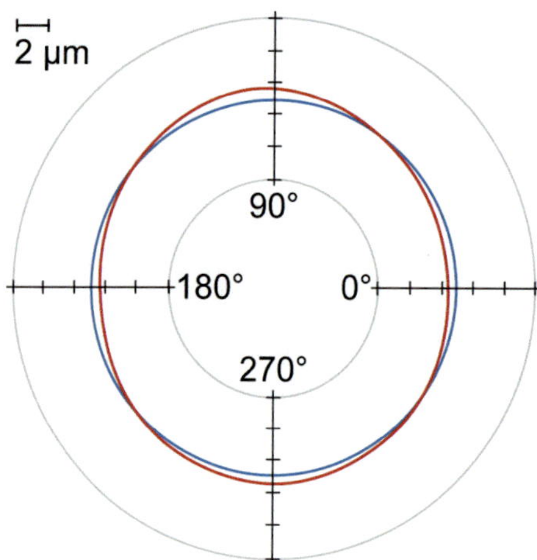

Figure 7.77:
Measurement results
of roundness of piston
pin in used condition

In the "as delivered" state, piston pins are nearly perfectly round and cylindrical. As is evident in the measurement record, the piston pin shows permanent ovalization in the used condition. Residual bending, in turn, is evident from the mantle curve measurements in **Figure 7.78** and **Figure 7.79**. Also evident in the area of the mantle curve at 180° is a significantly different wear pattern than with the other measured mantle curves. This shows, without a doubt, that the pin was no longer turning for some time prior to being removed. The wear pattern would otherwise have the same structure at every point around the circumference.

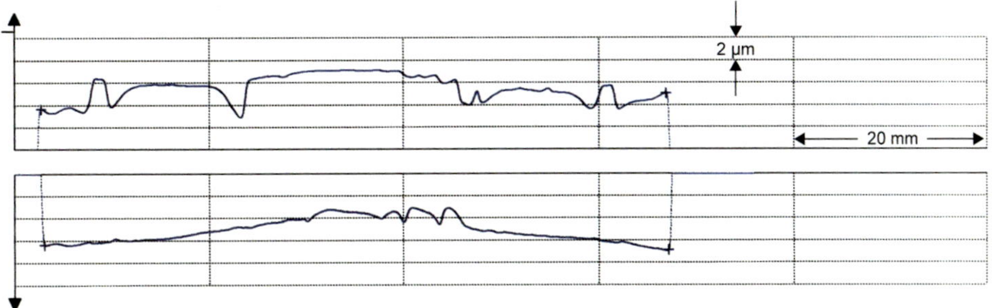

Figure 7.78: Measurement results of the mantle curve measurement of a piston pin in used condition, in the direction of 0°–180°

Top: mantle curve at 180°

Below: mantle curve at 0°

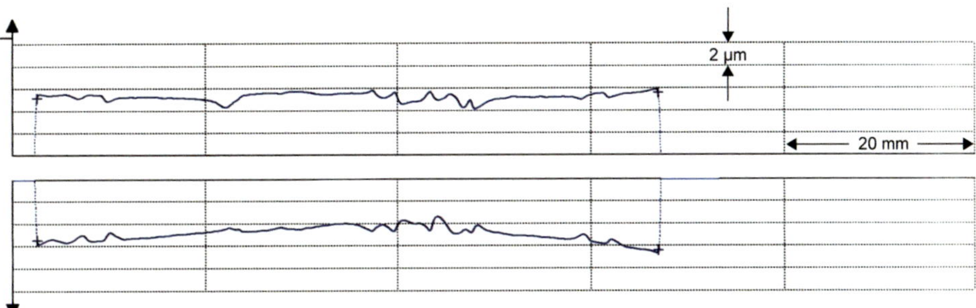

Figure 7.79: Measurement results of the generating line measurement of a piston pin in used condition, in the direction of 90°–270°

Top: mantle curve at 270°

Below: mantle curve at 90°

Figure 7.80:
Photo of used piston pin

The individual contact zones on the pin can be seen in **Figure 7.80**. The engagement area of the small end bore is located in the center. A narrow range with no contact can be seen to the right and left. The brownish circumferential stripes are due to oil varnish buildup. The two ends of the pin are located in the piston pin boss.

7.4.4.2 Piston pin boss

The dynamic deformation of the piston and the bending of the pin that occur at peak cylinder pressure change the shape and the alignment of the pin bore. For this reason, pin bores are often designed as so-called shaped pin bores. They deliberately deviate from a precise cylindrical shape and extend the diameter toward the ends of the bore (crowning). This can be done both internally and externally on the pin bore, in order to prevent an increase in local wear. **Figure 7.81** shows the dimensioning of the support points of such a shaped pin bore.

The values B and C indicated in **Figure 7.81** define the important node points for manufacturing and measuring the crown.

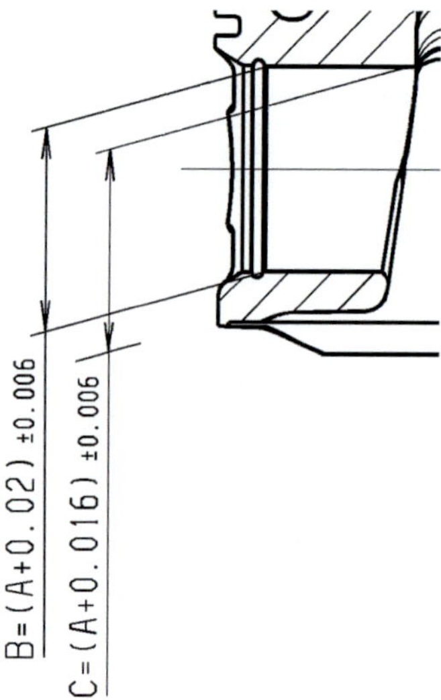

Figure 7.81:
Dimensions of support points of a shaped pin bore with crowning

A: nominal diameter of the crowned pin bore
B: definition of the support dimension of the pin bore in the indicated area, important for machining
C: definition of the support dimension of the pin bore at the inner extension of the pin bore, important for machining

Figure 7.82 shows the measurement of the shape of the pin bore of an unused piston. The inner and outer areas of the bores exhibit a defined crown.

The measurement of a pin bore of a used piston in **Figure 7.83** shows that the bore shape has not changed significantly, even after a long running time.

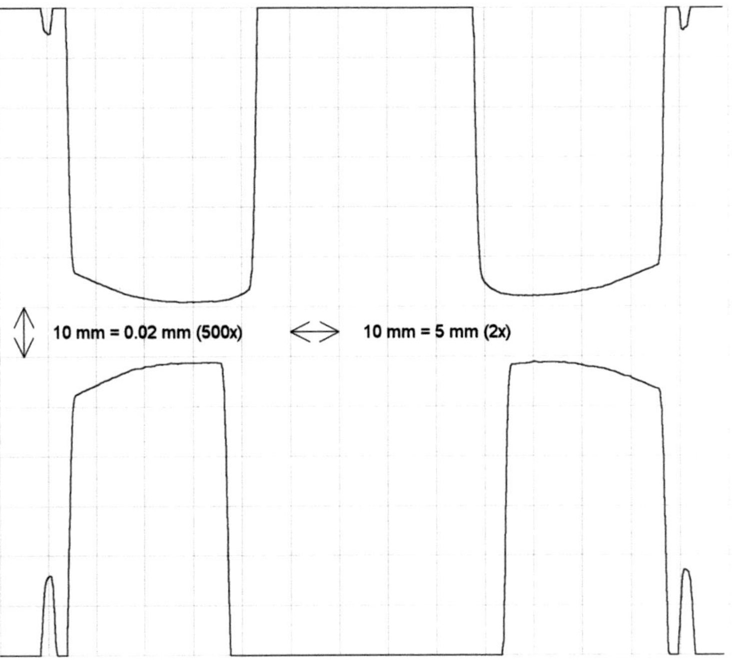

Figure 7.82: Measurement of the shaped pin bore of a piston pin boss in "as delivered" state

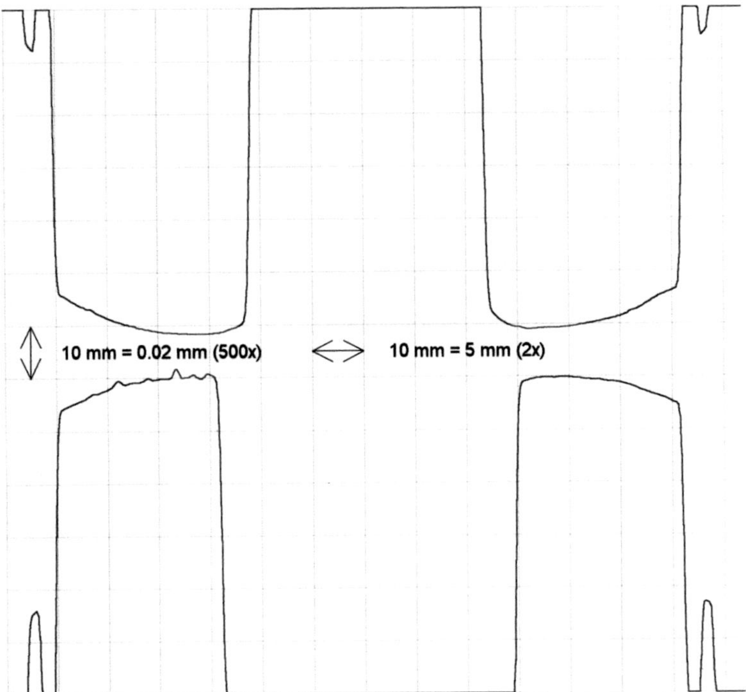

Figure 7.83: Measurement of the shaped pin bore of a piston pin boss in used condition

7.4.5 Circlip and circlip groove

The task of the circlip is to limit the motion of the piston pin in the axial direction in the piston. Theoretically, there are no axial forces acting on the piston pin in a symmetrically designed crank mechanism. Any type of asymmetry, however, such as a tilt in the orientation of the piston bore or the conrod bore, or due to shifting of the connecting rod shank, can give rise to an axial force component. The associated motion behavior of the connecting rod in the longitudinal axis of the engine can lead to what is known as pin displacement. In some cases, wear and even damage can occur in the area of the circlip groove.

Figure 7.84 shows the dimensioning of a circlip groove in an excerpt from a piston drawing.

In the simplest case, the associated circlip consists of an annular, bent spring steel wire with a round cross section and a diameter that is slightly less than the groove width.

Figure 7.85 shows a cross section through the circlip groove according to **Figure 7.84**, which is widened significantly as a result of the axial forces acting on the piston pin. The dashed-line circle shows the original machining radius and location. The circlip groove is plastically deformed in the outward piston direction by the circlip, which is under load, to a width of over 2 mm. The deformation can also be seen in the material that is built up above the right side face.

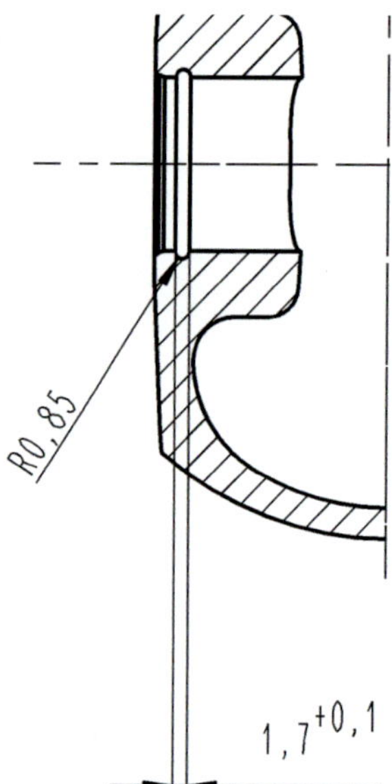

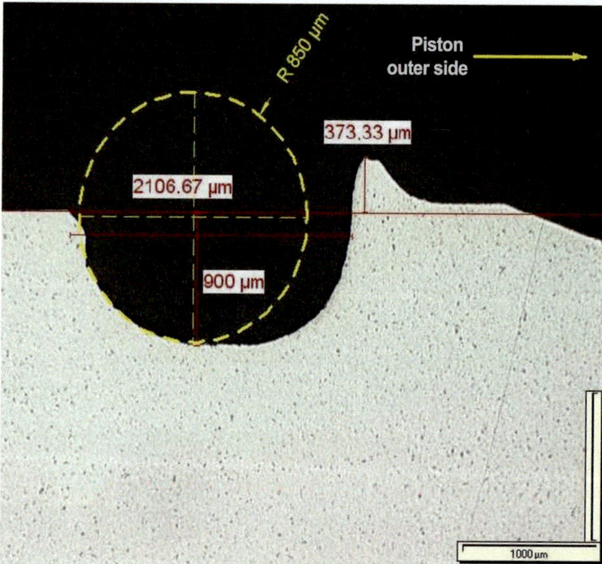

Figure 7.84: Dimensioning of circlip groove

Figure 7.85: Cross section through circlip groove

7.5 Piston loading due to knocking combustion

In the context of more severe exhaust gas legislation, with simultaneous demand for improved fuel consumption, modern gasoline engines increasingly reach the limits of thermal and mechanical load carrying capacity. Downsizing, greater compression, and optimized combustion lead to higher loads. The components in the combustion chamber are particularly affected. The use of supercharging technologies such as turbocharging is indispensable in the development of downsizing engines. Improved cylinder charging with optimized combustion leads to increased combustion pressures. There is also a trend toward increasing compression ratios and boost pressures in turbocharged engines. These efficiency optimizations are limited by the occurrence of knocking combustion in gasoline engines. In knocking combustion, the residual mixture after regular combustion autoignites as a result of the increase in pressure and high local temperatures. This residual mixture combusts explosively in the local area, i.e., at a very high velocity. The resulting rapid local pressure increase causes pressure waves that can propagate through the combustion chamber at sound velocity. The gas force load overlapping regular combustion can cause mechanical damage to the components that bound the combustion chamber.

High compression enables optimized consumption values in the partial-load range, but requires additional measures in the full-load range, such as dynamic retardation of the ignition angle through knock control. Modern measurement methods, such as the MAHLE KI meter [15, 16], can help with tuning.

Specific power output, also a measure of engine loading, has increased significantly in recent years. **Figure 7.86** shows trends in the development of gasoline engines over the last 35 years. The significant increases in cylinder pressure and specific power output are largely due to the increase in turbocharging of gasoline engines in conjunction with downsizing.

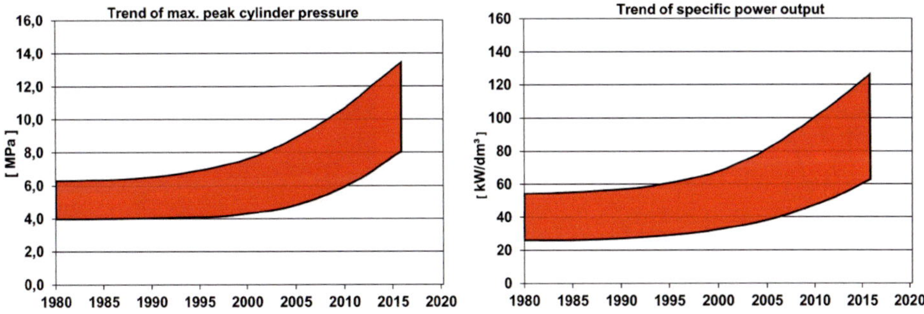

Figure 7.86: Change in maximum peak cylinder pressure and specific power output for gasoline engines (1980–2015)

7.5.1 Knock damage and damage evaluation

Figure 7.87 shows knock damage on the top land and in the top ring groove. The erosion damage on the top land shows that knocking combustion has occurred here over a long period of time.

The shape measurement of the top ring groove, shown in flat projection, shows significant local wear on the upper and lower flanks. This is known as "hammering out" the piston ring groove. During knocking combustion, the first piston ring is excited by the high pressure amplitudes such that plastic deformation of the piston material occurs at the upper and lower flanks.

The greatest widening of the groove often exhibits areas of outward deformation as well, which can make contact with the cylinder wall. This hammering of the groove due to knocking combustion is also sometimes seen without the erosion damage to the top land that is typical of knocking. It can occur over a relatively short running time because of singular high knock amplitudes. This damage is mechanical in nature. The component temperatures are in an uncritical range, according to measurements. The fact that this damage is of a mechanical nature is demonstrated by a test, in which the area of damage is bounded. A relief groove, 1 mm deep, was made in a piston at the lower area of the top land.

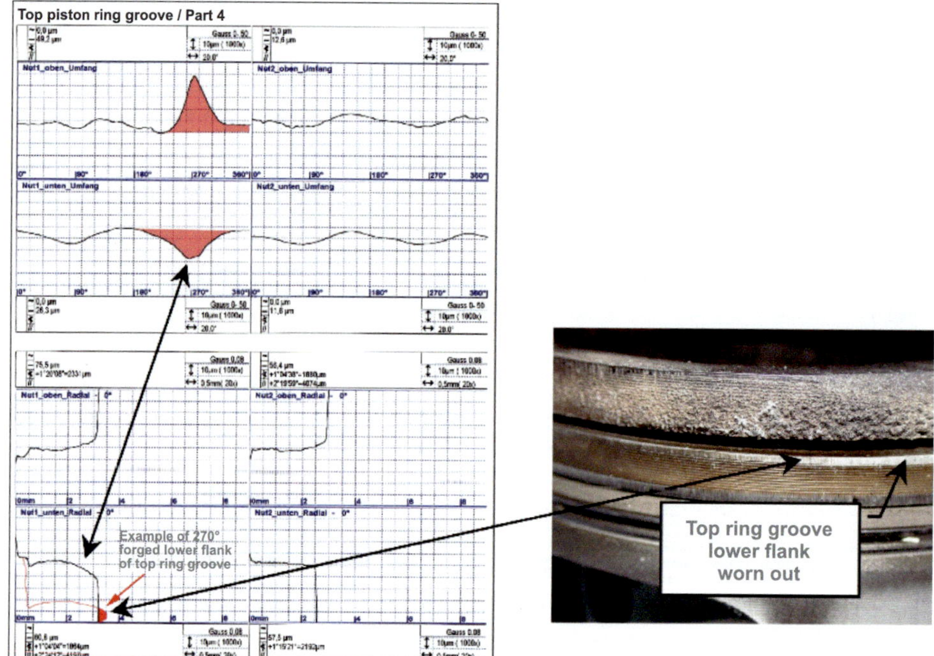

Figure 7.87: Deformation and wear of the top ring groove due to knocking combustion

	Top land with no relief groove	Top land with relief groove
Knock damage	Fine knock damage over a large area, up to the upper surface of the top piston ring groove (no coating on groove)	Knock damage on the top land, only above the relief groove (no coating on groove)

Figure 7.88: Knock damage on the top land, with and without relief groove

Figure 7.88 shows the piston after 105 h of long-term knocking under full load. Both pistons were operated under continuous knocking, with maximum knock amplitudes of about 25 bar.

The piston on the left exhibits knock damage that extends to the lower edge of the top land. The damage to the top land of the piston on the right extends only as far as the relief groove. No knock damage is detected below the relief groove. The pressure wave is dissipated in the relief groove, so that no damage can occur below the relief groove.

This measure for limiting the damage is restricted by the design of modern gasoline engine pistons with short top land width. The relief groove also loses effectiveness in normal operation if it is clogged by carbon deposit.

The damage shown thus far is relatively easy to identify as knock damage. It is more problematic to evaluate damage that occurs as a result of individual knocks of very high amplitude, which does not leave the erosion damage that is typical of knocking.

Figure 7.89 shows four pistons from an engine with ring land fractures. In all four pistons, the fracture area is opposite the spark plug, in the direction of the exhaust valve. Appropriate measurements show that the very high knock amplitudes of over 100 bar are overlaid over the normal cylinder pressure. A single one of these high knock amplitudes can be sufficient to cause such damage.

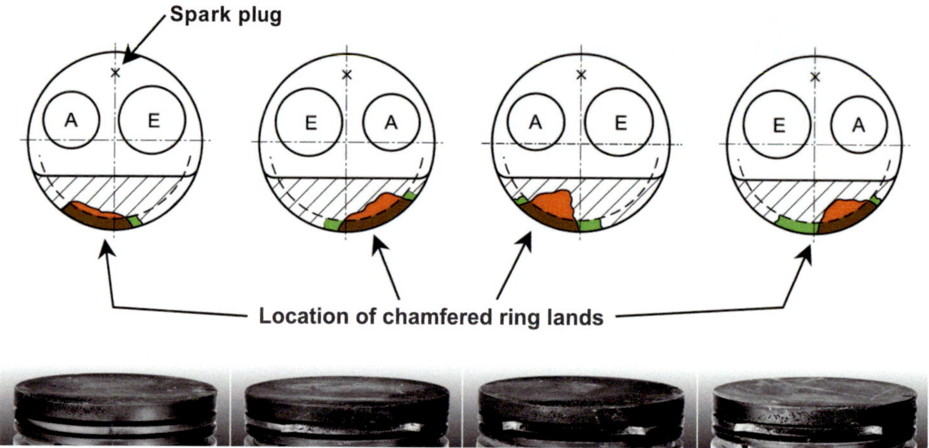

Figure 7.89: Ring land fractures due to knocking combustion

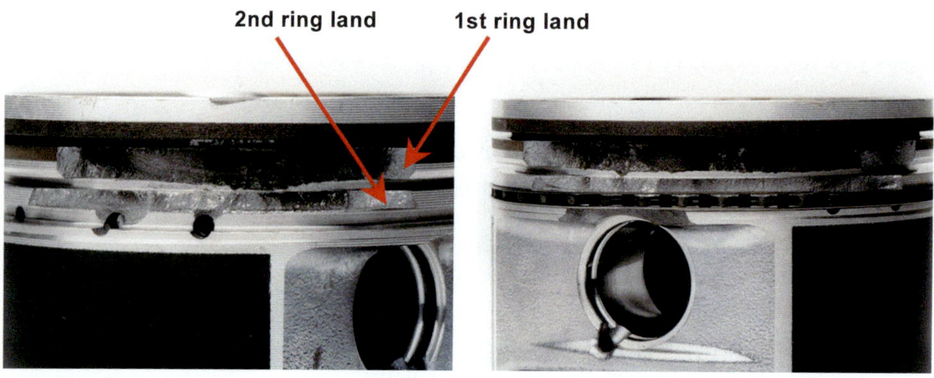

Figure 7.90: Ring land fracture at ring land 1 and ring land 2–caused by severe knocking combustion

Figure 7.90 shows two pistons with fractures at the first and second ring land after only a brief engine running time. The support load on the second ring land was so high that it could not withstand the stress. The support for the third piston ring also broke off from one piston. This was caused by individual, extremely high knock amplitudes. Other typical signs of knock damage are therefore not evident on the piston.

7.5.2 Knock measurement and the MAHLE KI meter

Our objective was to develop a measurement method for the quantitative evaluation of the knock strength. Knocking occurs rather stochastically. The measurement method must thus be able to capture a large number of operating cycles, evaluate them statistically, and make them available in real time. The maximum, positively superimposed knock amplitude

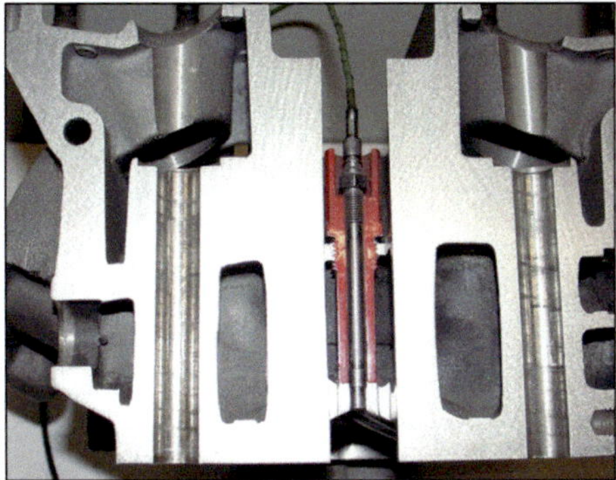

Figure 7.91: Pressure transducer in the cylinder head–installation examples (Kistler 6001 or 601 pressure transducer, installed flush with combustion chamber)

is derived as an indicator of damage; **Figure 7.93**. The KI meter (KI = Knock Intensity), developed by MAHLE, fulfills these requirements.

Precise indexing signals are needed in order to determine the exact knock amplitude level. A prerequisite for precise indexing is the correct positioning of the pressure transducer in the combustion chamber; **Figure 7.91**. It is particularly critical that it is properly bonded to the combustion chamber, in order to prevent connection channels. They would corrupt the signal with whistling vibrations, so that high knock amplitudes, in particular, would not be recorded at the correct level.

The KI meter, **Figure 7.92**, is capable of displaying measured signals for each operating cycle for up to eight cylinders, in real time. The measurement system can capture the ignition angle for each cylinder separately for each operating cycle, without needing to access the standard controller. Further signals can be captured on two additional channels per cylinder (e.g., acceleration signals), which can be displayed precisely relative to the operating cycle.

Automatic analysis allows this information to be used to determine for each operating cycle whether a measured knock amplitude was detected by the knock control system and whether the ignition angle was reduced; **Figure 7.93**. After a measurement, it is thus possible

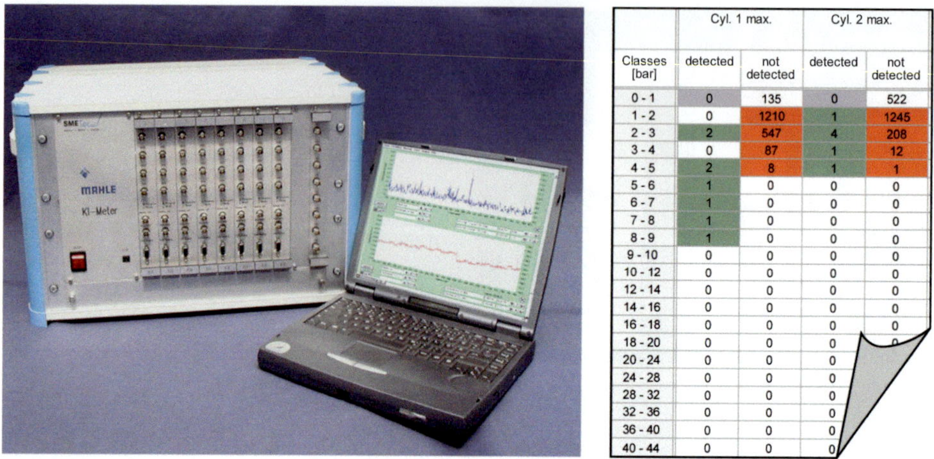

	Cyl. 1 max.		Cyl. 2 max.	
Classes [bar]	detected	not detected	detected	not detected
0 - 1	0	135	0	522
1 - 2	0	1210	1	1245
2 - 3	2	547	4	208
3 - 4	0	87	1	12
4 - 5	2	8	1	1
5 - 6	1	0	0	0
6 - 7	1	0	0	0
7 - 8	1	0	0	0
8 - 9	1	0	0	0
9 - 10	0	0	0	0
10 - 12	0	0	0	0
12 - 14	0	0	0	0
14 - 16	0	0	0	0
16 - 18	0	0	0	0
18 - 20	0	0	0	
20 - 24	0	0	0	
24 - 28	0	0	0	
28 - 32	0	0	0	
32 - 36	0	0	0	
36 - 40	0	0	0	
40 - 44	0	0	0	

Figure 7.92: The MAHLE KI meter [16] and the statistical analysis of the knocking operating cycles detected and not detected by the knock control system

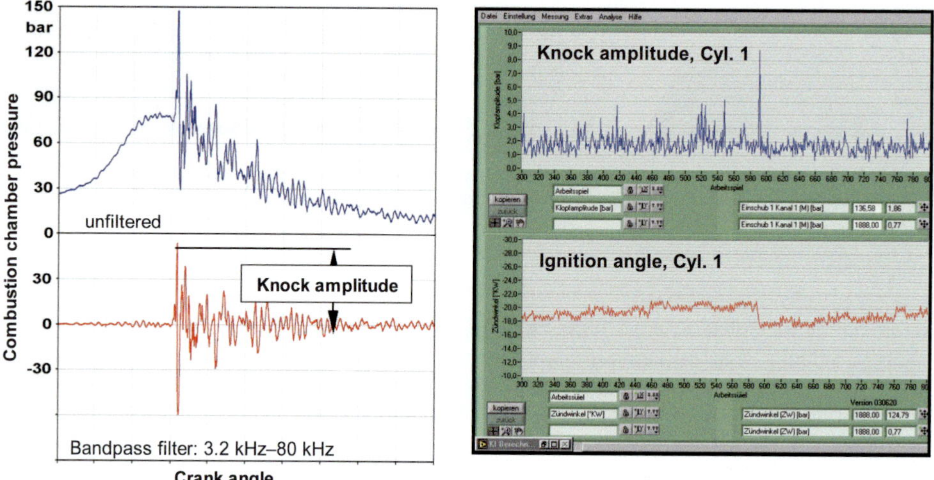

Figure 7.93: Analysis of positively superimposed knock amplitudes and control process of the knock control system, in online mode

to capture the knock amplitudes that were and were not detected by the knock control (KC) system in tabular form.

In order to analyze the quantitative knock strength, a value must be created that reflects the quantity and level of knock amplitudes. A knock intensity (KI) factor is defined for this purpose.

The calculation of the knock intensity factor KI uses the summation formula shown in **Table 7.6** for a predetermined number of four-stroke cycles to be measured. The example shows the knock amplitudes measured with increasing advanced ignition and the calculated KI factor.

Table 7.6: Calculation of the MAHLE knock intensity factor KI

Window position: 0 degrees

Window width: 150 degrees

Engine speed: 5,000 rpm, full load

Load cycles: 200

Class	Evaluation factor	Pressure range [bar]	Variation in ignition angle [°CA]					
			22°	23°	25°	27°	30°	35°
			Number of four-stroke cycles					
0	0	0–2	199	199	190	145	82	9
1	0.5	2–4	1	1	10	48	69	15
2	1	4–8	0	0	0	7	41	70
3	2	8–12	0	0	0	0	6	49
4	3	12–16	0	0	0	0	2	25
5	4	16–20	0	0	0	0	0	9
6	5	20–24	0	0	0	0	0	8
7	6	24–28	0	0	0	0	0	5
8	7	28–32	0	0	0	0	0	4
9	8	32–36	0	0	0	0	0	4
10	9	36–40	0	0	0	0	0	0
11	10	40–44	0	0	0	0	0	0
12	11	44–48	0	0	0	0	0	2
13	12	48–52	0	0	0	0	0	0
14	13	52–56	0	0	0	0	0	0
15	14	56–60	0	0	0	0	0	0
16	15	60–64	0	0	0	0	0	0
Overrun		> 64	0	0	0	0	0	0
Knock intensity factor KI			0.05	0.05	0.5	3.1	9.35	43.85

$$KI = N \cdot \frac{\sum\limits_{k=0}^{m} (n_k \cdot f_k)}{c}$$

KI: knock intensity factor

k: class number

m: maximum number of classes

n_k: number of cycles with pressure amplitudes in a class

f_k: weighting factor of a class

c: number of four-stroke cycles measured

N: Standardization constant

Using the functions of the KI meter as described above, it is possible, for example, to run a performance check on the knock control. This can expose errors in the control strategy or show a level of detection sensitivity for each cylinder.

Using an automated analysis system, the antiknock property of each individual operating cycle can be monitored during an endurance test and displayed graphically in real time. For example, individual severe knocking events that occur very rarely (mega-knocks) can be detected. In addition, the system shown can perform systematic investigations of knocking behavior, such as the influence of knocking combustion on the piston temperature or the influence of the air-fuel mixture on the knock limit.

7.5.3 Examples of measurement results

Figure 7.94 shows an example of the result of an analysis using the measurement method described, together with a piston temperature measurement.

As is evident from the figure, the KI factors for cylinders 2 and 3 increase relative to the ignition angle much earlier and more quickly than for cylinders 1 and 4. At this operating point, the test engine has a nonuniform mixture distribution between the inner (#2 & 3) and outer (#1 & 4) cylinders. Cylinders 2 and 3 are running a relatively lean mixture, so that knocking occurs more intensely under high loads.

From **Figure 7.94**, it can also be seen that the increase in piston temperatures rises linearly up to the knock limit, at about 2.5°C/°CA. A disproportional increase in piston temperature is evident only after severe knocking has occurred.

In another series of tests, the influence of the air-fuel mixture (lambda) on the piston temperature and knock limit was determined.

Figure 7.95 shows a shift in the knock limit for a rich mixture of about 3° CA in the direction of an advanced ignition angle. The piston temperature is lower for the rich mixture, but the temperature at the knock limit (now 3° CA earlier) is greater than for the lean mixture.

Both figures show that the ignition angle (IA) fundamentally has a significant influence on component temperatures, even without knocking. An additional increase in temperature due to knocking combustion, in contrast, is evident only at very high knocking intensities.

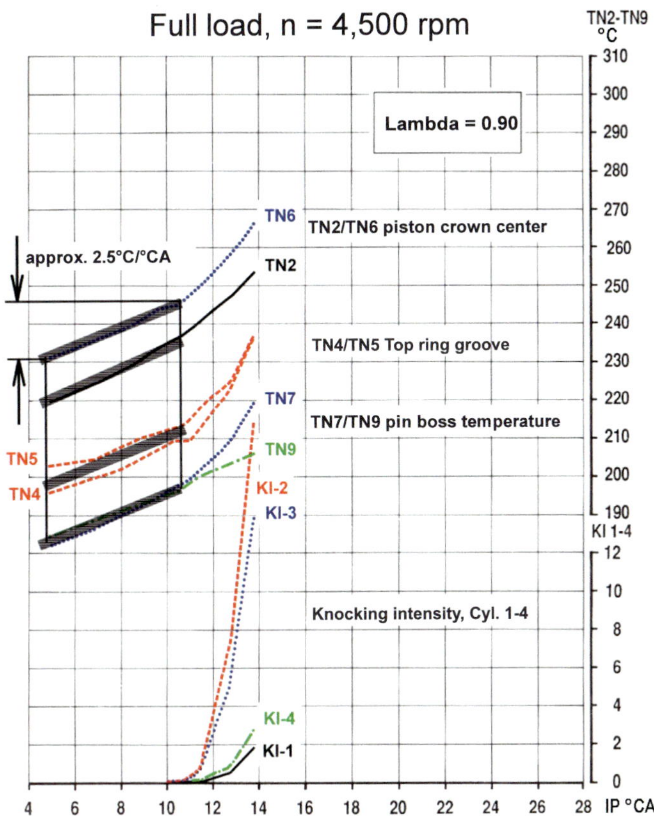

Figure 7.94: Piston temperatures and knocking intensity as a function of the ignition angle (comparison of all four cylinders) for a 2.0-liter, 4-valve, naturally aspirated engine

Operating conditions:

Full load, n = 4,500 rpm

T_{Oil} = 100°C

T_{Wa} = 90°C

Fuel: Reference fuel SUPER ROZ 96

Temperature descriptions:

TN2 Center of piston recess, piston 1,

TN4 Top ring groove, ATS, piston 2

TN5 Top ring groove, TS, piston 2

TN6 Center of piston crown, piston 3

TN7 Vent side boss, piston 3

TN9 Flywheel side boss, piston 4

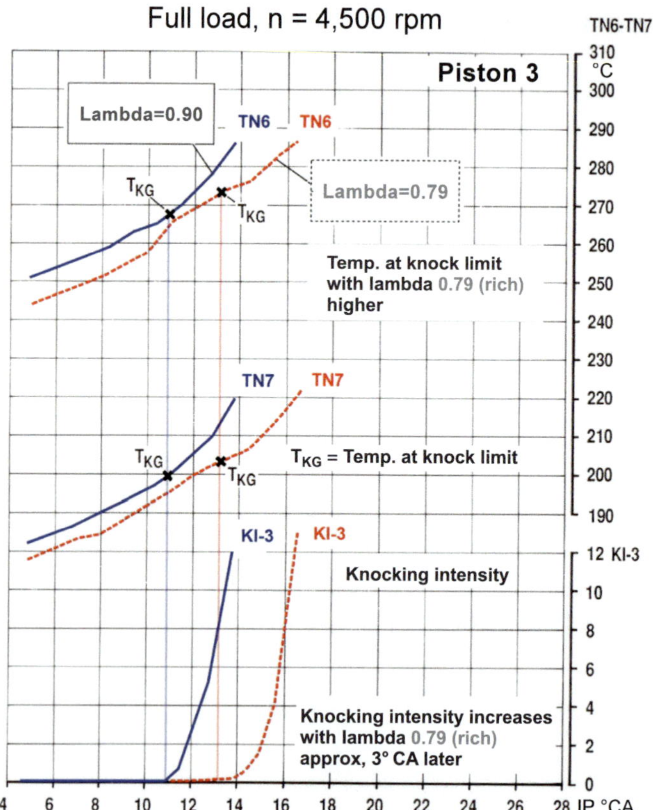

Figure 7.95: Piston temperatures and knocking intensity as a function of the ignition angle, at $\lambda = 0.90$ (lean) and $\lambda = 0.79$ (rich)

Operating conditions:

Full load, n = 4,500 rpm

T_{Oil} = 100°C

T   = 90°C

Fuel: Reference fuel SUPER ROZ 96

Temperature descriptions:

TN6 Temperature at center of piston crown, piston 3

TN7 Temperature at vent side boss, piston 1

T_{KG} Temperature at knock limit

7.5.4 Detection quality of knock control systems

Knock damage, such as axial deformation of the top ring groove, **Figure 7.87**, can be attributed to deficient knock-detecting functionality of the knock control system, among other factors.It was demonstrated that even relatively slight increases in knock amplitudes can cause such damage.

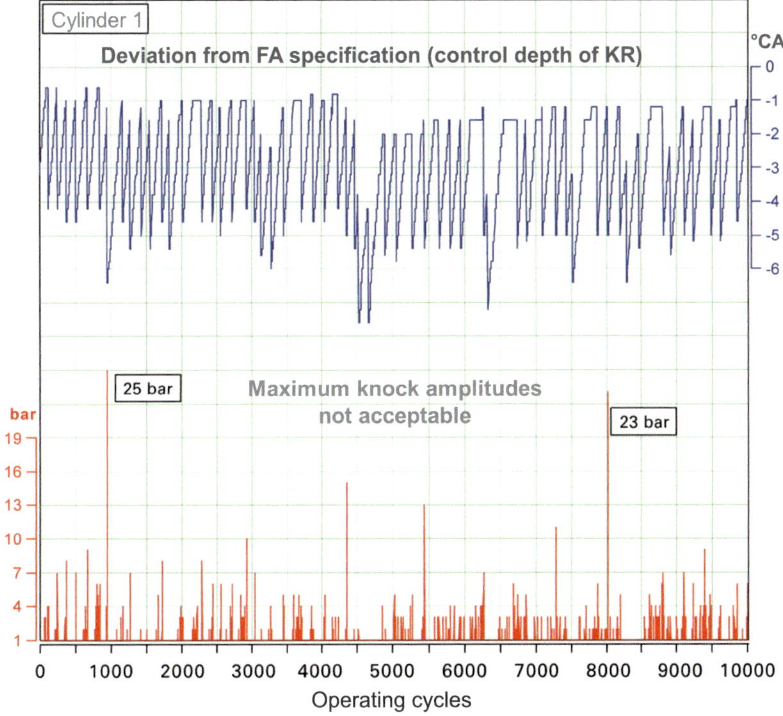

Figure 7.96: Maximum knock amplitude and associated change in ignition angle for 10,000 operating cycles, full load, 5,600 rpm, with ROZ 95, 1.8-liter turbocharged engine

Such knocking behavior is shown in **Figure 7.96**. For each operating cycle, these are the maximum knock amplitudes and the cylinder-specific retardation angle of the knock control system for cylinder 1. For the ROZ 95 fuel that was used, a so-called average control drop of about 3° CA was determined.

Knock amplitudes of up to 25 bar were measured in cylinder 1. With this knock control setting and ROZ 95 fuel quality, the knock damage indicated previously was caused in a full-load endurance test. With a fuel quality of ROZ 99, in contrast, comparable full-load endurance testing showed no knock damage whatsoever; **Figure 7.97**. The maximum knock amplitudes dropped to about 5 to 6 bar, and the average control drop is much lower.

The level and quantity of the maximum permissible knock amplitudes cannot be generalized, but must be determined by endurance testing for each application. Knock amplitudes of greater than 8 to 10 bar have already caused problems in a wide range of cases and should be avoided.

It is evident from **Figure 7.98** that the previous example is a cylinder-specific problem. The measured maximum knock amplitudes for all engine cylinders are summarized in a table.

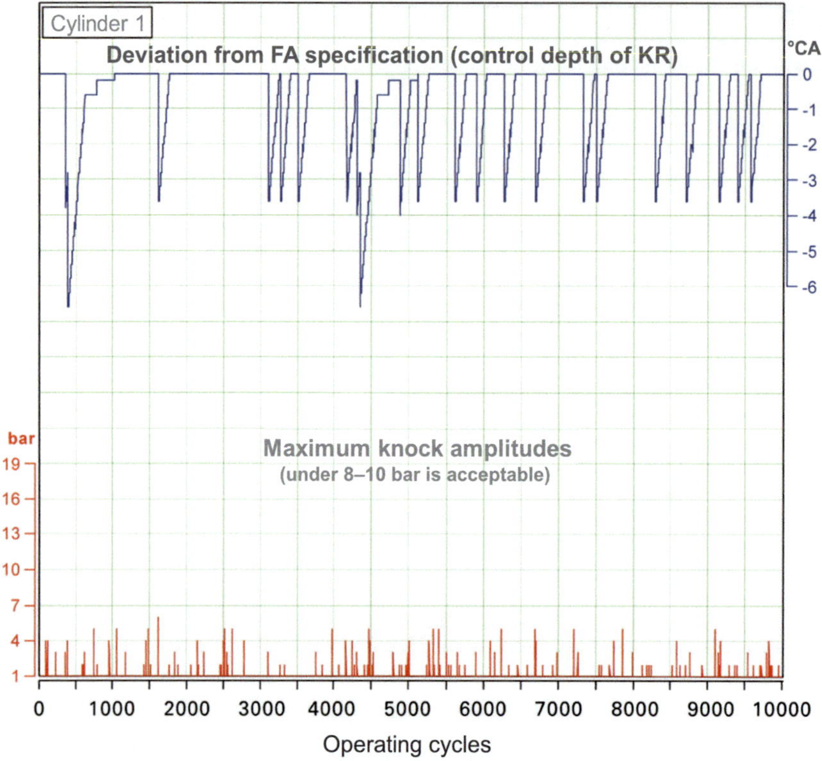

Figure 7.97: Maximum knock amplitude and associated change in ignition angle for 10,000 operating cycles, full load, 5,600 rpm, with ROZ 99, 1.8-liter turbocharged engine

Knock amplitudes that were and were not detected by the knock control system are differentiated. The left part of the figure (results with ROZ 95 fuel quality) shows that the problem indicated earlier affects only cylinder 1. The detection quality of the knock control system is not satisfactory for this cylinder. It failed to detect knock amplitudes of 8 to 9 bar. The poor detection quality for cylinder 1 was evident for ROZ 99 fuel quality as well.

Deficient detection quality can have many causes. The causes often cannot be addressed by design changes, because the engine development has already progressed too far. Software applications can provide a remedy in this connection.

One method is to back off the ignition angle values in the operating map, thus moving away from the knock limit and reducing the control drop of the knock control system. Typically, this can only be implemented for all cylinders at once. The result, however, is a degradation in engine efficiency, even in operating ranges where the knock limit is not reached.

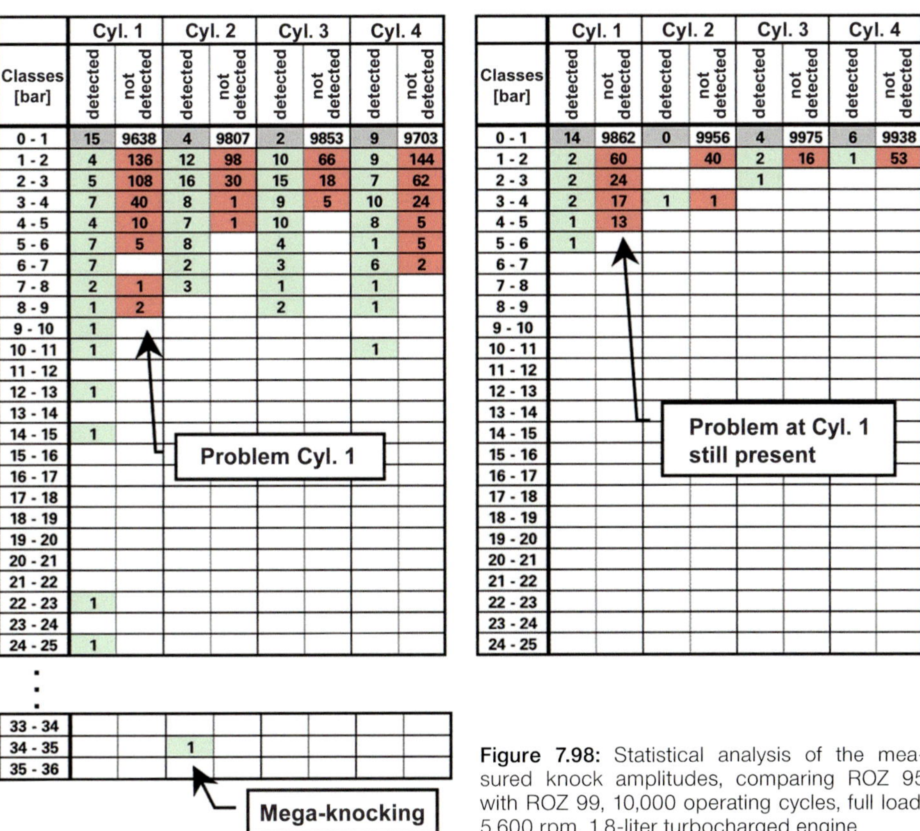

Classes [bar]	Cyl. 1 detected	Cyl. 1 not detected	Cyl. 2 detected	Cyl. 2 not detected	Cyl. 3 detected	Cyl. 3 not detected	Cyl. 4 detected	Cyl. 4 not detected
0 - 1	15	9638	4	9807	2	9853	9	9703
1 - 2	4	136	12	98	10	66	9	144
2 - 3	5	108	16	30	15	18	7	62
3 - 4	7	40	8	1	9	5	10	24
4 - 5	4	10	7	1	10		8	5
5 - 6	7	5	8		4		1	5
6 - 7	7		2		3		6	2
7 - 8	2	1	3		1		1	
8 - 9	1	2			2		1	
9 - 10	1							
10 - 11	1						1	
11 - 12								
12 - 13	1							
13 - 14								
14 - 15	1							
15 - 16								
16 - 17								
17 - 18								
18 - 19								
19 - 20								
20 - 21								
21 - 22								
22 - 23	1							
23 - 24								
24 - 25	1							
⋮								
33 - 34								
34 - 35			1					
35 - 36								

Classes [bar]	Cyl. 1 detected	Cyl. 1 not detected	Cyl. 2 detected	Cyl. 2 not detected	Cyl. 3 detected	Cyl. 3 not detected	Cyl. 4 detected	Cyl. 4 not detected
0 - 1	14	9862	0	9956	4	9975	6	9938
1 - 2	2	60		40	2	16	1	53
2 - 3	2	24			1			
3 - 4	2	17	1	1				
4 - 5	1	13						
5 - 6	1							
6 - 7								
7 - 8								
8 - 9								
9 - 10								
10 - 11								
11 - 12								
12 - 13								
13 - 14								
14 - 15								
15 - 16								
16 - 17								
17 - 18								
18 - 19								
19 - 20								
20 - 21								
21 - 22								
22 - 23								
23 - 24								
24 - 25								

Figure 7.98: Statistical analysis of the measured knock amplitudes, comparing ROZ 95 with ROZ 99, 10,000 operating cycles, full load, 5,600 rpm, 1.8-liter turbocharged engine

7.5.5 Mega-knocks and premature ignition

A mega-knock is an extremely high knock amplitude that can be up to 100 bar or greater, which occurs very rarely. As shown in Chapter 7.5, one single knock amplitude at this level can destroy the top land or the ring land.

Figure 7.99 shows a typical mega-knock detected in an eight-cylinder naturally aspirated engine in full-load/partial-load operation. The maximum knock amplitude and ignition angle trace for cylinder 1 are shown for 10,000 operating cycles, as well as the engine load at a constant speed of n = 6,000 rpm. The knock control system works perfectly under these operating conditions. In operating cycle no. 4747, an extreme knocking combustion event took place, with a maximum knock amplitude of 60 bar.

This operating cycle was recorded online and is shown in **Figure 7.100**. At the start of the pressure curve, normal combustion is evident, which then leads to an extreme knocking event, without any apparent cause.

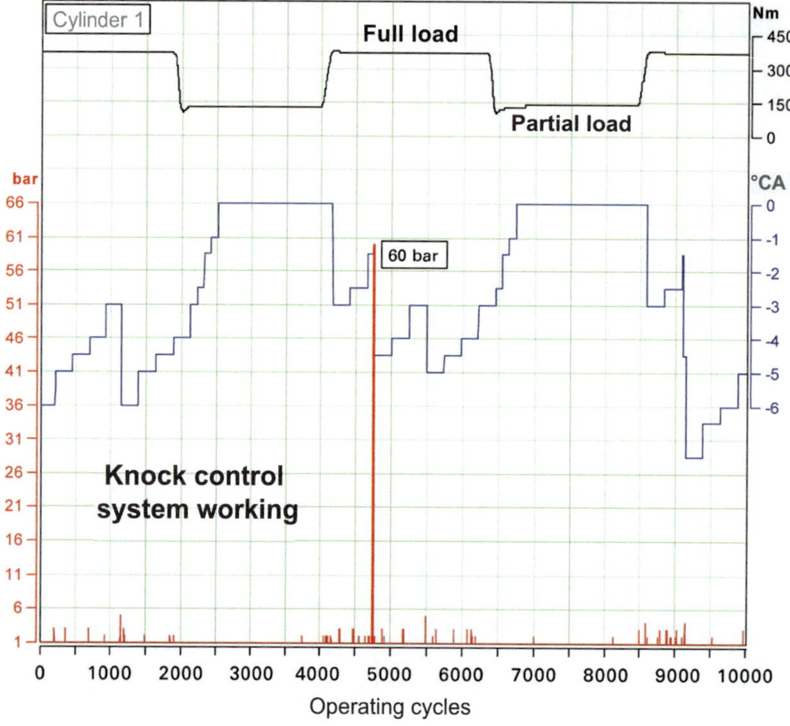

Figure 7.99: Mega-knock in cylinder 1, full-load/partial-load operation at n = 6,000 rpm, ROZ 95 fuel, 8-cylinder naturally aspirated engine

Mega-knock in cyl. 1

Engine: 8 cyl. naturally aspirated

Fuel: ROZ 95

Classes [bar]	Cyl. 1		Cyl. 2		Cyl. 3		Cyl. 4		Cyl. 5		Cyl. 6		Cyl. 7		Cyl. 8	
	detected	not detected	detected	not detected	detected	not detected	detected	not detected	detected	not detected	detected	not detected	detected	not detected	detected	not detected
0 - 1	0	9928	0	9949	1	9932	9	9987	3	9651	1	9921	4	9985	0	9959
1 - 2	2	40	2	42	2	55		2	3	29	2	64	3	5		26
2 - 3		23	2		4	2				4	2	3	1		2	3
3 - 4	2		1		1				2	4	1	1			1	3
4 - 5	2		1		1						1				3	
5 - 6			1							1						
6 - 7									1							
7 - 8											1					
8 - 9															1	
9 - 10											1					
10 - 11																
11 - 12																
12 - 13																
13 - 14																
14 - 15																
15 - 16																
16 - 17																
17 - 18																
18 - 19																
19 - 20																
⋮																
58 - 59																
59 - 60	1															
60 - 61																

Figure 7.100: Measurement of 10,000 operating cycles, with mega-knock in cylinder 1

The table with the data from all eight cylinders shows nothing else unusual. The mega-knocks do not occur only in cylinder 1 in this engine. No particular cylinder seems to be preferred. The running program–alternating load in this case–was likewise not a determining factor. It was established, however, that the mega-knocks occurred only at high speeds and under full load.

Potential defects in ignition can be ruled out by monitoring the ignition signal.

Mega-knocks are known not only in naturally aspirated engines, but also in turbocharged engines with indirect and direct injection. In most cases, the main initiator of such mega-knocks seems to be nonhomogeneous mixture formation in the combustion chamber. Hotspots can form in which the air-fuel mixture leads to severely knocking combustion [17].

The mega-knock phenomenon should not be confused with that of premature ignition. Premature ignition occurs occasionally in the development of highly stressed turbocharged engines. In this case, the air-fuel mixture ignites prior to actual induced ignition by the spark plug [18]. Premature ignition is caused by an extreme increase in the peak cylinder pressure in the compression phase. This greatly increased pressure curve can also be superimposed with additional knock amplitudes. Cylinder pressures of greater than 300 bar have been measured.

These phenomena, mega-knocks and premature ignition, cannot be seen in conjunction with the function of the knock control system.

7.6 Piston noise and transverse motion

7.6.1 Procedure for systematically minimizing piston noise

In order to optimize the acoustics of a combustion engine, the individual sources of sound and their contribution to the overall engine noise must be identified as precisely as possible. This is the only way to demonstrate which combination of measures will have the best effect, with the minimum number of engine tests.

One of the noise-generating mechanisms in a reciprocating piston engine is the motion of the piston, owing to its clearance perpendicular to its normal running direction. Under the influence of the gas and inertial forces, referred to hereafter as inertia forces, the connecting rod angle and the necessary installation clearance inevitably cause transverse motion of the piston. When the piston contacts the cylinder wall, it causes impacts that excite structure-borne noise in the engine structure. This is the cause of piston noise.

Piston noise contributes a significant component of the mechanically caused crankshaft drive noise. It is therefore important to optimize piston transverse motion in order to minimize engine noise in the summation of the sound level, but also in the degree of subjective unpleasantness. The tendency to use pistons with large installation clearances in conjunction with small piston pin offsets in order to reduce friction power loss in the engine and thus CO_2 emissions will make acoustic optimization more difficult in the future.

The following boundary conditions should be met in order to quickly and successfully achieve the optimization:

- An occurring piston noise should be clearly identifiable and assessable as such, at least for gasoline engines.
  This prevents expensive changes to components or clearances without any detectable improvement in the auditory impression.
- Each individual type of piston noise should be objectively quantifiable.
  This allows a quantitative analysis of the effects on noise excitation of changes made to the piston.
- For each individual type of piston noise, the characteristic motion sequence of the piston must be known.
  Visualization of the piston transverse motion measured on a loud piston shows the mechanism of the motion that must be prevented for the particular piston noise.
- The influence of various piston design measures on piston transverse motion must be precisely identified.
  Systematic parameter studies assist in acquiring the appropriate expert knowledge in order to keep the number of engine tests to a minimum.

In order to meet these requirements, the assessment and systematic minimization of piston noise follows four different principle approaches, as listed in **Figure 7.101**. These four approaches to the investigation of piston noise and its causes require completely different levels of time and resources. Depending on the problem, task definition, and objective, they are pursued at different intensity levels and combined as needed. Common to all procedures is the overall goal of systematically expanding knowledge and insight in order to be able to perform future optimization efforts more quickly and reliably.

Subjective assessment and evaluation of engine noise must be considered a necessary step at the start and finish of every optimization process. At the start of optimization, the auditory impression provides an initial diagnosis and helps to qualitatively evaluate the initial state as quickly as possible. A defined airborne noise level recording allows subsequent auditory comparisons. After completing the optimization, subjective assessments and a direct comparison to the initial state serve as verification that the improvements made have brought about a subjectively improved auditory impression. Such subjective assessment of engine noise can be performed either in the vehicle or on a test bench.

If subjective assessments and specifically prepared auditory comparisons are not sufficient as diagnostic tools, which is frequently the case with diesel engines, then reproducible airborne and structure-borne sound measurements must be taken on the test bench. The analysis

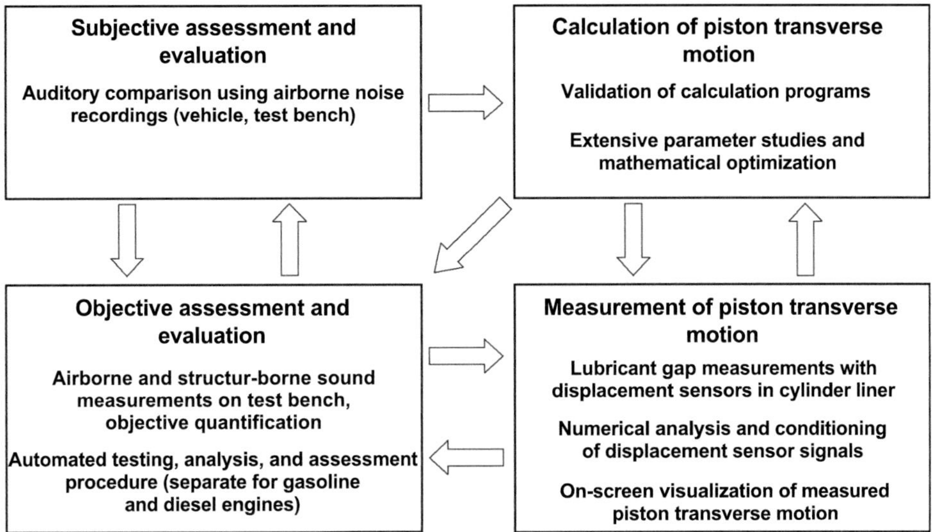

Figure 7.101: Procedure for systematically minimizing piston noise

of signals measured with different pistons, comparing them with one another, can provide an objective quantification of the piston noise. Standardization and partial automation of the analysis process allows the necessary experimental variations to be performed quickly.

A significant goal of systematic test planning is to minimize the number of variations of experiments, and therefore the number of engine test runs, while obtaining as much information as possible. The existing wealth of experience and the ability to make predictive computations contribute greatly to this goal.

Using numerical simulations of piston transverse motion, not only can the acoustically objectionable piston motions be recorded and analyzed, but the appropriate solution alternatives can also be qualitatively assessed, thereby reducing the time needed to implement them. This assumes that appropriate parameter studies have been completed.

In order to validate simulation programs and in support of basic development work, additional measurements of piston transverse motion are carried out. The effort needed to equip an engine is considerable, in terms of both time and resources. It is more economical to prepare a few typical representatives as experimental engines for such experiments–e.g., a conventional gasoline engine, one with direct injection, IDI and DI diesel engines, and commercial vehicle diesel engines. Using a measurement method with displacement sensors mounted on the cylinder [19], a number of critical parameters and their influence on piston transverse motion can be investigated on these engines with little retooling effort. The analysis and visualization procedures are largely automated. The valuable results support the targeted selection of test variants and the preparation of action plans.

7.6.2 Piston noise in gasoline engines

7.6.2.1 Subjective noise assessment

In gasoline engines, piston noise can be identified as such on the basis of the subjective auditory impression. Several different types of noise can even be distinguished this way. A practiced listener can determine the actual excitation mechanism in the engine on the basis of the sound pattern and the operating conditions.

The individual types of piston noise are classified by cause, location of excitation on the piston or cylinder, crank angle, and range of speed, load, and temperature. On the basis of the auditory impressions, "sound-rich" descriptions have been developed over time for these various types of excitation, a list of which is shown in **Figure 7.102**, without making a claim as to its completeness.

Several recordings of auditory examples for each type of piston noise are now available to experienced development engineers, which can be used as comparison samples for assessing the acoustics of an engine. In many cases, the cause of a noise can be diagnosed right away in the first subjective assessment. This means that the best possible test variants can be proposed early in the process for any subsequent engine testing.

A grading scale from 1 to 10 is typically used in the automotive industry for subjectively assessing engine noise, as documented in **Figure 7.103**.

If several piston variants are run in comparison during acoustic optimization, then this type of subjective assessment is often not sufficient for precise quantification. An objective noise analysis and quantification of the noises must then be performed.

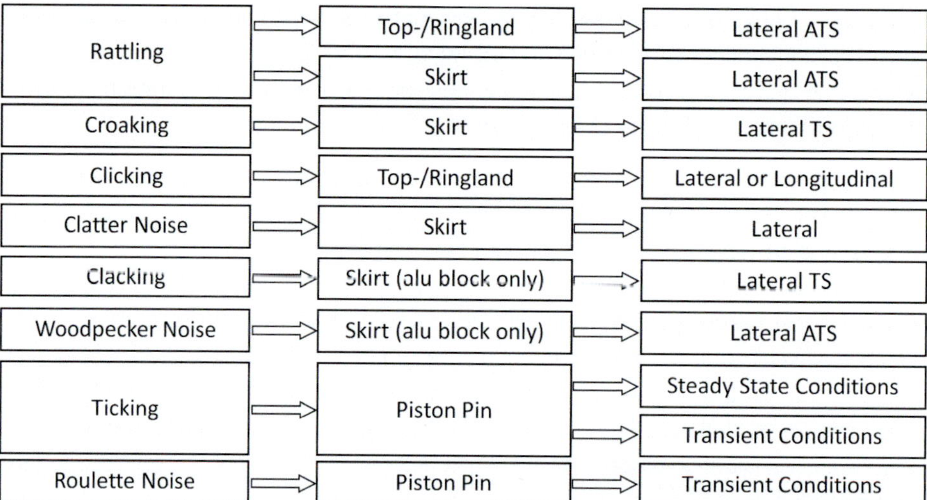

Figure 7.102: Piston noises in gasoline engines

Grade	Evaluation	Description	Comments
10	Excellent	Cannot be detected, even by practiced evaluators	Marketable
9	Very good	Can only be detected by practiced evaluators	
8	Good	Can only be detected by critical customers	
7	Satisfactory	Can be detected by all customers	
6	Acceptable	Considered disturbing by some customers	
5	Unsatisfactory	Considered disturbing by all customers	Not marketable
4	Deficient	Considered defective by all customers	
3	Insufficient	Claimed as a severe defect by all customers	
2	Poor	Only conditionally functional	
1	Very poor	No longer functional	

Figure 7.103: Grading scale for subjective assessment of piston noise

7.6.2.2 Objective noise assessment and quantification

The most common piston noises in a gasoline engine are as follows:

- So-called "rattling"–impact of the piston crown or, depending on the piston geometry, the rigid upper part of the piston skirt against the antithrust side in the crank angle range immediately prior to or at the TDC fired (TDC in the expansion stroke). The excitation mechanism is based on an inertia-induced traverse of the clearance, from the thrust side to the antithrust side, superimposed by a rotary motion component of the piston crown toward the antithrust side.

- So-called "croaking"–the impact and deformation of the elastic piston skirt on the thrust side of the cylinder, in the crank angle range after TDC fired. The excitation mechanism is based on a rotary motion, induced by the gas force, about a center of rotation in the lower skirt area.

The motion mechanisms described here are shown schematically in **Figure 7.104**. The most critical geometric parameters that can affect piston noise behavior with a reasonable amount of effort are piston pin offset, piston crown offset, cylinder distortion, piston shaping (contour and ovality), and clearance. Note that the minimum permissible installation clearance is largely determined by the profile of the piston skirt.

Both noises are very typical of gasoline engines and occur almost exclusively in a cold engine under low to medium load.

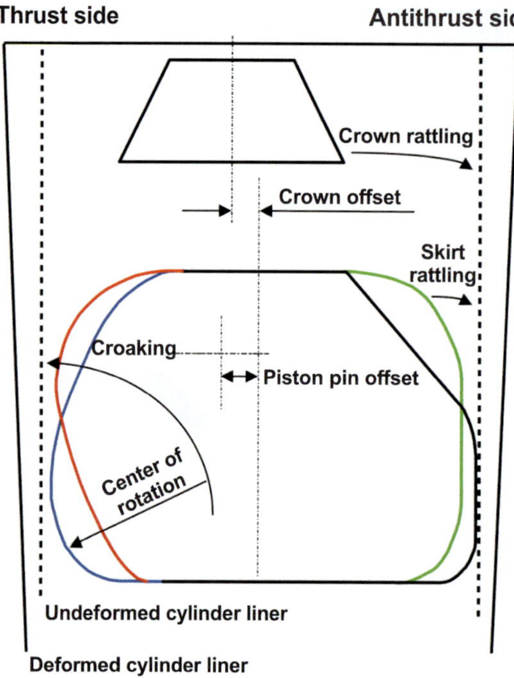

Figure 7.104:
Relationship between piston design and piston noise in gasoline engines

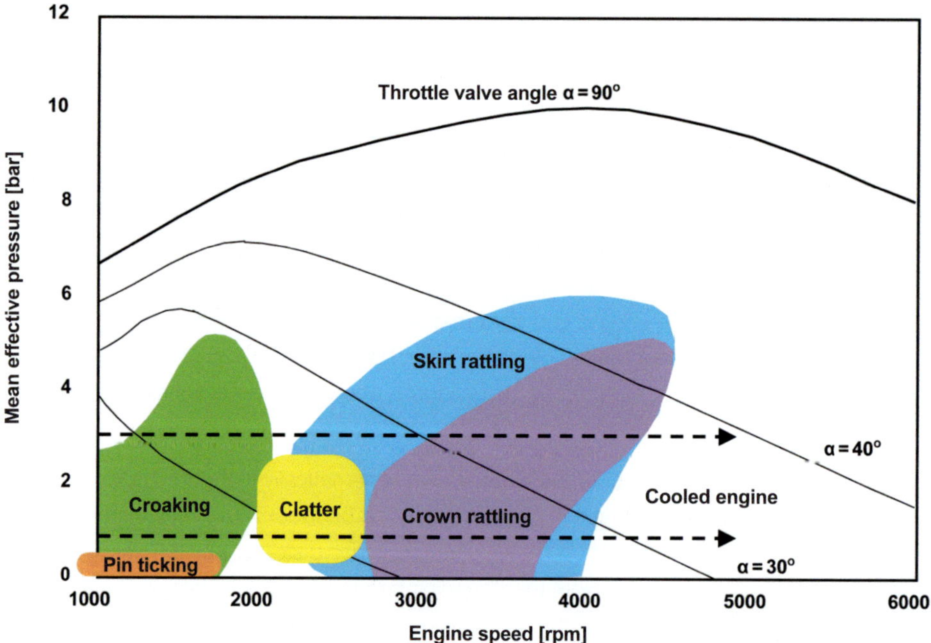

Figure 7.105: Gasoline engine operating map with typical operating ranges for piston noises and recommended measurement program

"Croaking," similar to the dark tone of a frog's call, mostly occurs at low engine speeds of less than 2,000 rpm. The brighter, sharp "crown rattling," on the other hand, is generally observed at speeds of greater than 3,000 rpm. If "skirt rattling" occurs because of the piston geometry, then it may be observed at lower speeds of around 2,500 rpm.

The gasoline engine operating map in **Figure 7.105** shows the operating ranges in which the piston noises can occur in principle.

If an engine is to be checked for the piston noises of "croaking" and "rattling" using measurement technology, then continuous measurement of the airborne and structure-borne noise should be applied during an increase in engine speed, under a constant, low load. The standard measurement program suggested for two different load cases in a cooled engine (coolant and oil temperature near −20°C) is indicated by the dashed-line arrows in **Figure 7.105**.

Figure 7.106 shows an example of the spectral distribution of the airborne noise signal measured during a run-up. As would be expected from the subjective auditory impression, the "croaking" and "rattling" piston noises have very different frequency distributions. The "croaking" caused by the piston skirt shows up as a relatively narrow band in the frequency range between 1 and 2 kHz. "Rattling," however, occurs as a more wide-band excitation because of the high stiffness of the impacting components. Depending on the impact intensity and the engine structure, it can propagate in the range from 2 to well over 5 kHz.

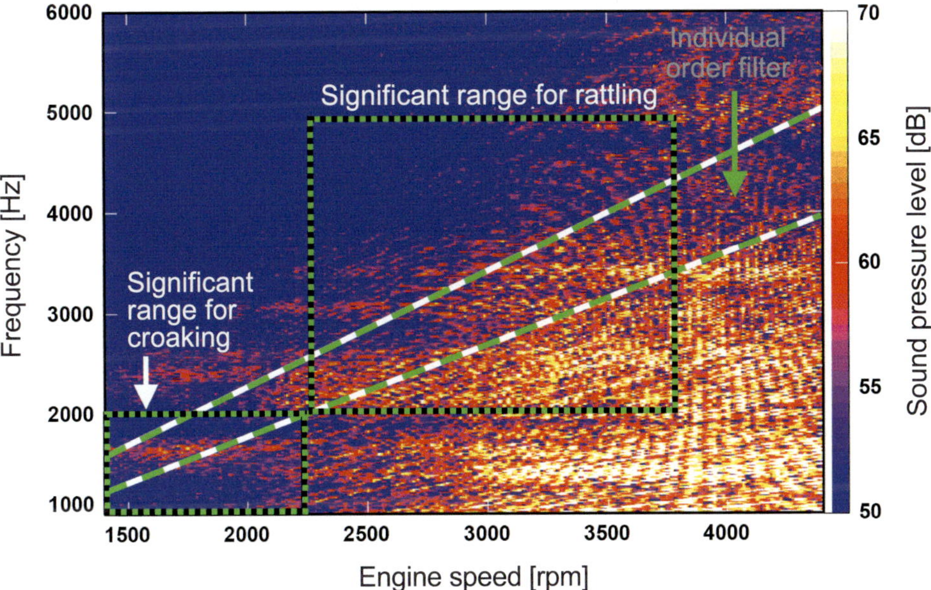

Figure 7.106: Spectral distribution of the airborne noise measured during a run-up in engine speed, showing "croaking" and "rattling"

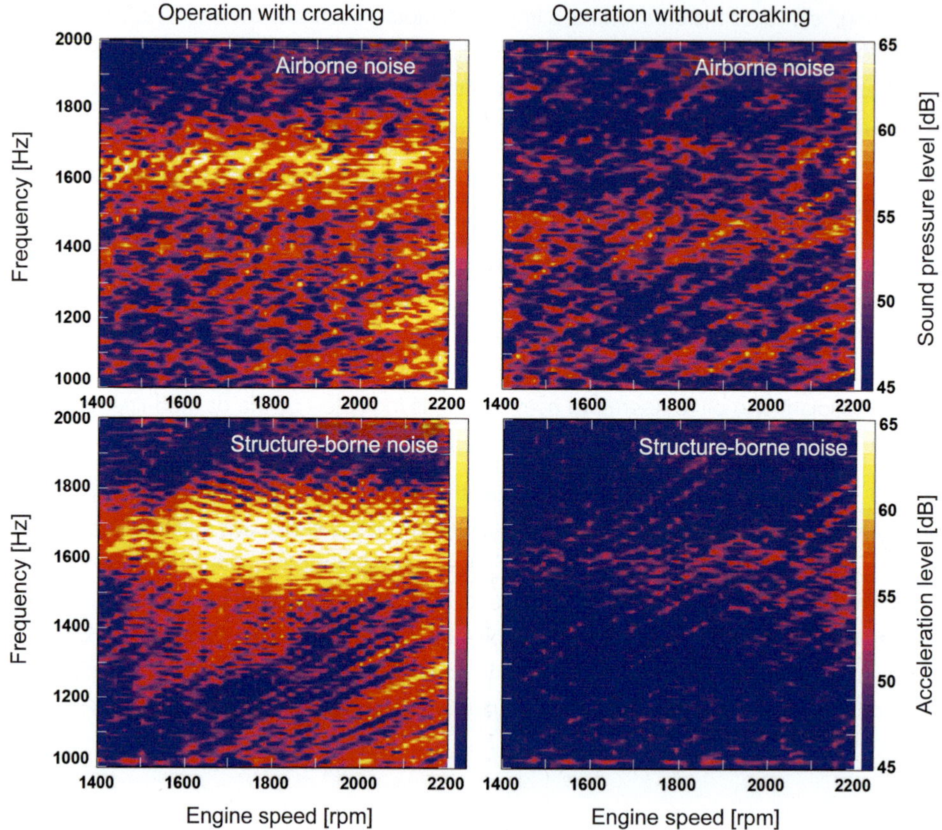

Figure 7.107: Partial airborne noise spectra (far field, thrust side) and structure-borne sound spectra (cylinder block, thrust side) for operation with and without "croaking"

In addition to engine speed, another important aspect for using measurement technology to differentiate the two types of noise has thus been identified. Both criteria together allow selective assessment of the signals, using defined frequency-speed windows, as shown in **Figure 7.106**.

In order to ensure as part of such a procedure that the objectionable noises lie within the selected frequency-speed windows, a comparison of the measurement data with and without the piston noise should be performed at all times. An estimate of the potential, comparing a piston version that has been deliberately designed to be acoustically problematic with one that has been acoustically optimized with very little installation clearance, can be helpful in this regard. Often, no optimized piston is available and the engine is not allowed to be reconfigured, so measurements from cold- and hot-engine operation can be compared instead. Because not all of the differences detected in the airborne noise spectrum can be attributed to piston noise, however, the ranges with the piston as the predominant cause of excitation must be verified using structure-borne sound signals.

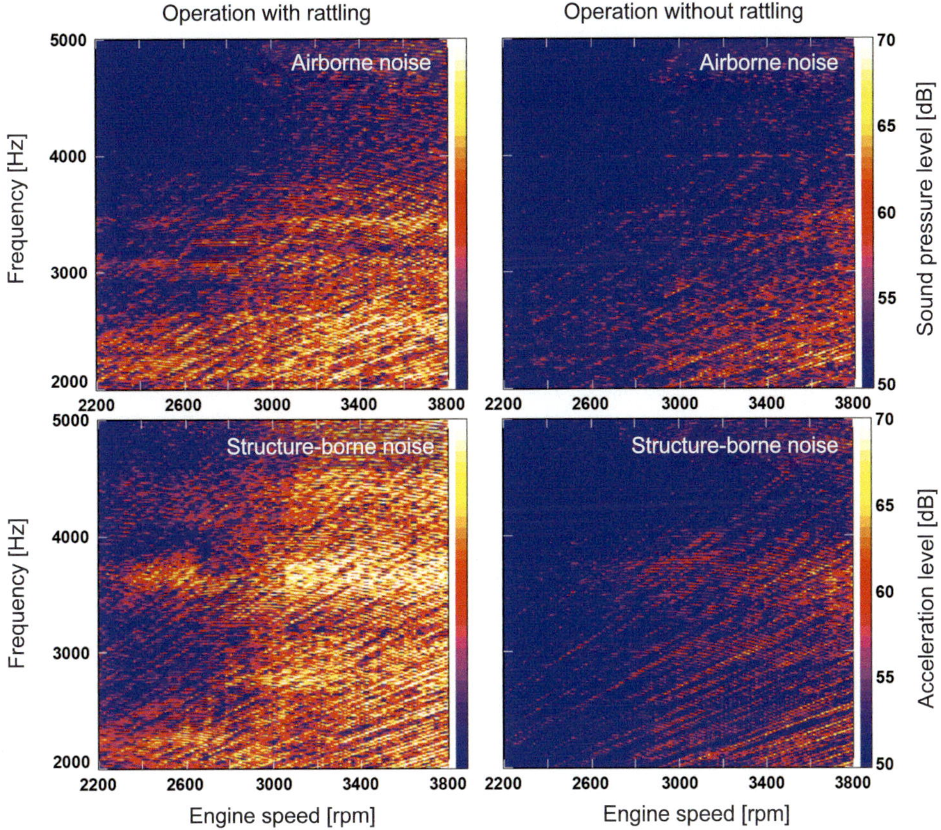

Figure 7.108: Partial airborne noise spectra (far field, antithrust side) and structure-borne sound spectra (cylinder block, antithrust side) for operation with and without "rattling"

This is done by analyzing simultaneous measurements of the accelerations on the cylinder block, which represent the piston noise to a large degree. **Figure 7.107** shows a comparison of airborne and structure-borne sound spectra for a cold engine (with "croaking") and a hot engine (no "croaking").

The airborne and structure-borne sound spectra in **Figure 7.108** indicate that this type of representation also can visualize wide-band "rattling" and that the point of impact (crown or skirt) can also be differentiated in the frequency range. It is notable that the excitations that occur at a higher speed also affect a higher frequency range. This is due in part to the stiffness of the components undergoing impact, and in part to the rising gas and inertia forces at higher speeds.

Depending on the location of the piston noises identified within the frequency-speed windows, an individual filter band can be defined as a function of the speed. This band covers the ranges of interest for piston analysis and masks the less informative ranges. A simple narrow band-pass filter, as is indicated in **Figure 7.106**, can often be used for this purpose as well. If

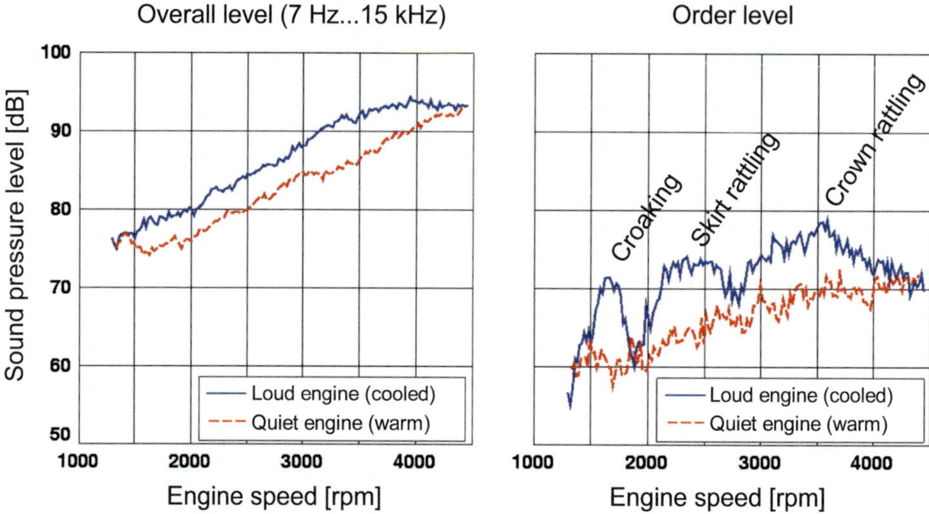

Figure 7.109: Overall levels and order levels as functions of engine speed, calculated from identical airborne noise signals

the level of only that portion of the signal within the filter band is calculated, the piston noises present in the range are indicated by significant peaks. **Figure 7.109** shows the curves of these order levels in comparison to the calculated overall level of the airborne noise, as measured for cold- and hot-engine operation.

If different piston variants are run under essentially identical operating conditions during the optimization process, then the differences in sound levels can be used to quantitatively analyze the individual piston noises. An analogous procedure can also be used for the measured structure-borne sound signals, allowing additional cylinder-specific analysis, depending on the location and number of the measurement points.

7.6.2.3 Piston transverse motion and influence parameters in gasoline engines

The impact excitation caused by the piston can be minimized only by optimizing the piston transverse motion. The overall goal should therefore be to "model" the piston transverse motion for all operating points such that only a minimum amount of the impact energy is transmitted to the engine structure upon contact alteration.

In order to change the motion mechanism of the piston at the point in time that contact is made with the cylinder wall, the preceding phase of motion must be targeted. Precise knowledge of the motion behavior of the piston, as a function of the speed, load, and temperature, is therefore just as critical as the knowledge of the influence of individual design parameters on piston transverse motion.

Measuring the piston transverse motion is therefore an indispensable means for obtaining basic knowledge, and also serves to validate simulation models and parameter identification for computations.

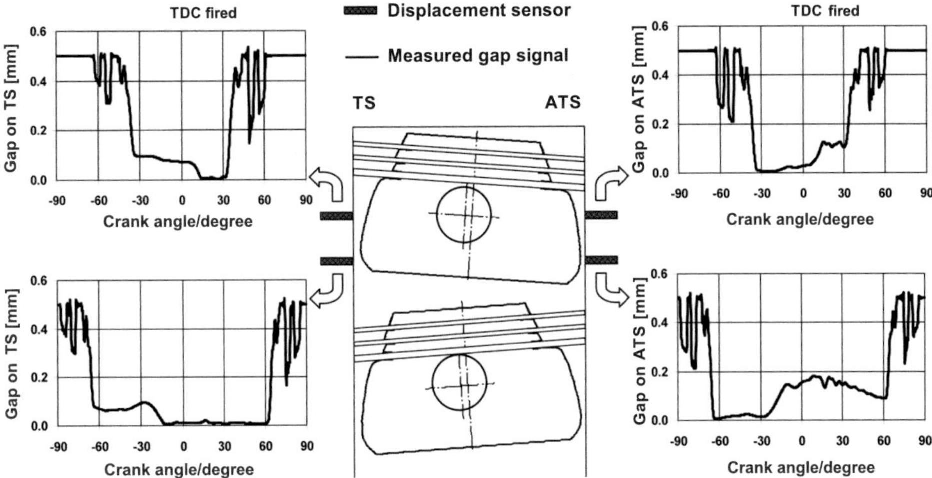

Figure 7.110: Lubricant gap signals between the piston surface and cylinder wall

When applying a measurement method with displacement sensors [19] mounted in the cylinder, a moving measurement linkage is not needed for data transmission. The advantage of this method is a quick and simple changeout of the test variants, without requiring setup of the measurement equipment on the piston. A disadvantage is the increased programming effort for the numerical analysis of the individual lubricant gap signals measured between the piston and the cylinder wall. The tilt angle curve for the piston axis and the translational displacement of the piston pin must be calculated from the usable individual segments of such lubricant gap curves as shown in **Figure 7.110**. Signals caused by the piston rings must be ignored. The signal range that can be interpreted may be limited to the crank angle range about the top dead center and depends on the piston skirt length and the application to the cylinder.

In order to reduce the effort for setup, in terms of both time and resources, a few typical engine variants have been prepared as test carrier. A number of critical parameters and their influence on piston transverse motion and structure-borne sound excitation can be investigated using these setups.

Selected examples will clarify the motion mechanisms that are associated with the noises known as "croaking" and "rattling." A few typical displacement signals for the motion curves are observed, along with the accelerations measured simultaneously at the cylinder.

Figure 7.111 (left side) shows the typical signal curves for operation with "croaking."

At lower speeds, the gas force is predominant, even in the early compression phase, thereby inducing contact on the antithrust side as the piston is thrust upward. Not until the crank angle range of 30° to 10° before TDC fired does the bottom end of the skirt change contact from the antithrust side to the thrust side. The associated impact of the skirt, which is very elastic at this height, generally does not lead to any significant structure-borne sound excitation. The resulting diagonal orientation remains until after TDC fired. Under the influence of gas force, the piston then straightens up, and the more rigid upper skirt area strikes the thrust side of the cylinder. The piston skirt is deformed over a large area, so the bottom end of the skirt lifts only slightly off the cylinder wall. The thrust-side structure-borne sound excitation generally occurs in the crank angle range between 10° and 25° after TDC fired.

Measures with a positive effect include

- greater barrel shape of the piston contour;
- greater tapering of the lower end of the skirt;
- greater stiffness in the middle of the skirt;
- greater piston pin offset toward the thrust side;
- reduced installation clearance.

A piston transverse motion under similar operating conditions, but without measurable piston noise, can serve as a model for an optimized, "quiet" sequence of motion. Direct comparison of such data often reveals the cause for the structure-borne sound excitation. The diagrams on the right side in **Figure 7.111** show that "croaking" does not occur for measurements at lower gas force, and thus lower lateral force, despite similar behavior of the piston transverse motion. In the present case, this indicates insufficient skirt stiffness as the cause of the "croaking."

An optimized profile and stiffness of the piston skirt reduces the severity of the skirt deformation under load. The improved shape stability promotes continuous rolling of the piston skirt on the thrust side, thus preventing the impacts that lead to "croaking," as has already been determined for operation with lower lateral force loads. A characteristic of improved rolling motion is a more obvious lifting of the bottom end of the skirt at the point of the maximum tilt angle.

Figure 7.112 shows the typical signal curves for operation with "rattling." At high speeds, the piston is located on the thrust side during the upward stroke, because of the influence of inertia force. The increasing influence of the gas force causes the piston crown to move from the thrust side to the antithrust side well before TDC fired. If the top land or the first ring land of the piston crown strikes the antithrust side, this is known as "crown rattling." If the impact occurs at the rigid upper skirt area, then this is logically known as skirt rattling. The antithrust-side structure borne sound excitation takes place between 15° and 5° after TDC fired.

Measures with a positive effect include

- crown offset toward the thrust side;
- greater tapering of the upper end of the skirt;
- adapted top land and ring land clearances;
- lesser piston pin offset toward the thrust side;
- reduced installation clearance;
- reduced cylinder distortion.

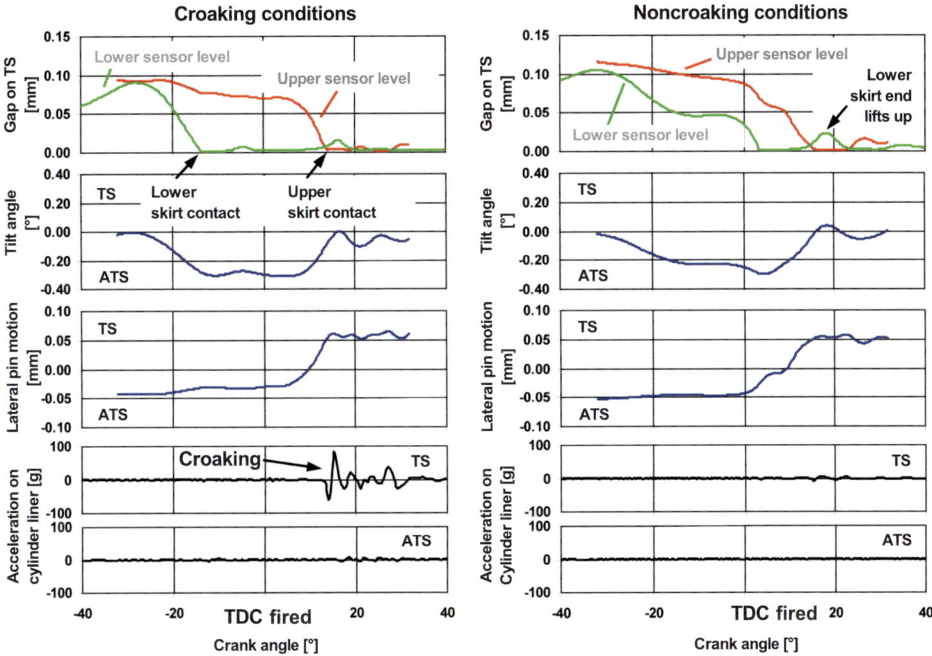

Figure 7.111: Relationship between measured lubricant gap, piston axis motion, and structure-borne sound excitation during operation with "croaking" (left, low speed, low load) and without "croaking" (right, low speed, no load)

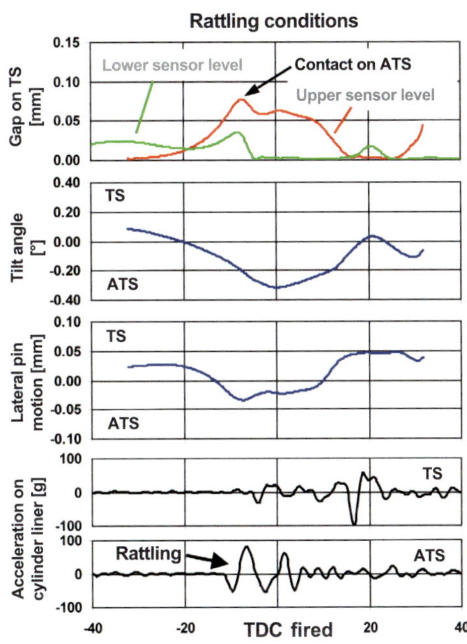

Figure 7.112:
Relationship between measured lubricant gap, piston axis motion, and structure-borne sound excitation during operation with "rattling" (high speed, low load)

Because of the diagonal piston orientation arising in the region of the TDC fired, the sequence of motion described above for "croaking" typically follows this effect. A second structure-borne sound excitation on the opposite thrust side, as can be seen in **Figure 7.112**, is therefore often detected, but is not subjectively perceived as additional "croaking" at the higher speed.

As an adjunct to manual analysis and interpretation of the individual diagrams, the measured displacement signals can be used as input variables for computer animation of the piston transverse motion. Only a clear visualization and comparison of a number of suitable test variants can provide the insight that the engineer needs for quick and reliable development.

7.6.3 Piston noise in passenger car diesel engines

7.6.3.1 Subjective noise assessment

The acceptance of the diesel engine as a comfortable vehicle drive has increased greatly since the introduction of the high-torque direct-injection passenger car engine. For this reason, the acoustic optimization of diesel engine combustion is considered by vehicle manufacturers to be one of the most critical future tasks, besides the reduction of consumption and exhaust gas emissions.

Modern injection systems, which allow pre-injection based on the operating point, or modeling of the injection sequence, contribute greatly to this end. This and the current pump concepts with reduced hydraulic noise lead to a reduction in noise level, which has caused mechanically induced noise to play a more prominent role again. Systematic minimization of piston noise is therefore more and more important in diesel engines as well.

New piston concepts such as the steel piston, which has friction power loss advantages over aluminum pistons under warm operating conditions due to its low thermal expansion, present new challenges for acoustic optimization on account of their large running clearances and short skirt heights.

In contrast to the gasoline engine [20, 21], diesel engines [22] have no typical types of piston noise that can be identified by their sound alone. The physical boundary conditions that prevail in diesel engines, however, do not support the conclusion that less piston noise should be expected than in gasoline engines. Higher gas forces, more rigid pistons with greater mass, and greater installation clearances indicate that greater structure-borne sound excitation is more likely.

Nevertheless, for many drivers, the degree of sensitivity to piston noise appears to be less and the general tolerance of mechanically induced noises to be greater. Both tendencies are certainly due to the dominant combustion noise that is common in diesel engines.

The acoustic similarity of the combustion noise and piston noise often makes a subjective assessment of piston noise difficult. A direct comparison of recorded signals, however, allows the human ear to detect even slight differences very well. When variants are compared, how-

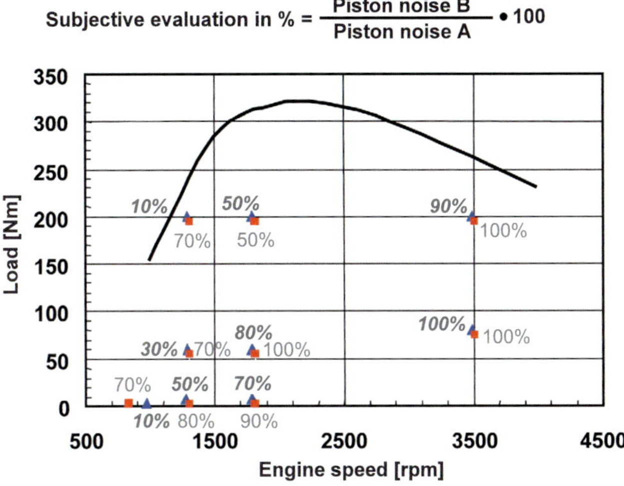

Operating point	Idle 0 Nm	1300 rpm 0 Nm	1300 rpm 60 Nm	1300 rpm 200 Nm
Warm ▪				
Piston A	100%	100%	100%	100%
Piston B	70%	80%	80%	70%
Cooled ▲				
Piston A	*100%*	*100%*	*100%*	*100%*
Piston B	*10%*	*50%*	*30%*	*10%*

Figure 7.113: Engine operating map of a DI passenger car diesel engine, with subjective assessment of the piston noise of two different pistons, assessed in direct listening comparison

ever, because these differences are attributed to the change in the running characteristic of the piston, scrupulous attention must be paid to the fact that in such a comparison the combustion sequence remains unchanged and must not appear here as a second parameter. In order to ensure this, the combustion chamber pressure curve must be recorded and appropriately controlled, even in the case of signal recording for subjective noise assessment.

In order to determine the subjective, acoustically relevant operating ranges in the diesel engine, an estimate of potential is helpful. The engine noises are compared between a selected loud piston (piston A: without piston pin offset, with acoustically unfavorable piston profile, and large installation clearance) and a piston designed specifically to be quiet (piston B: with computationally optimized piston pin offset, acoustically favorable piston profile, and very small installation clearance). The most important results of such a comparative, subjective noise assessment are shown in **Figure 7.113**.

Piston A is defined as 100% for each operating point. The result for the piston noise detected from the quiet piston B is therefore a lower value.

While the subjectively perceived difference is very low at high speeds–although this is precisely where a high proportion of mechanical noise is present–a significant influence on the overall noise by the piston excitation can be detected at low speeds. This applies particularly to speeds of less than 2,000 rpm, at very low or high loads. In the lower partial-load range, where combustion causes high pressure rise values, however, the combustion noise is dominant, so that the piston noise that must surely also be present in this range plays a secondary role.

The influence of piston design on the piston noise is fundamentally greater in a cold engine than in a hot engine. Nevertheless, an improvement can be perceived subjectively, even at operating temperature.

Therefore, it is proposed that a measurement program be standardized that includes both the cooled engine (symbol △, water and oil temperature around −20°C) and the engine at operating temperature (symbol □). **Figure 7.114** again shows the operating map of a diesel engine, now with the measurement program resulting from the above conclusions, consisting of run-up curves above the load at low speed and for different engine temperatures (symbolized by the vertical arrow), and a few selected stable measurement points from the ranges that are particularly relevant from an acoustical standpoint. Previous experience has shown that these operating conditions are not only representative of subjective assessments, but are also very well-suited for computational optimization processes and as measuring points for objective noise assessment.

In the crank angle range around the TDC fired, the contact side of the piston in the cylinder changes from the antithrust side to the thrust side. For pistons with no piston pin offset or with an offset toward the antithrust side, this is a largely translational motion sequence with small tilt angles about the piston axis. This can give rise to a free motion phase for the piston, which causes the piston to accelerate toward the thrust side. The piston then strikes the thrust-side cylinder wall, causing structure-borne sound excitation in the engine structure.

If the piston is offset toward the thrust side, however, then the contact alteration starts with the bottom end of the skirt, at an earlier crank angle, and the motion of the piston crown toward the thrust side is slowed. The result is greater tilt angles of the piston axis and a temporary diagonal orientation in the cylinder. Owing to the supporting lateral forces and the greater deformations of the piston skirt, the structure-borne sound excitation can be reduced [23]. In current DI engines, the optimum piston pin offset is generally between 0.3 and 0.6 mm toward the thrust side. Therefore, a starting value of 0.5 mm is suggested for the experimental development phase.

Seizure resistance must be considered in the optimization of the barrel-shaped, oval piston profile. Because this has a direct interrelationship with the installation clearance, only an overall analysis makes sense. This circumstance can be clarified with the help of two piston skirt profiles; **Figure 7.115**.

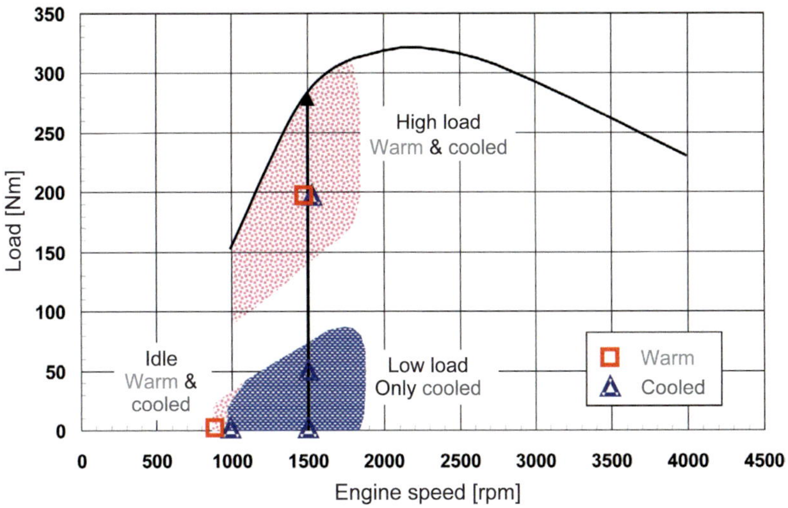

Figure 7.114: Noise operating map of a DI passenger car diesel engine and recommended measurement program, based on subjective auditory impression, for subjective and objective noise assessment

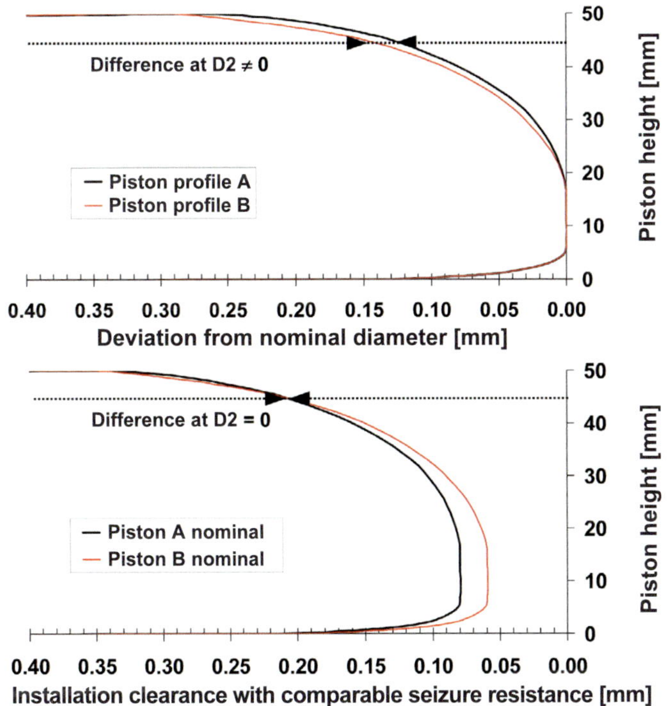

Figure 7.115: Interrelationship between the piston profile, installation clearance, and seizure resistance
D2: Piston diameter at a fixed piston height in the upper skirt area

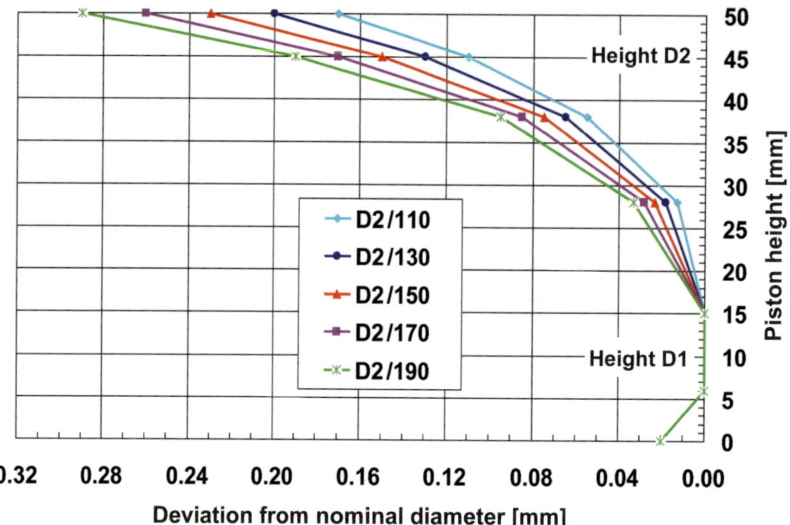

Figure 7.116: Piston skirt geometries used for calculations, with different implementations of seizure resistance (height D1: piston height at which the piston has the greatest diameter D1)

If a piston has less deviation in the upper skirt area (piston profile A)–indicated here as height D2, according to the piston drawings–then the guidance of the piston in the cylinder appears to be better, and less structure-borne sound excitation is to be expected. If, however, the reduction in seizure resistance is taken into consideration, then the piston must be installed with a nominal installation clearance that is greater by this amount; **Figure 7.115**, lower diagram. In this case, piston B appears to provide better guidance in the cylinder, and therefore less structure-borne sound excitation. The two influence parameters, piston profile and installation clearance, can be clarified by a fast and rough projection to determine which design will tend to lead to a running characteristic that is more favorable from an acoustical standpoint.

The key parameters for five piston skirt profiles are sketched in **Figure 7.116**, differentiating the upper skirt taper at height D2 between 110 and 190 μm in steps of 20 μm. The lower skirt taper remains unchanged. The analysis of the piston transverse motion is made for different assumptions of installation clearance, such as the steps of 20 μm.

If the maximum contact force that occurs on the thrust side is again taken as an approximate criterion for structure-borne sound excitation, then the result is the lines drawn in **Figure 7.117** for the individual piston profiles, which indicate smaller installation clearances, as expected. Also, as assumed, pistons with increasingly large upper skirt taper cause a shift toward greater maximum forces.

If points with identical running clearance at height D2 (deviation of the piston geometry at height D2 plus installation clearance) are connected to one another, then the result is the connection line, as a line of comparable seizure resistance. This is additionally shown in **Figure 7.117**, using the example of 190 μm. Its progression shows clearly that for such

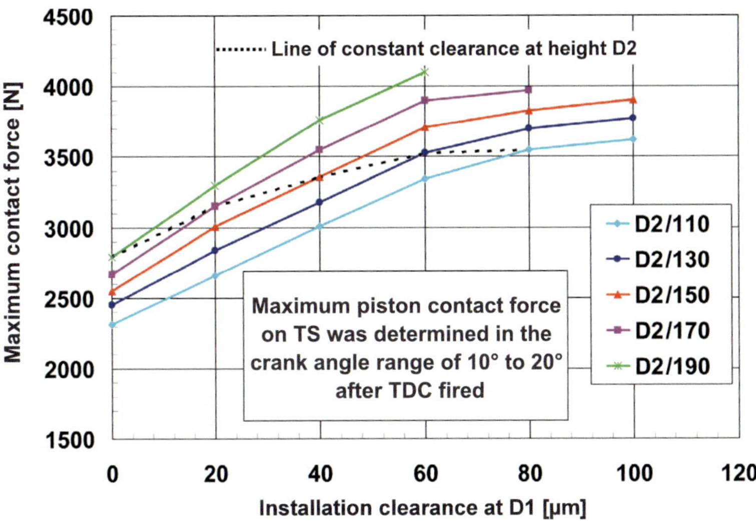

Figure 7.117: Maximum contact force as a function of piston profile and installation clearance, calculated for low speed and low load

engines with comparable seizure resistance, the least possible installation clearance with the correspondingly necessary large upper skirt taper provides the best acoustic solution.

This potential for improving the acoustics often justifies the implementation of small installation clearances, using a graphite coating on the piston skirt to increase seizure resistance. The positive influence of reducing the installation clearance and the negative influence of the greater upper skirt taper cancel each other out only for large installation clearances greater than 60 µm.

Owing to the high stiffness of the skirt in diesel engine pistons, compared with gasoline engine pistons, the typical ovality values in series production diesel engine pistons are considerably less dispersed. A computational pre-estimate is therefore not absolutely necessary for a first test. The final ovality values arise largely from the optimization of the wear pattern on a piston that has been run. Particularly for pistons with optimized profiles and small installation clearances, the acoustic influence of the ovality of the piston skirt is of secondary importance.

7.6.3.2 Objective noise assessment and quantification

Noise analyses and a subsequent objective noise assessment have the goal of identifying an existing piston noise as such and quantifying it, so that piston variants can be made objectively comparable.

In order to determine the absolute value of the noise component caused by the piston, a systematic partial sound source analysis must be performed with the goal of quantitatively separating the noise excitation from combustion, the injection system, the valve train system, the crank mechanism, and various auxiliary systems [24]. Such extensive tests are primarily suited for basic research work, but not for the regular optimization work leading up to series production.

A method is used, therefore, that allows piston variants to be objectively compared with one another with relatively little effort. The portion identified as piston noise, however, can be quantified only relative to other piston variants. An initial basis for such an A/B comparison is often the potential estimate suggested previously, with selected loud and quiet piston designs. Every other test variant can be arranged between them. The improvement that has been achieved between the initial state and the optimized piston can then be documented, as shown in the following example.

If a noise excitation due to piston impact occurs in a diesel engine, then it can be assumed that this mechanical impact excitation will be found to be much stronger in a structure-borne noise signal measured in the upper area of the cylinder block, on account of the dominant external structure-borne noise transfer path, than in a signal measured at the main bearing cover of the associated cylinder [25]. The force excitation due to combustion and the impact excitation of the crankshaft in the bearing, on the other hand, are greater in the area of the main bearing than at the exterior of the crankcase. The internal structure-borne noise transfer path is dominant.

If such structure-borne sound measurements are performed with piston variants that are very different acoustically and the signal levels measured at the main bearing do not significantly deviate from one another, then the difference that can be measured in the upper crankcase area and in the airborne noise can only be attributed to excitation by the piston. A prerequisite, however, is that each measured operating point has been checked using the combustion chamber pressure curves captured at the same time, to determine whether operation with both piston variants and identical combustion has taken place.

Figure 7.118 shows such a measurement result, using overall levels (up to 10 kHz) over the load, derived for the sound pressure at a distance of 130 cm from the engine, for the structure-borne noise on the thrust side of the crankcase (centered on the selected cylinder), and for the structure-borne noise at an associated bolt for the main bearing cover.

The acoustically optimized piston B differs from the acoustically objectionable piston A by a piston pin offset optimized by computation and experimentation, by a modified skirt profile having greater upper skirt taper, and by minimized installation clearance. The skirt ovality and the piston ring and top land clearances are adapted to the modified running characteristic with the help of visual inspection of the pistons that have been run. All prescribed seizure tests have been passed previously.

The reduction in sound level achieved by these measures can be as much as 5 dB. The significant reduction in the structure-borne sound level on the thrust side is effective only at very low loads and high loads—as is sketched schematically in **Figure 7.113**. The accelerations measured at the main bearing cover do not exhibit these blatant differences. It can be assumed, therefore, that the difference in noise excitation in this case can be attributed solely

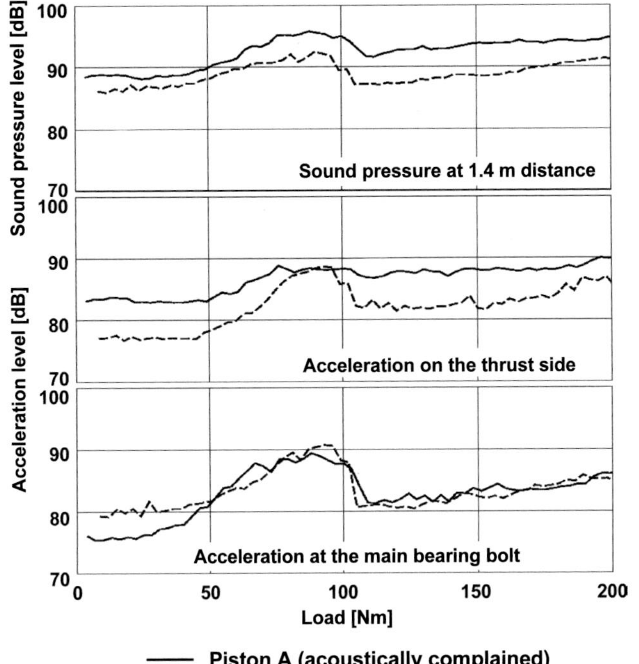

Figure 7.118:
Sound pressure and acceleration level as a function of the load, measured for a DI passenger car diesel engine, engine speed 1,500 rpm, cooled engine (total level over 0–10,000 Hz)

to the piston. In the lower partial-load range, in contrast (cf. load range from 50 to 100 Nm), the dominant excitation is due to very rapidly advancing combustion caused by the considerable ignition delay, especially in the cooled engine. In this load range, the combustion noise determines the sound level. In the lower partial-load range, therefore, neither subjective nor objective assessment of the piston noise is recommended.

The airborne and structure-borne sound levels depicted in **Figure 7.118** are analyzed more closely in order to quantify the noise components caused by the piston. The comparison of the two Campbell diagrams for airborne noise shown in **Figure 7.119** demonstrates clearly that the frequency range around 1,850 Hz is emitted more strongly with the acoustically objectionable piston over the entire load range.

The spectral representation of the accelerations measured on the thrust side, shown in **Figure 7.120**, confirms that the piston is the cause of this partial increase in sound level. The accelerations measured at the main bearing, in contrast, would not show this increased excitation in the case of piston impact, which is also evident in **Figure 7.121**.

The frequency content of an impact excitation caused by the piston has been empirically found to have a broader band than is evident in the diagrams in **Figure 7.119** and **Figure 7.120**. The intense increase in just a narrow range is surely not attributable to the excitation characteristic; rather, it is due to the operational vibration behavior of this special engine structure. In many diesel engines, however, this has been observed in a similar manner when

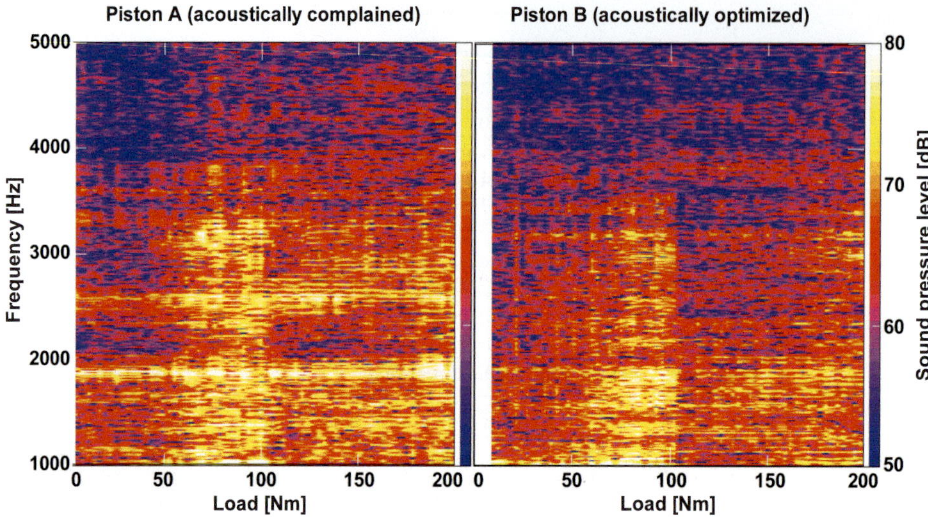

Figure 7.119: Sound pressure spectra as a function of load, measured for a DI passenger car diesel engine, engine speed 1,500 rpm, cooled engine

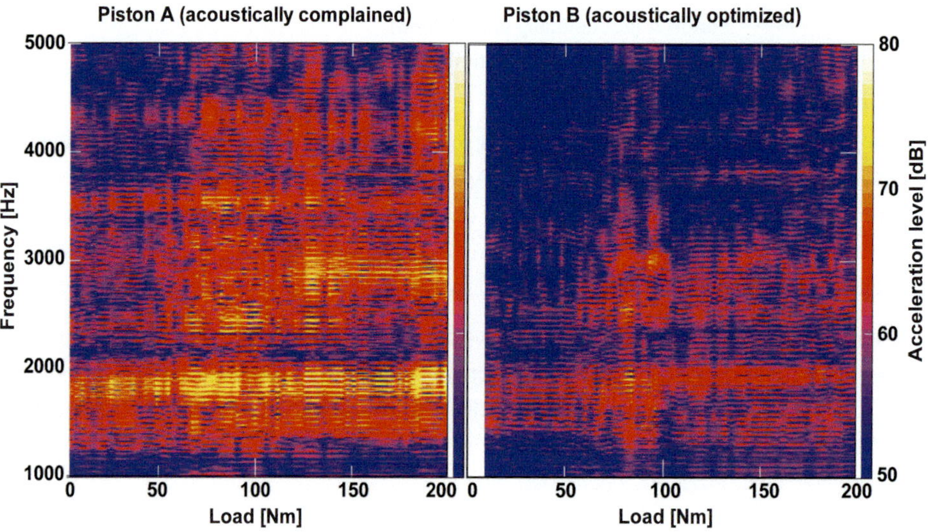

Figure 7.120: Acceleration spectra (thrust side) as a function of load, measured for a DI passenger car diesel engine, engine speed 1,500 rpm, cooled engine

piston impact occurs. The band-pass levels measured in the narrow raised frequency range, as shown in **Figure 7.122**, can be used as a comparative criterion for objective analysis of the individual piston variants. The acceleration measured on the thrust side of the engine block represents the influence of the piston on the noise excitation much more specifically than the airborne noise signal measured at a distance from the engine and is therefore also more appropriate for such a quantitative analysis in most cases.

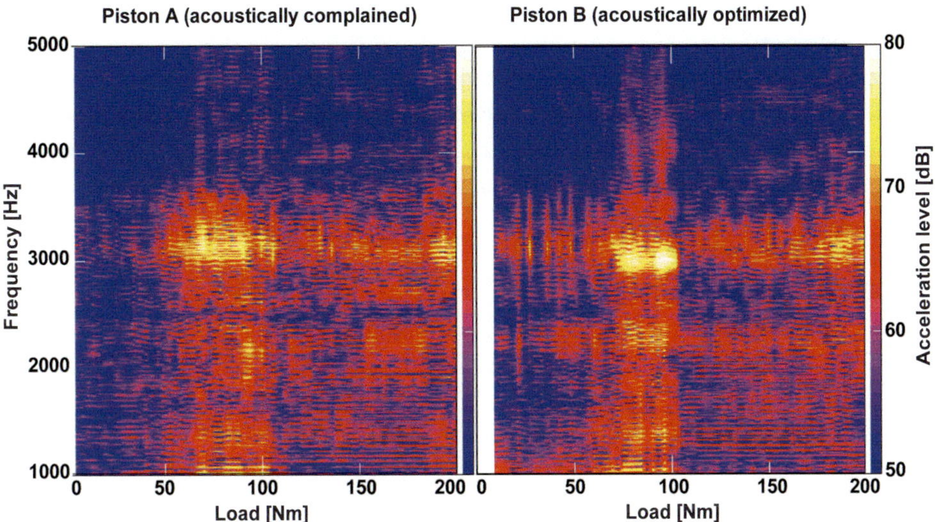

Figure 7.121: Acceleration spectra (main bearing cover bolt) as a function of load, measured for a DI passenger car diesel engine, engine speed 1,500 rpm, cooled engine

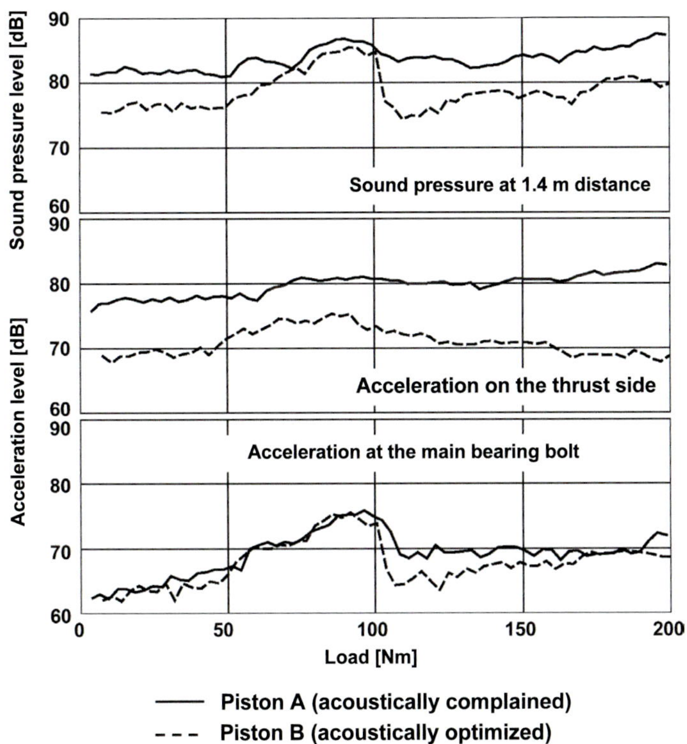

Figure 7.122: Sound pressure and acceleration level as a function of the load, measured for a DI passenger car diesel engine, engine speed 1,500 rpm, cooled engine (band-pass filter frequency 1,750–1,950 Hz)

7.6.3.3 Piston transverse motion and influence parameters in passenger car diesel engines

Analogous to the procedure for gasoline engines, a characteristic behavior of the piston transverse motion is to be depicted for the diesel engine. **Figure 7.123** shows an analysis of such superimposed signals together with the pressure curve measured simultaneously in the combustion chamber and the structure-borne noise measured at the cylinder.

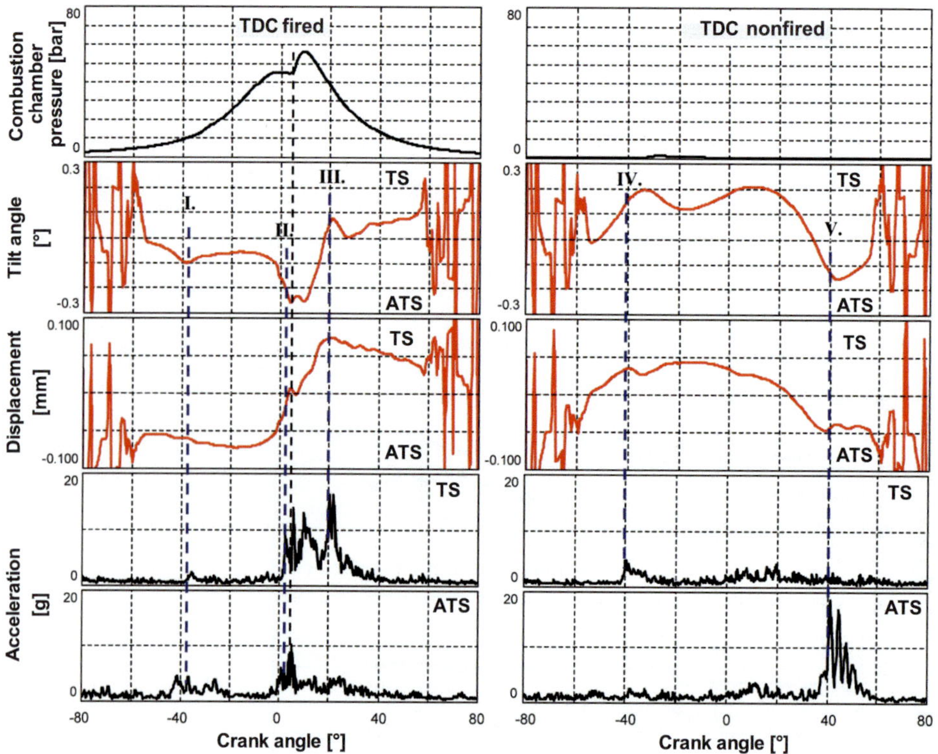

Figure 7.123: Mechanisms for structure-borne sound excitation in diesel engines

The example shown is typical of the sequence of motion of a diesel engine piston with a slight offset toward the thrust side, immediately after cold start. Owing to the very large installation clearance selected for this measurement, five excitation mechanisms that may occur in the diesel engine can be depicted using a single operating point. Assigning the translational and rotational piston transverse motion to the corresponding structure-borne sound excitation, in terms of both time and location, defines the mode of action of mechanisms I through V unambiguously.

If a piston noise is identified as such in the diesel engine, and if a known mechanism of motion can be determined as the cause on the basis of experience, then the number of engine tests needed for optimization depends only on the correct decision of how many test variants to run.

In order to increase the understanding of concrete motion sequences and of the measures to be taken, measured motion information is prepared so that the entire piston transverse motion, including the deformations of the piston skirt, can be visualized quickly and easily. This allows the engineer to find a solution in a quick and precise manner.

Using the insight drawn from several DI passenger car diesel engines, a few helpful tips can be formulated for the design of an acoustically favorable diesel engine piston:

- The distinct barrel shape of the upper area of the skirt, together with sufficient stiffness in the middle area of the skirt, supports rolling of the piston against the cylinder wall, thus helping to reduce impact excitations.
- The skirt end taper with a hydrodynamically optimized design should act over a circumferential angle of about 60 degrees, in order to ensure seizure resistance even at low installation clearances.
- The installation clearance, to be defined in the lower skirt area, should be as small as possible. In case of insufficient seizure resistance in the area of the thrust-side or antithrust-side mantle curve of the piston skirt, the upper skirt taper should without fail be increased before increasing the installation clearance. The use of a graphite coating on the piston skirt can provide a double enhancement in this case.
- The ovality of the skirt can be used exclusively for optimizing the width of the wear pattern and helps to ensure the necessary seizure resistance in the more rigid side areas of the piston skirt as well. The direct influence of skirt ovality on the acoustic behavior is of secondary importance in a piston design that has been optimized in all other aspects.
- For diesel engine pistons, a small piston pin offset of 0.3 to 0.6 mm toward the thrust side is preferred over a design with no piston pin offset. A potential increase in carbon buildup, which can lead to "bore polishing" under certain circumstances and thus to increased oil consumption, can be counteracted with suitable measures, such as an offset of the piston crown or adaptation of the conicity and clearance at the top land or ring land.

7.7   Piston pin noise

7.7.1   Causes of noise

As has been described in Chapter 7.6, the subjectively objectionable noises that can be attributed to the piston and its periphery in a gasoline engine have sound characteristics that are very different from one another. In addition to the piston noises that have already been analyzed, in some cases a harsh, metallic, impulse noise can occur erratically at idle-running speed or increased speed (up to about 2,000 rpm), only at zero load. It is subjectively perceived to be very objectionable.

This noise, known as "pin ticking," is induced as the pin impacts against the bore wall after passing through the clearance of the piston pin in the pin bore. Noise induced by the pin in the small end bore, which can also cause a comparable ticking noise, and is therefore also sometimes called pin ticking, is not the subject of this discussion. As has been shown by experience, however, the procedure described for measuring and analyzing the pin noise in the boss is also suitable for quantifying the excitation caused in the small end bore.

Complaints due to pin noises are unknown for the diesel engine, as these experience completely different pressure ratios in the compression phase. All further considerations, therefore, deal with gasoline engines exclusively.

In a gasoline engine, in the affected operating ranges, the cylinder is under vacuum in the early compression phase. Combined with the arising inertia forces, this forces the pin to make contact in the lower boss area. As the compression pressure increases, the sign of the resultant vector for the gas and inertia forces changes in a crank angle range well before TDC fired. At the point when this resultant force changes sign, the contact of the piston pin changes from the bottom to the zenith of the boss. If this change occurs along a more or less direct path to the opposite wall rather than as a rolling or sliding motion along the pin bore, then this can cause noise excitation after passing through the clearance [21].

While the piston noises of "croaking" and "rattling" are generally audible at low engine temperatures, and decrease as the engine warms up, piston pin noises in the pin bore occur predominantly during the warm-up phase or in a warm engine. One reason for this phenomenon is the difference in thermal expansion coefficients between the aluminum piston and the steel pin. Particularly during the warm-up phase, but also in the warm operating condition, the thermally induced running clearance between the pin and piston can increase relative to the installation clearance, which is measured at 20°C. This promotes the generation of this noise.

7.7.2 Structure-borne noise transfer paths and measurement program

If a pin noise is to be objectively analyzed, then the best and simplest way to capture it with measuring equipment must first be determined. **Figure 7.124** shows the locations of structure-borne sound excitation for the piston noises of croaking and rattling, as well as for pin ticking. For the first two noises, the cylinder wall is excited directly by the piston on the thrust side (TS) or antithrust side (ATS). Following the outer structure-borne noise transfer path, the acceleration measurement is applied directly to the surface of the block. For the pin noise, the structure-borne noise transfer primarily takes the inner transfer path, via the connecting rod, the crank pin, and the crankshaft, into the main crankshaft bearing. The main crankshaft bearing cover, or one of the two mounting screws for the cover, is thus ideal as a measurement position for an accelerometer [26].

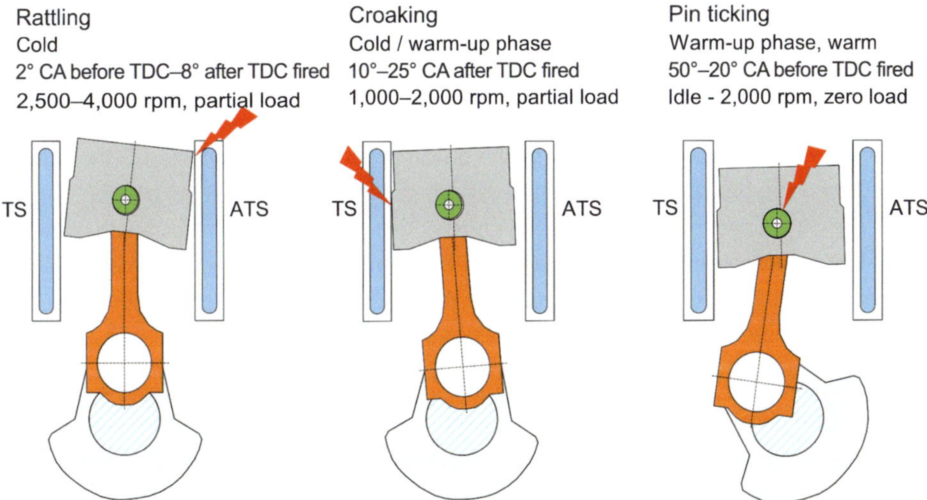

Figure 7.124: Locations of structure-borne sound excitation for typical piston and pin noises in a gasoline engine

Figure 7.125 shows the time signal traces of various acceleration sensors and a microphone at an operating point with a clearly audible pin noise. Five operating cycles are shown. A piston with an oversized boss diameter is installed only in the test cylinder, resulting in oversized pin clearance. All the other cylinders in the four-cylinder engine are equipped with pistons that allow pin clearance at the lower tolerance limit, and therefore verifiably do not produce pin noises.

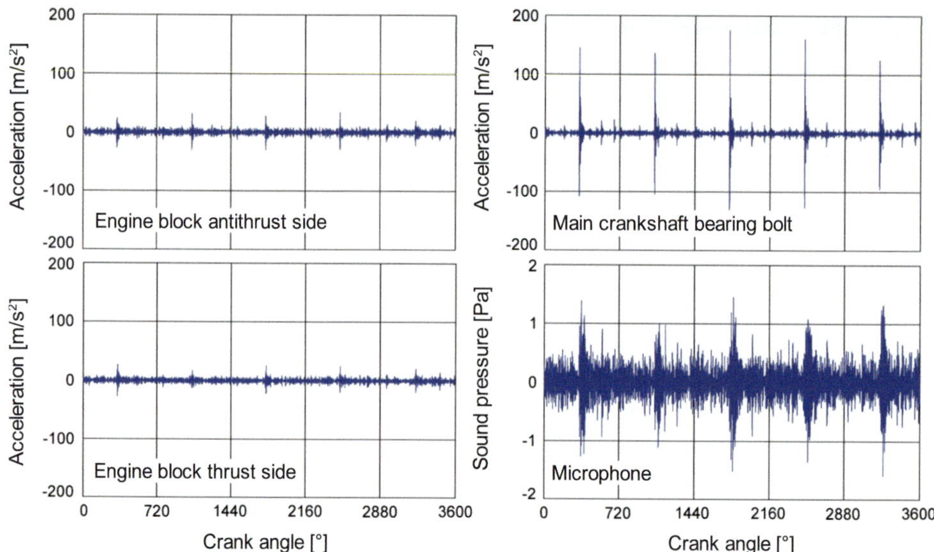

Figure 7.125: Typical time signal traces for a given piston pin noise

The structure-borne sound excitation by the piston pin can also be detected in the signals from the sensors on the thrust and antithrust sides of the engine block surface, but significantly greater spacing between the useful signal and the background noise can be obtained at the screw location on the main crankshaft bearing cover. Therefore, this measurement position is ideal for a refined analysis of the different variants that affect the pin noise. The microphone position is selected at the location where the noise of interest is subjectively most clearly audible.

The pin noise behavior of different engines will normally vary greatly. Therefore, a standardized measurement program has been developed, which allows any pin noises to be detected. It consists of various running programs at constant speed, and dynamic run-ups followed by a drop in speed. The water temperature is varied through the range from 30 to 100°C for each run. The temperature range with the most acoustic activity is then measured more precisely at a constant water temperature using dynamic variations in speed.

The measurements are taken continuously over the entire rise in temperature. A large amount of data is produced, because the measurement period lasts for several minutes. In order to extract the important information from the measurement signals, the method described in the next section is used. The structure-borne noise at a main bearing cover bolt, which corresponds very well to the airborne noise, is used exclusively for this analysis.

7.7.3 Method of analysis in the time domain

The amplitude of the excitation determines the intensity of the perceived pin noise. Therefore, only the maximum acceleration amplitude that is excited by the piston pin at the structure-borne sound measurement point on the main bearing cover bolt is used. If the structure-borne sound excitation is as dominant as in the example in **Figure 7.125**, the maxima can simply be sought out in the individual operating cycles. For less obvious pin noises, and if other noise sources are also present, as is often the case, then the results can easily be misinterpreted. Therefore, the maxima should be sought out only for the crank angle range in which pin noises might occur. In this example (see **Figure 7.126**), a crank angle range from 50° to 20° prior to TDC fired was selected for a maximum value search. All operating cycles occurring during the measurement period are analyzed for maximum amplitude in the crank angle range indicated. These maximal amplitudes are then plotted against temperature and time.

In this example, no remarkable events can be seen during the temperature run-up at idle speed. The test engine used does not have any pin noise in this operating state, despite its large pin clearance. Experience shows, however, that other engines can generate clearly audible pin noises at idle speed. At first, when measured at a constant speed of 1,500 rpm, pin excitation still cannot be detected. Above a water temperature of approximately 83°C, however, a significant increase in the amplitude becomes evident. The temperature run-up with dynamic increase in speed, from idle-running speed to 2,000 rpm, shows a similar behavior. Here the first signs of noise excitation by the piston pin can already be detected at a low temperature. The greatest amplitudes occur at a water temperature of about 95°C. For

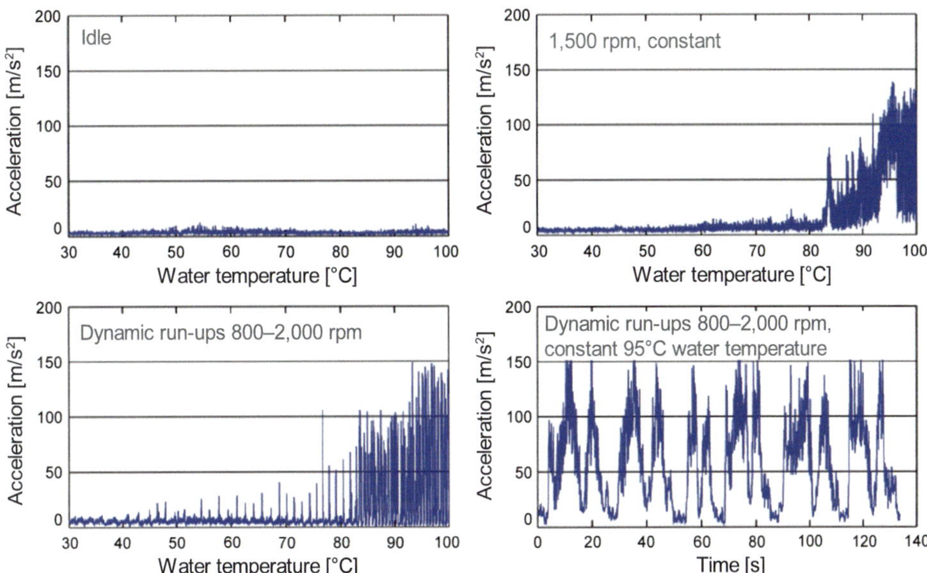

Figure 7.126: Influence of various operating conditions on piston pin noise using the standardized measurement program

this reason, the measurement at constant water temperature with a dynamic speed increase will also be run at 95°C. This measurement shows the very characteristic behavior of the pin noise in the test engine, and, after further refinement of the analysis method, helps to evaluate various parameters that affect the pin noise.

The speed, the maximum acceleration amplitudes determined for a crank angle window prior to TDC fired, and the crank angle at which the individual amplitudes occur are all plotted against the measurement time in **Figure 7.127**.

At both rising and falling speeds, the greatest amplitudes are observed in the range between 1,600 and 1,800 rpm. The crank angle values at which the maximum amplitudes occur are between 35° and 20° crank angle before the TDC fired.

In order to reliably evaluate and classify the results in a parameter study, a large number of such dynamic acceleration and deceleration runs must be incorporated in the analysis. Using a special type of averaging, reproducible and meaningful results can be obtained. Examples are included in the following section, using concrete parameter studies. Including a sufficiently high number of run-ups provides sufficiently stable results.

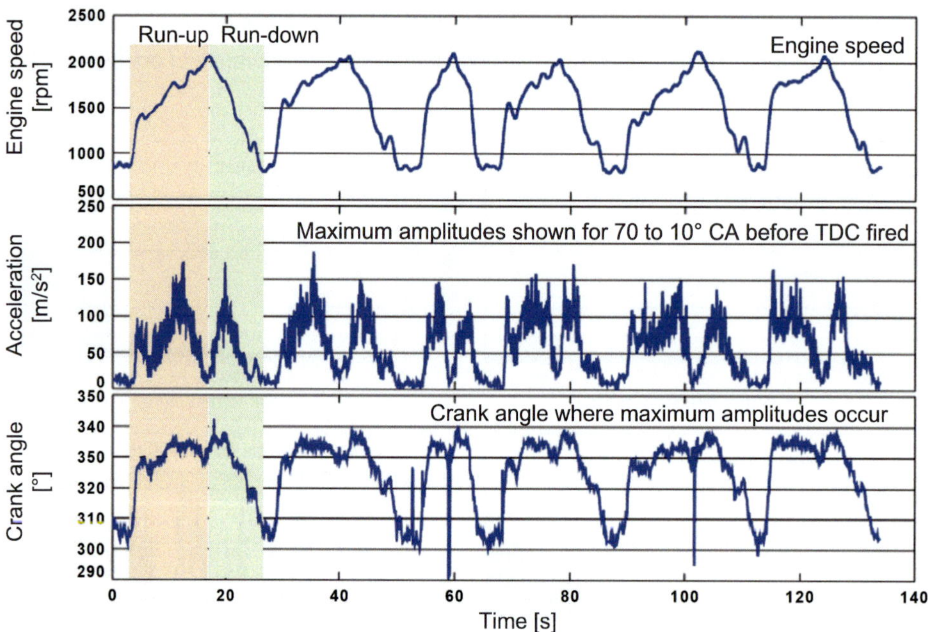

Figure 7.127: Maximum amplitudes, measured at the main bearing cover bolt and corresponding crank angle, at varying speed

7.7.4 Results of parameter studies

7.7.4.1 Influence of piston pin clearance

One parameter that has a great influence on piston pin noise is the piston pin clearance in the boss. In a floating piston pin bearing, the typical design clearance is between 2 and 12 µm. These values refer to a cold engine. During the warm-up phase, and in the warm condition, the operating clearance is greater because of the thermal expansion coefficient of the aluminum piston material, which is nearly double that of the steel used for the pin.

Figure 7.128 shows the influence of different pin installation clearances on the maximum structure-borne sound excitation, from 50° to 20° crank angle before TDC fired, measured at the main bearing cover bolt. Pin clearances were varied over a range from 5 µm to 24 µm. As expected, increasing pin clearance is associated with a continual increase in the maximum accelerations. In order to evaluate the amplitude above which the structure-borne noise measured at the main bearing can be heard as airborne noise, and therefore perceived as objectionable, the results of a subjective auditory impression are used. The variations in clearances differ significantly not only in the intensity of the pin noise, but also in the speed range at which a pin noise first becomes audible. While no pin noise can be heard at 5 µm and 8 µm clearance, an audible noise occurs at 12 µm clearance between 1,600 and 1,800 rpm. At 24 µm clearance, the speed range with audible pin ticking starts just above idle-running speed, at about 900 rpm, and ends again at about 1,800 rpm.

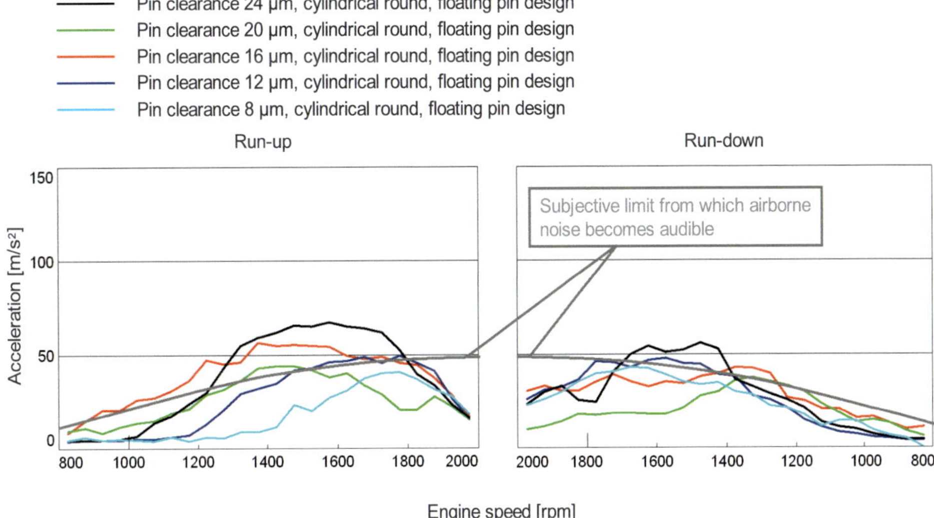

Figure 7.128: Influence of piston pin clearance on the structure-borne sound excitation in a round cylindrical pin bore

On the basis of this observation, the diagram in **Figure 7.128** for speed increases and decreases includes a curve showing the level at which the structure-borne noise becomes audible as objectionable airborne noise. This audible limit is included in all the other diagrams without further explanation.

7.7.4.2 Influence of pin boss geometry

There are many common versions of pin boss geometry designs. Deviations from the round cylindrical pin boss shape, however, are typically not related to the acoustic behavior of the engine; rather, they are intended to reduce local component stresses, thereby increasing service life. Other measures have the goal of improving lubrication in the boss, thus reducing wear and the risk of seizing.

7.7.4.2.1 Oil pockets and circumferential oil groove

The oil pockets (slots) are continuous recesses running along the pin bore, which improve the oil supply (see sketch in **Figure 7.129**). Circumferential oil grooves running perpendicular to the pin bore serve the same purpose.

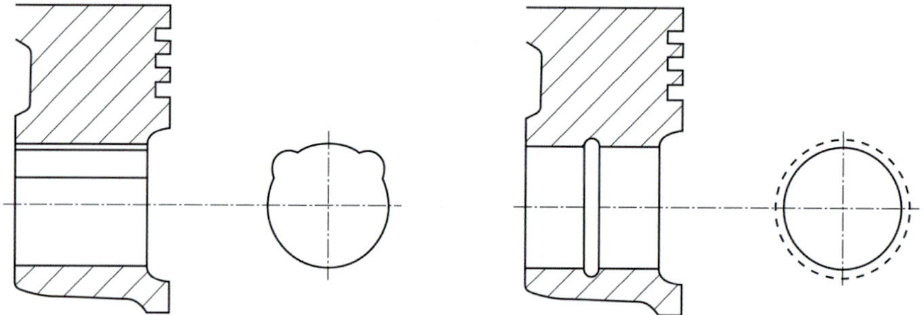

Figure 7.129: Design principle of a pin bore with oil pockets (slots, left) and circumferential oil grooves (right)

Both measures improve lubrication in the boss and reduce the structure-borne sound excitation considerably below an acoustically critical level; **Figure 7.130**. For the variant with a round cylindrical boss combined with a circumferential oil groove, even pin clearances of up to 24 µm were implemented without any acoustic events.

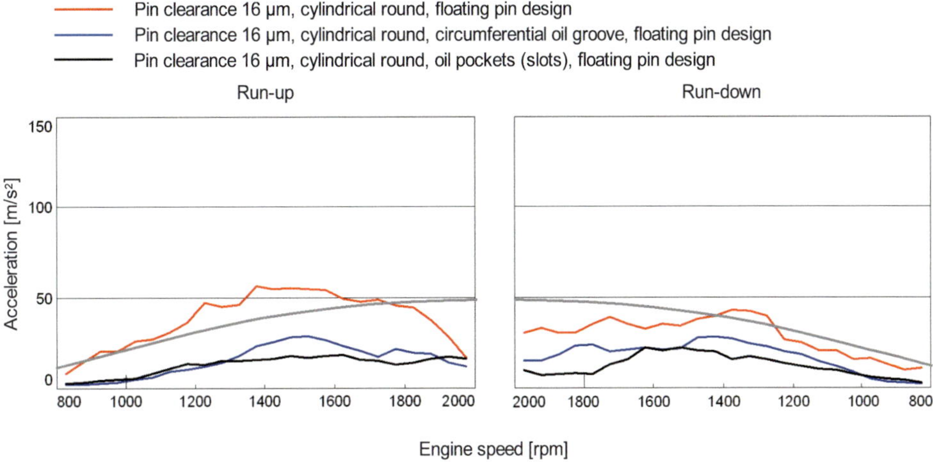

Figure 7.130: Influence of oil pockets and circumferential oil groove on the structure-borne sound excitation

7.7.4.2.2 Transverse oval pin bore and pin bore relief

Both measures help to reduce stresses in the pin boss support for piston pins that deform into an oval shape under high gas forces. For designs, see sketches in **Figure 7.131**.

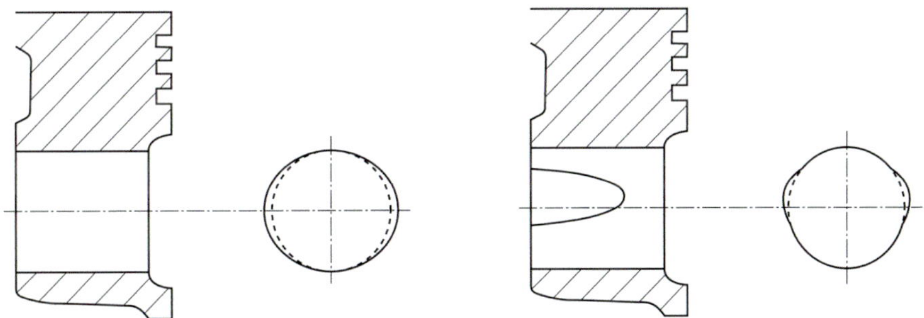

Figure 7.131: Design principle of a transverse oval pin bore (left) and a round pin bore with pin bore relief (side relief, right)

Figure 7.132 shows a comparison between round and transverse oval pin bore designs, as well as pin bores with pin bore relief (side reliefs), with 16 μm basic installation clearance for each pin. Both measures reduce the structure-borne sound excitation by the piston pin. The version with pin bore relief, however, is of limited value. The transverse oval pin bore achieves a significant improvement over the round cylindrical pin bore. It has significantly lower acceleration values, which do not cause any audible pin noise. With transverse pin

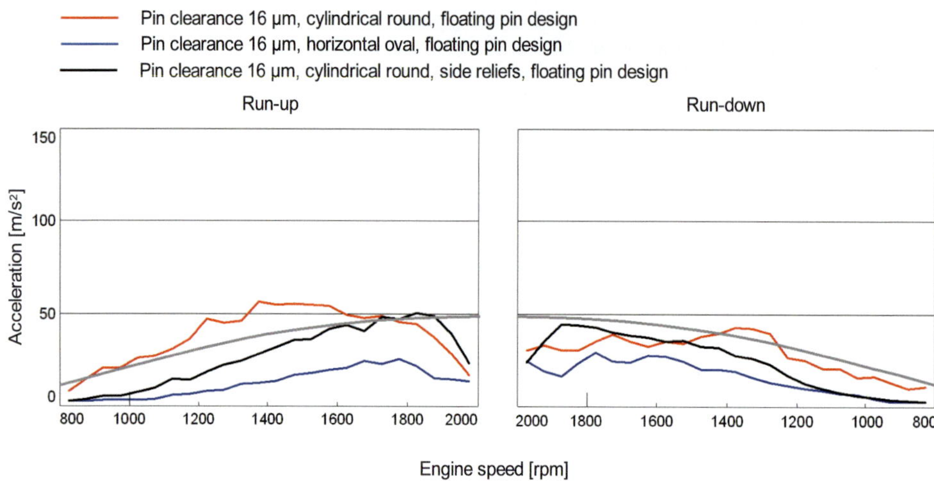

Figure 7.132: Influence of transverse oval pin bore and pin bore relief (side relief) on the structure-borne sound excitation

boss geometry, even a very large basic installation clearance can be implemented without causing perceptible noise. The transverse oval design of the pin bore is thus a very effective measure to reduce pin noises.

7.7.4.2.3 Single-sided vertical oval pin bore

In very highly stressed gasoline engines, the high gas forces induce large bending moments in the piston about the longitudinal axis of the pin. This is often associated with high stresses on the surface of the piston, which bounds the combustion chamber in the transverse direction of the engine. The vertical oval design of the upper half of the pin bore helps to reduce these stresses significantly; **Figure 7.133**. This measure, intended to increase service life, is also accompanied by an increased pin clearance in the vertical direction.

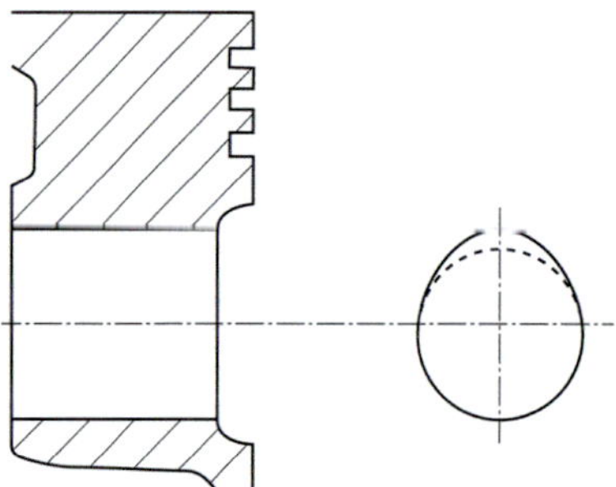

Figure 7.133:
Design principle of a pin bore with single-sided vertical ovality

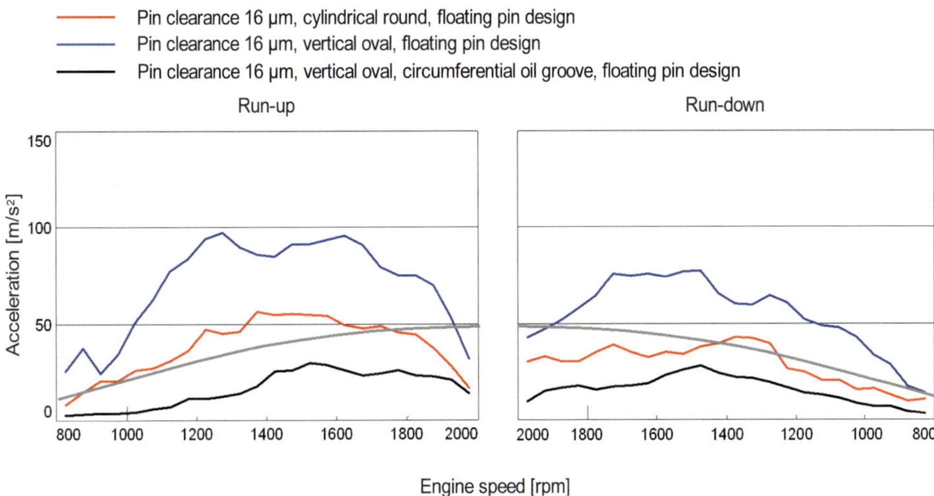

Figure 7.134: Influence of piston pin clearance on the structure-borne sound excitation in a single-sided vertical oval pin bore, with and without circumferential oil groove

Figure 7.134 shows the influence of a single-sided vertical oval pin bore, with and without a circumferential oil groove, as a potential acoustic improvement action. In this case, with 16 µm basic installation clearance, a significant increase in structure-borne sound excitation is observed for the single-sided vertical oval boss relative to the round pin bore. Other clearance variations show that only a very small basic clearance can provide acceptable acoustic behavior with regard to pin noise. Starting at a basic installation clearance of 8 µm, large accelerations are already evident at the structure-borne sound measurement point on the main bearing cover bolt, along with very clearly audible pin noise.

When the single-sided vertical oval boss is combined with a circumferential oil groove, the test engine used showed a significant improvement in structure-borne sound excitation well below the threshold at which audible pin ticking occurs.

The vertical oval boss is a very effective means for achieving sufficient service life under high loads and can be implemented with no acoustic disadvantages whatsoever, as long as it is combined with suitable design measures.

7.7.4.2.4 Shaped pin bore

A shaped pin bore is generally a trumpet-shaped expansion of the pin bore diameter, extending toward the inner or outer pin boss; **Figure 7.135**. This measure adapts the boss to the piston pin as it bends under the gas force. This reduces the maximum edge pressure in the boss area.

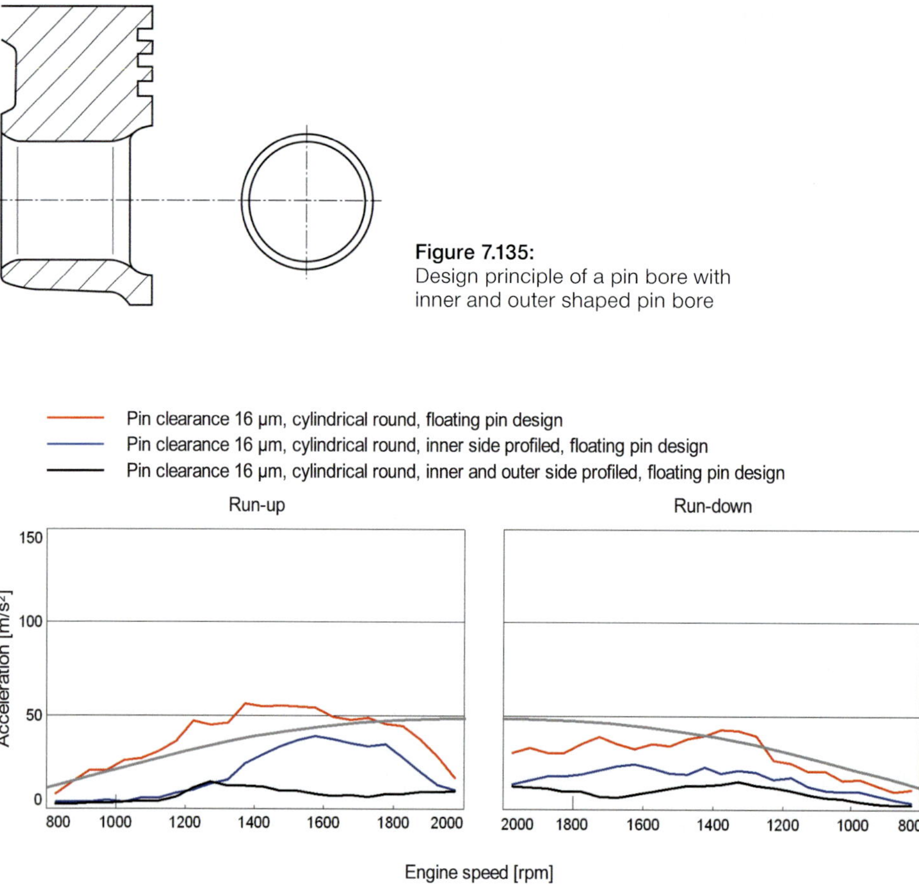

Figure 7.135:
Design principle of a pin bore with
inner and outer shaped pin bore

Figure 7.136: Influence of profiled pin bores on the structure-borne sound excitation

During noise testing, variants with such shaped pin bores were tested. The results for this variation are shown in **Figure 7.136** for round pin bores with no ovality. The basic installation clearance is a uniform 16 µm.

The inner shaped pin bore already shows a reduction in structure-borne noise compared with the round cylindrical boss, and no pin noise can be heard. The combination of inner and outer shaped pin bores ultimately has the least noise excitation, comparable to that of a 5 µm installation clearance in a round cylindrical pin bore.

This also explains why increased clearance due to wear does not necessarily have to lead to increased noise, because the ovality and form of the bore can also change, thus providing the positive effects described here.

7.8 Cavitation in wet cylinder liners of commercial vehicle diesel engines

In the physical sense, cavitation means the formation, growth, and sudden implosion of vapor bubbles in fluids.

Strictly speaking, cavitation refers to the phenomenon in which vapor bubbles form as a result of a drop in pressure below the vapor pressure curve at a constant temperature; **Figure 7.137**. In contrast, boiling refers to the phenomenon in which vapor bubbles form as a result of an increase in temperature at a constant pressure. This strict view applies only to chemically pure fluids, which are never encountered in practice.

Cavitation is a complex phenomenon that is influenced by many factors. Its complexity is illustrated by the fact that physical processes from hydrodynamics, thermodynamics, chemistry, plasma physics, and optics are all involved in the cavitation phenomenon [27].

Precise knowledge of the causes and effects of cavitation is a basic prerequisite for understanding complex cavitation phenomena in wet cylinder liners of commercial vehicle diesel engines. The two main causes for cavitation on the outside of the cylinder liners, which are surrounded by coolant, are the potentially adverse local flow conditions for the coolant and the vibrations of the cylinder wall caused by gas and inertia forces during a four-stroke cycle.

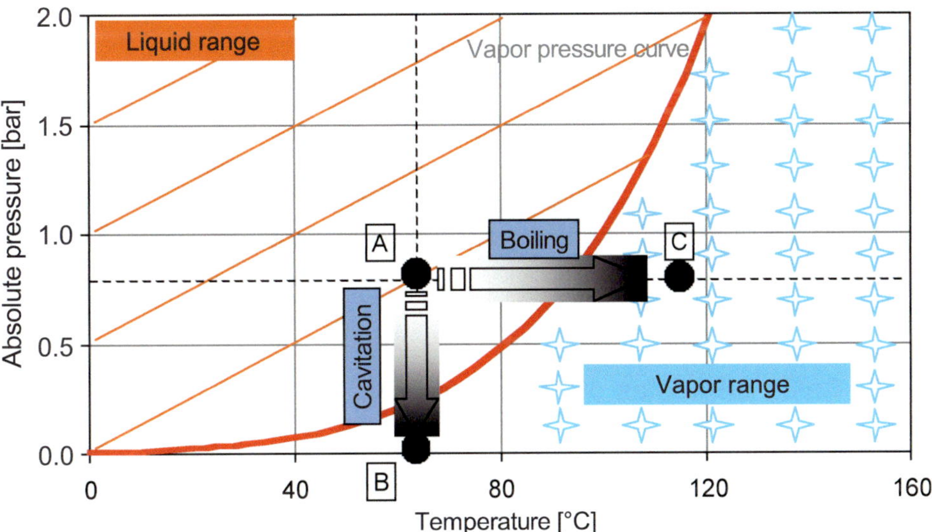

Figure 7.137: Definition of cavitation

Adverse flow conditions are found mainly around very narrow gaps in the coolant channels between the cylinder liner and engine block. Very high local flow velocities occur here, and locally cause both high dynamic pressure and low static pressure in the coolant. This promotes a tendency toward cavitation at the cylinder liner or the engine block.

During an operating cycle, the piston performs various secondary motion. In particular, these involve multiple contact alterations on the thrust and antithrust side, with corresponding impacts on the cylinder liner. This excites undesired, high-frequency vibrations in the cylinder liner [28].

The coolant surrounding the cylinder liner cannot keep up with these vibrations. The local static coolant pressure drops below the coolant vapor pressure. When this occurs, cavitation vapor bubbles form. If the coolant pressure rises above the vapor pressure again, then the cavitation vapor bubbles implode suddenly. Their collapsing causes physical and chemical effects, such as high pressure impulses, pressure waves, temperatures, high local flow velocities, and light effects (sonoluminescence). This results in the destruction of the material on the exterior of the cylinder liner where the coolant flows around it. This phenomenon is known as cavitation or pitting.

Repeated occurrence of the cavitation process in each operating cycle leads to erosion of the cylinder liner. Depending on how long the engine is operated and the intensity of the cavitation process, it is possible that eroded holes can completely penetrate the cylinder liner; **Figure 7.142**. Coolant can then flow into the combustion chamber or the crankcase through these holes and can cause severe engine damage. Depending on their intensity, visible damage to the cylinder liner can occur as a result of cavitation after only a few operating hours [29].

Cavitation can occur at any point on the cylinder liner that makes contact with the coolant if the local coolant pressure drops below the coolant vapor pressure.

Two typical forms of appearance are characteristic of cavitation on cylinder liners:
- A typical pattern of damage occurs on the thrust side–or, in rare cases, on the antithrust side–in the area of piston contact alteration at the TDC fired.
- Cavitation damage occurs in the area of very narrow gaps in the coolant channels between the cylinder liner and the engine block [30].

7.8.1 Basic principles of cavitation

From a theoretical point of view, cavitation occurs when the static pressure in a fluid drops below the vapor pressure associated with the ambient temperature. For flowing media, the fundamental equation for cavitation is derived from the Bernoulli equation. Bernoulli's law for flowing media states that the static pressure is reduced in a constriction, while the dynamic pressure and the associated flow velocity are increased. It follows that, for the dynamic pressure of incident flow at a constriction:

$$p_d = p_{tot} - p_{st} = \frac{1}{2} \cdot \rho \cdot v^2$$

p_d: dynamic pressure

p_{tot}:   total pressure

p_{st}:   static pressure

ρ: density of the fluid

v: velocity of the fluid

The characteristic factor for cavitation is the dimensionless cavitation number sigma, also known as the cavitation coefficient. It is calculated as the difference between the static pressure and the vapor pressure associated with the ambient temperature, divided by the dynamic pressure of the incident flow. If the cavitation number is negative, then cavitation will occur in a flow [31].

$$\sigma = \frac{p_{st} - p_v}{\frac{1}{2} \cdot \rho \cdot v^2}$$

σ: cavitation number

p_v: vapor pressure as a function of the ambient temperature

7.8.2 The physical phenomenon of cavitation

As mentioned previously, the precise physical definition of the cavitation phenomenon–vapor bubble formation due to a pressure drop below the vapor pressure of the fluid at a constant temperature–applies, strictly speaking, only to chemically pure fluids. In practice, however, chemically pure fluids are not available. Every fluid, including engine coolant, contains weak points (known as nuclei), which are the source of all cavitation events. The formation of vapor bubbles due to a pressure drop thus occurs at such weak points, because the fluid breaks down preferentially at these points [32].

They include:
- gas bubbles in the fluid;
- gas dissolved in the fluid;
- hydrophobic solid particles, known as pore seeds;
- gas formation when flowing past fine peaks of surface roughness (bubble eddies);
- particles with gas inclusions, which often occur in recesses.

7.8.3 Types of cavitation

Cavitation events depend on the behavior of the weak points. Therefore, in real fluids, the following categories of cavitation are differentiated; **Figure 7.138**:

- Gas cavitation
- Pseudocavitation
- Vapor cavitation
- Cavitation in real flows.

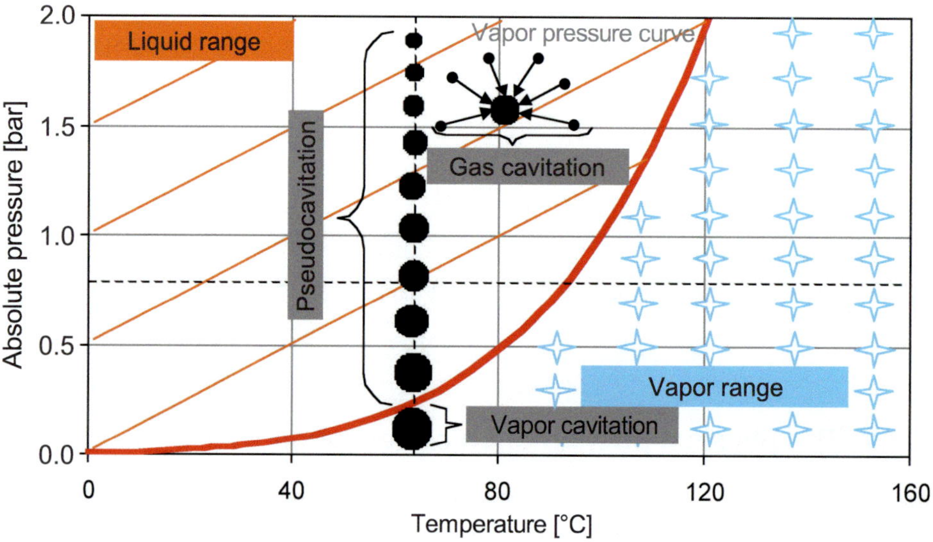

Figure 7.138: Types of cavitation

7.8.3.1 Gas cavitation

Gas cavitation is a phenomenon in which a gas dissolved in the fluid undergoes a transition to an undissolved state as a result of a pressure drop; **Figure 7.138**. The bubbles that form contain the gas components, which were previously dissolved in the fluid (typically air). If the pressure rises again, the bubbles collapse and the gas components dissolve back into the fluid.

In comparison with pseudocavitation, Chapter 7.8.3.2, and vapor cavitation, Chapter 7.8.3.3, gas cavitation is a relatively slow process that actually dampens the implosion of cavitation bubbles occurring with vapor cavitation. In order to minimize this dampening effect when measuring cavitation, the engine coolant is degassed in a degassing run prior to measuring. Gas cavitation can occur above the vapor pressure curve as well.

7.8.3.2 Pseudocavitation

The expansion of gas bubbles present in the fluid due to a pressure drop is called pseudo-cavitation; **Figure 7.138**. Like gas cavitation, pseudocavitation can also occur above the vapor pressure curve.

7.8.3.3 Vapor cavitation

Vapor cavitation, unlike pseudocavitation or gas cavitation, occurs only if the pressure drops rapidly below the vapor pressure. The cavitation bubbles that form are filled with the vapor of the surrounding fluid. If the pressure rises again, then these cavitation bubbles collapse again with an implosive reduction in volume. This phenomenon is called bubble implosion or bubble collapse; **Figure 7.138**. Very high local pressures, shock waves, flow velocities, and, briefly, very high temperatures can occur as a result of this bubble implosion. If the bubble implosion also occurs directly adjacent to a material surface, such as a cylinder liner, then fluid jets (microjets) may also occur.

These effects cause material destruction in the vicinity of material surfaces. Vapor cavitation is therefore responsible for potential material damage to material surfaces. If the bubble implosion is severe enough that the bubbles break up into many smaller bubbles, this is known as transient cavitation [31].

7.8.3.4 Cavitation in real flows

In practice, cavitation occurs as a combination of gas, pseudo-, and vapor cavitation. As shown in **Figure 7.138**, bubbles first grow to a critical radius on the cavitation seeds as a result of gas cavitation and pseudocavitation, and when this radius is reached and the pressure correspondingly drops below the vapor pressure curve, this initiates vapor cavitation [33].

7.8.4 Cavitation bubble dynamics and cavitation bubble collapse

Cavitation bubbles form at weak points, which are distributed unevenly in the fluid, often in the form of an entire bubble cloud. The individual bubbles of the bubble cloud are of different sizes. After they form, the cavitation bubbles expand if the static pressure in the surrounding fluid drops further. Substances dissolved in the surrounding fluid can then diffuse into the cavitation bubbles. Many other factors, including thermal and inertia effects, mass distribution, and compression characteristics, as well as the roughness of material surfaces, affect the growth of the cavitation bubbles and bubble implosion [34]. If the static pressure of the surrounding fluid increases again, or if the cavitation bubbles flow into areas of higher pressure, then they collapse suddenly with an implosive reduction in volume. This phenomenon is referred to as cavitation bubble collapse or cavitation bubble implosion.

Depending on the distance between the cavitation bubbles and a solid material surface, cavitation bubble implosion can be classified as either radial (or spherical) or aspherical. During bubble implosion, various effects occur, which cause material destruction in the vicinity of material surfaces.

7.8.4.1 Spherical cavitation bubble implosion

If a cavitation bubble is not near a solid material surface, then the bubble implosion is spherical; **Figure 7.139**. This also requires constant density, constant dynamic viscosity, and uniform temperature distribution in the surrounding fluid. The temperature and pressure distribution in the bubble must also be constant. The cavitation bubble grows constantly from its initial radius. In this case, the pressure in the cavitation bubble is greater than that of the surrounding fluid. The cavitation bubble implosion starts when the maximum bubble radius R_M is reached. For a very brief moment, equilibrium exists between the internal bubble pressure and the external fluid pressure. If the external fluid pressure then increases above the internal bubble pressure, then the bubble volume decreases suddenly. The cavitation bubbles implode, and very high pressures and temperatures occur in them. The pressure that builds up in the surrounding fluid at the bubble boundary is emitted outward as a shock wave after passing through the minimum radius.

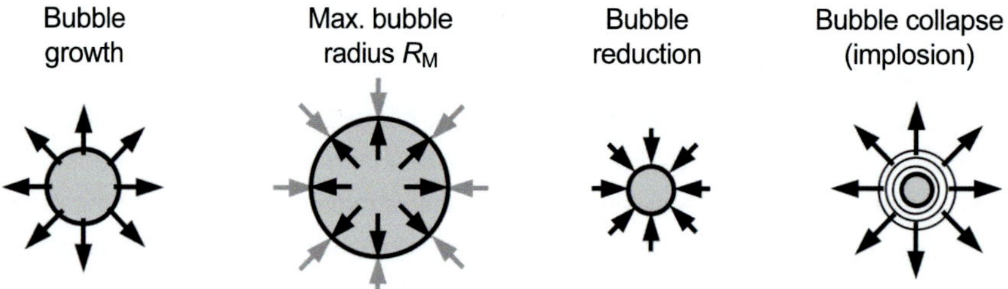

Figure 7.139: Spherical cavitation bubble implosion

7.8.4.2 Aspherical cavitation bubble implosion

If a cavitation bubble is near a solid material surface, or if there are inertia or thermal instabilities, then the cavitation bubble implosion is aspherical; **Figure 7.140** [34]. The cavitation bubble grows radially from an initial radius R_0 to the maximum bubble radius R_M. If the fluid flow in the area around the bubble is disturbed by a solid material surface, such as a wall, then the cavitation bubble becomes unstable. It implodes in a very specific manner, and creates a fluid jet. At the beginning of the implosion or bubble collapse, the side of the bubble furthest from the wall folds inward on itself. A fluid jet forms and shoots through the bubble, then impinges on the opposite bubble wall from the inside. The wall is deformed into a thin

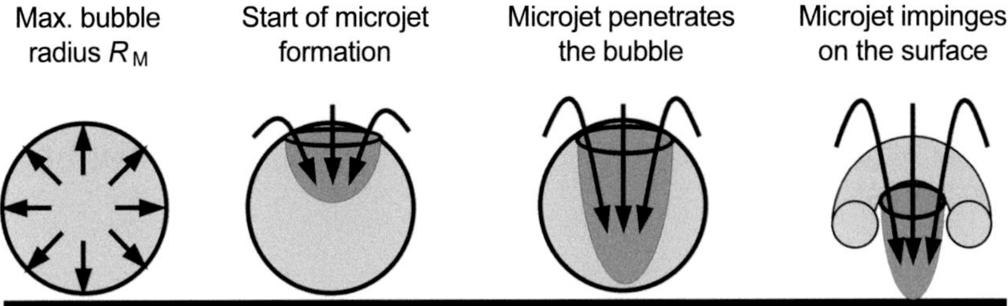

Figure 7.140: Aspherical cavitation bubble implosion

cone. After penetrating the opposite bubble wall, the fluid jet impinges on the adjacent material surface. The bubble, now deformed into a torus by the fluid jet, implodes further, also emitting shock waves.

If a cavitation bubble implodes in such a manner that the initial volume is reduced by a large multiple, or if the implosion is particularly severe, then the bubble can disintegrate into a whole cloud of smaller bubbles [35]. The individual bubbles in the cloud can each form fluid jets in the vicinity of the wall, and these jets can influence one another as well.

Fundamental knowledge of cavitation bubble dynamics in the vicinity of solid material surfaces is particularly crucial as they are the source of material damage [31]. For this reason, many investigations have been carried out in the field of cavitation bubble dynamics in recent years [36]. Scientific studies have shown that material damage is sure to occur if the dimensional distance parameter γ is less than two. It is defined as follows:

$$\gamma = \frac{s}{R_M}$$

γ: distance parameter

s: distance from center of cavitation bubble to material surface

R_M: maximum radius of cavitation bubble

The most severe damage occurs if the cavitation bubble makes direct contact with the material surface. In this case, the fluid jet reaches the material surface at full speed. The following factors are involved in material destruction by cavitation:

- Fluid jets, which impinge on the material surface with a large impulse value. Very high local pressures can occur briefly
- The high internal temperature and pressures in an imploding bubble, particularly when it implodes directly on the material surface
- Shock waves induced by the bubble implosion.

It is still not entirely clear which factor is dominant. New investigations have shown, however, that it is mainly the mechanical effects, such as high pressures resulting from the fluid jet, that are responsible for material damage [31]. In addition, the high temperature peaks–though they likely play a subordinate role–can contribute to material damage, despite their extremely short duration [27].

7.8.5 Cavitation damage in wet cylinder liners

Fundamentally, cavitation on wet cylinder liners can occur at any point that is in contact with the coolant if the local coolant pressure drops below the coolant vapor pressure.

With regard to the pattern of cavitation damage, two different characteristics have been determined:

- A more planar cavitation damage, often on the thrust side of the cylinder liner, but in rarer cases also on the antithrust side.
- A more linear cavitation damage, also known as gap cavitation, which often occurs where narrow cooling gallery cross sections are found between the cylinder liner and engine block. The linear cavitation damage is often identified slightly above the seal between the cooling gallery and the crankcase. In a more advanced state, it can extend over almost the entire circumference of the cylinder liner.

Figure 7.141 shows a complete cylinder liner, left, with planar and linear cavitation damage. In the right part of the figure, the different types of cavitation damage are shown under magnification. As can be seen, the holes resulting from planar cavitation damage have different diameters and are not uniformly distributed. The cavitation damage shown in **Figure 7.141** occurred after 900 engine operating hours.

The cross sectional view in **Figure 7.142** shows a magnified section through a 6 mm thick cylinder liner with cavitation damage. As is evident, the "pitting corrosion" caused by the cavitation does not propagate in a single direction; rather, the distribution is more crooked, with locally broadened areas. Therefore, the cavitation damage observed on the surface of the cylinder liner is often of only limited informational value with regard to cavitation damage.

Fundamentally, the propagation of "pitting corrosion" due to cavitation is determined by the stresses arising at the material surface, which are transferred to the components of the microstructure. The same applies to the residual stresses present in each material. Depending on the microstructure, grain size, and the hardness and toughness of the cylinder liner material, normal, shear, and residual stresses can arise with varying effects. For example, the microstructure of a cast iron liner contains brittle graphite and hard cementite components. The normal and shear stresses thus arising can then extend only as far as a boundary, such as a graphite layer. This results in a change in the course of the cavitation damage [37].

Figure 7.141:
Types of cavitation damage

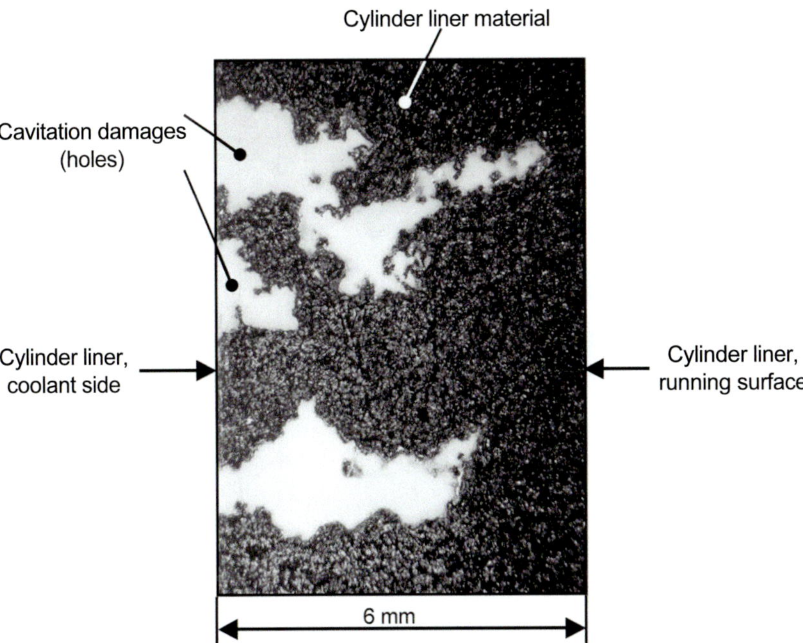

Figure 7.142: Cross section through a cylinder liner with cavitation damage

7.8.6 Cavitation measurement equipment

Cavitation is a complex and highly dynamic phenomenon, regardless of any other influences on engine operation. The entire cavitation process is therefore very difficult to detect for measurement purposes in a running combustion engine. Currently, only the effects of imploding cavitation bubbles in the coolant can be captured. The measurement method for detecting cavitation events in a running combustion engine is therefore based on the measurement of local changes in coolant pressure. Imploding cavitation bubbles lead to local high-frequency dynamic pressure peaks in the coolant. The intensity of these pressure amplitudes is correlated with the severity of the imploding cavitation bubbles. The associated pressure amplitudes can therefore be used as a measure of potential material damage resulting from cavitation.

Special pressure sensors are used for measuring high-frequency dynamic pressure peaks. Precisely determining the position of the potential cavitation damage is a prerequisite for the placement of the sensors. To this end, the engine is fitted with cylinder liners that are painted on the outside. The locations of potential cavitation damage on the cylinder liners can be determined precisely, using a special engine running program and specific cooling system conditions. Placing the sensors typically requires a great deal of machining on the engine block. Care must also be taken to ensure that the coolant flow is not significantly impaired. **Figure 7.143** shows a cross section through an installed pressure sensor with add-on parts.

Figure 7.144 shows two installed sensors from the interior of an engine (cooling gallery side), left, and from the exterior of the engine, right. The sensors are placed in the interior of an

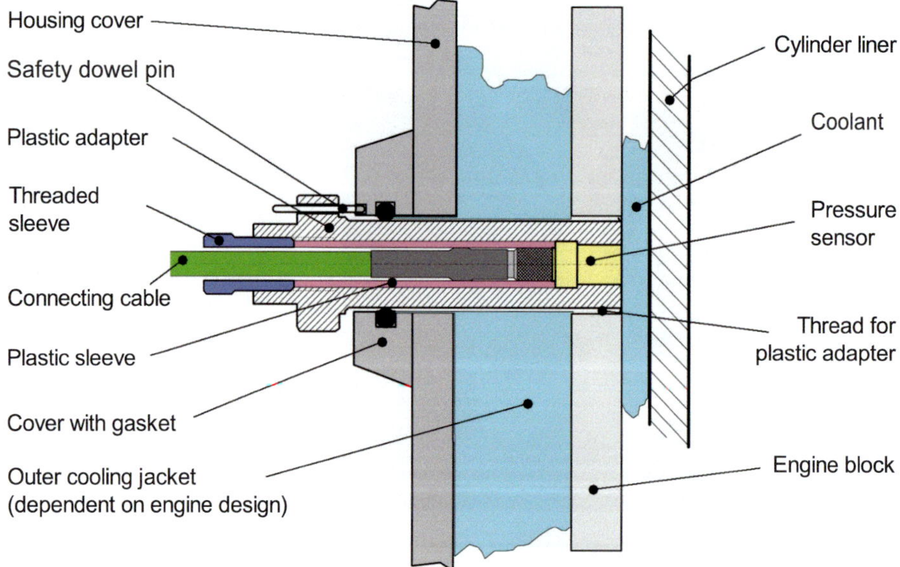

Figure 7.143: Schematic diagram of a section through an installed measurement point with sensor for measuring water pressure

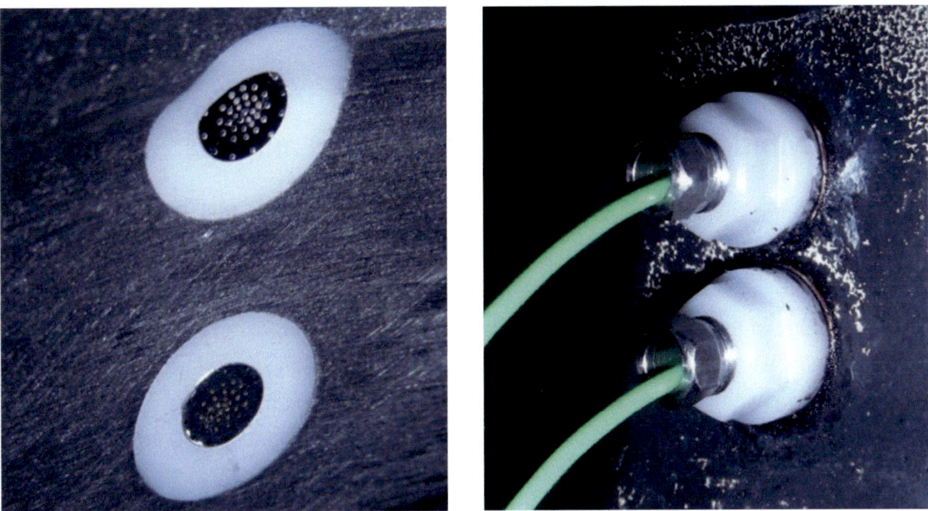

Figure 7.144: Installed pressure sensors on the engine interior where the coolant flows (left), and on the engine exterior (right)

engine so as to avoid influencing the flow of the coolant. The top sensor is used to measure gap cavitation here, and the bottom one is used to measure surface cavitation.

The measurement signals are recorded and numerically processed using a computer-aided data acquisition unit (KI meter) specially developed for solving cavitation problems. Further processing of the measurement data is done by special PC-based software. **Figure 7.145** shows the complete measurement chain. The data acquisition unit is able to process the

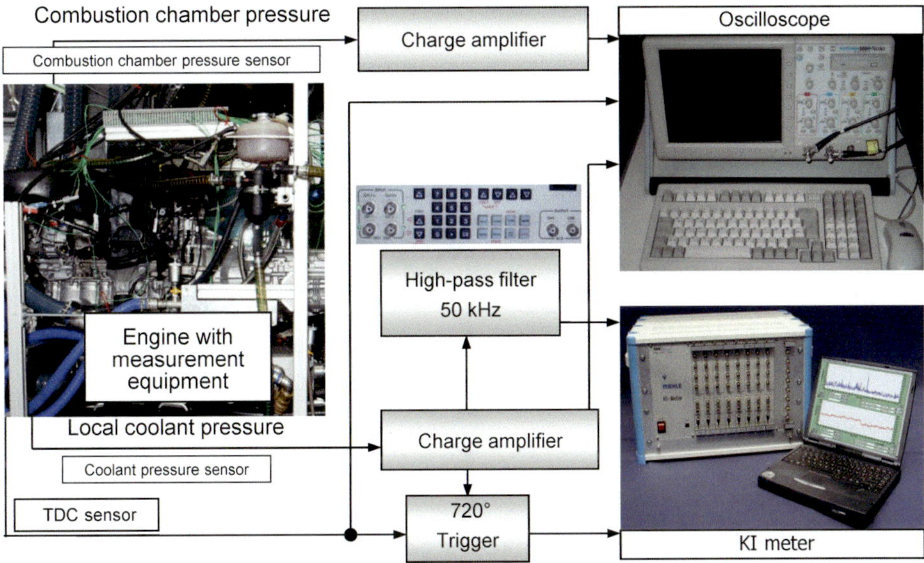

Figure 7.145: Setup of the entire measurement chain for measuring cavitation

data from eight cylinders at the same time. Up to 10,000 operating cycles can be captured for each cylinder.

7.8.7 Cavitation intensity factor and signal analysis

The data are validated on the basis of the analysis of the greatest positive high-frequency pressure amplitude, which is measured at a predetermined crank angle range in every operating cycle; **Figure 7.146**.

The initial value of the crank angle and the length of the measurement range are determined individually for each engine–depending on the angular position of the crankshaft at which the imploding cavitation bubbles occur. The local coolant pressure curve measured for a measurement window is first conditioned by a high-pass filter so that only the pressure amplitudes that are induced by cavitation bubble implosions ultimately determine the signal curve.

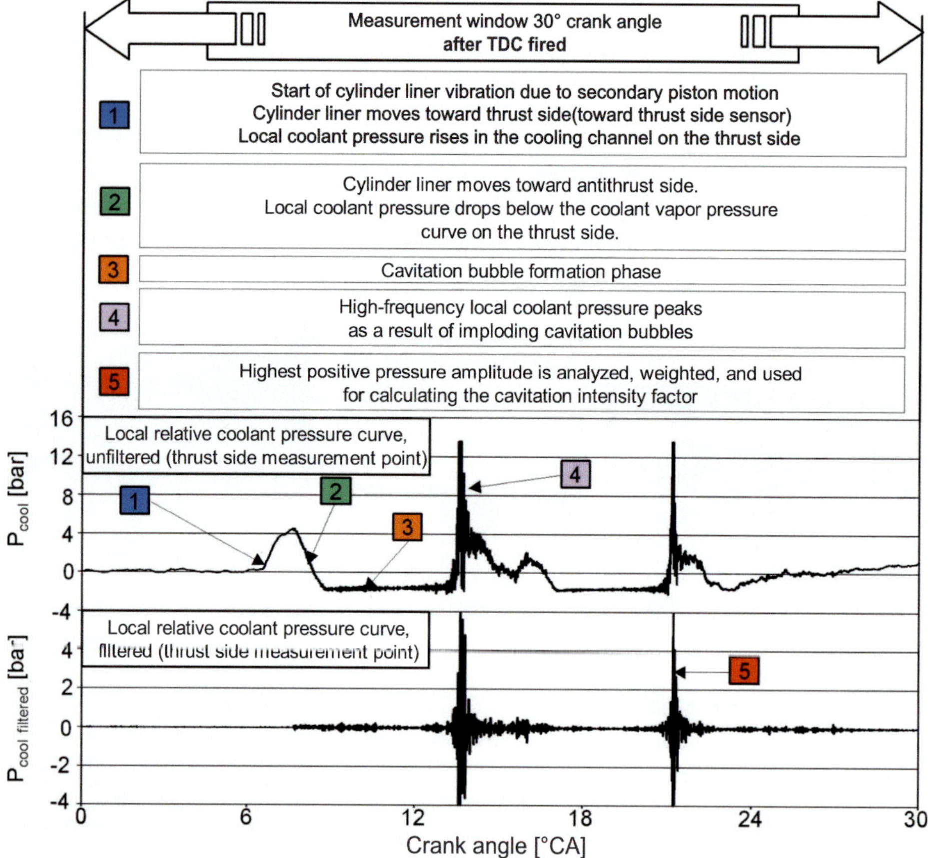

Figure 7.146: Signal analysis of the relative local coolant pressure curve

The greatest positive pressure amplitude in the high-pass filtered signal curve is captured and classified. The cavitation processes, and thus the severity of imploding cavitation bubbles, can be very different from operating cycle to operating cycle, even in steady-state engine operation. Therefore, several successive operating cycles must be analyzed in order to calculate the cavitation intensity factor, KI. Every measured pressure amplitude is assigned to a pressure class based on its level. The number of events in each class is then assigned a weighting factor. The classes that represent greater pressure amplitudes are assigned a greater weighting factor. In order to compare transient and steady-state measurements with each other, the cavitation intensity factor is uniformly standardized to N operating cycles (standardization constant). It is calculated as follows:

$$KI = N \cdot \frac{\sum_{k=0}^{m} (n_k \cdot f_k)}{c}$$

KI: cavitation intensity factor

k: class number

m: maximum number of classes

n_k: number of events per class

f_k: weighting factor

c: number of four-stroke cycles measured per KI factor

N: standardization constant

7.8.8 Test bench setup for cavitation measurements

Figure 7.147 shows the test bench setup for cavitation measurements on a six-cylinder in-line engine. Two pressure sensors are shown symbolically on cylinder 1 for measuring the local changes in coolant pressure. The cavitation process is influenced by many physical effects. Therefore, for a qualitative analysis of the results, it is necessary that other physical parameters be measured in addition to the local fluctuations in coolant pressure resulting from imploding cavitation bubbles. As **Figure 7.147** shows, the inlet and outlet temperatures of the coolant, the corresponding pressures, and the coolant pressure directly downstream of the coolant pump are captured, along with the coolant volume flow rate.

In order to influence the cooling system's system pressure and coolant volume flow, regardless of the operating condition of the engine, the test bench is also equipped with external regulators for the pressure in the cooling system. The coolant pump can be driven externally by a controlled electric motor. A coolant pump operation independent of the engine speed is therefore a prerequisite for determining various parameters.

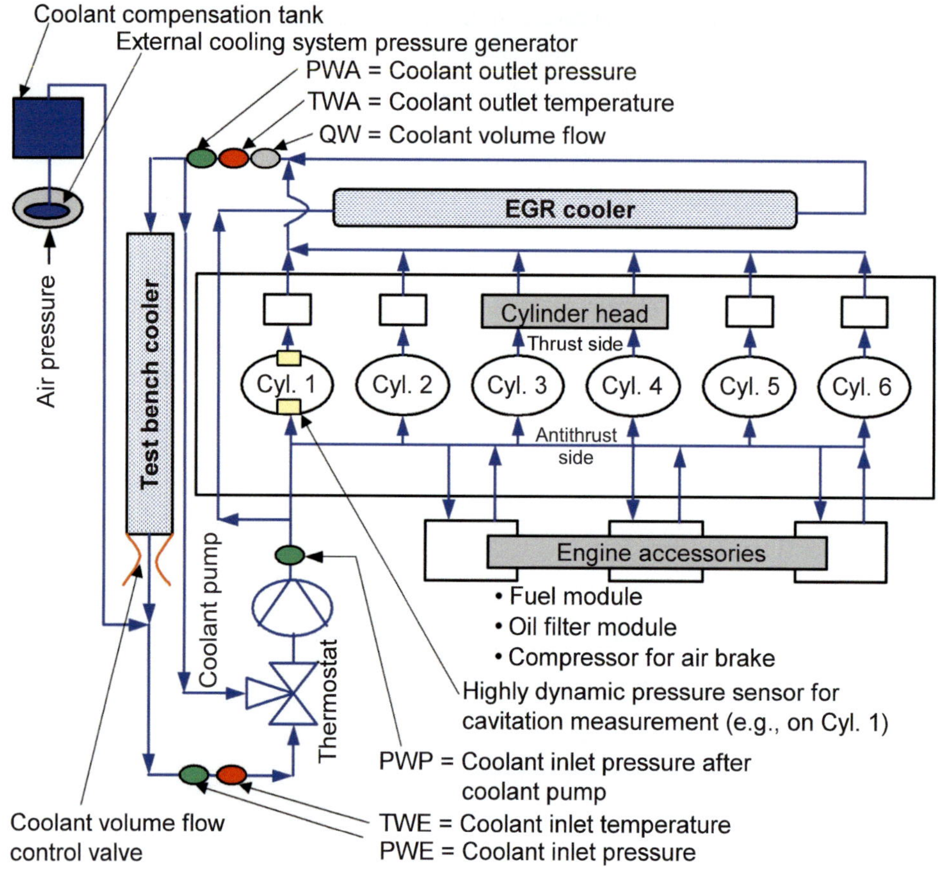

Figure 7.147: Test bench setup for cavitation measurements

The flow conditions can vary greatly within an engine from cylinder to cylinder, and depending on the type and number of engine auxiliaries through which the coolant flows.

7.8.9 Test run programs for cavitation measurements

Cavitation measurements are particularly helpful if the influence of individual parameters can be determined. The best results are obtained with:

- steady-state engine operation programs, if the influence of an individual parameter is to be determined; and
- transient cold-warm programs, if the goal is to determine the greatest cavitation intensity under the least favorable operating conditions. Operating conditions are generated

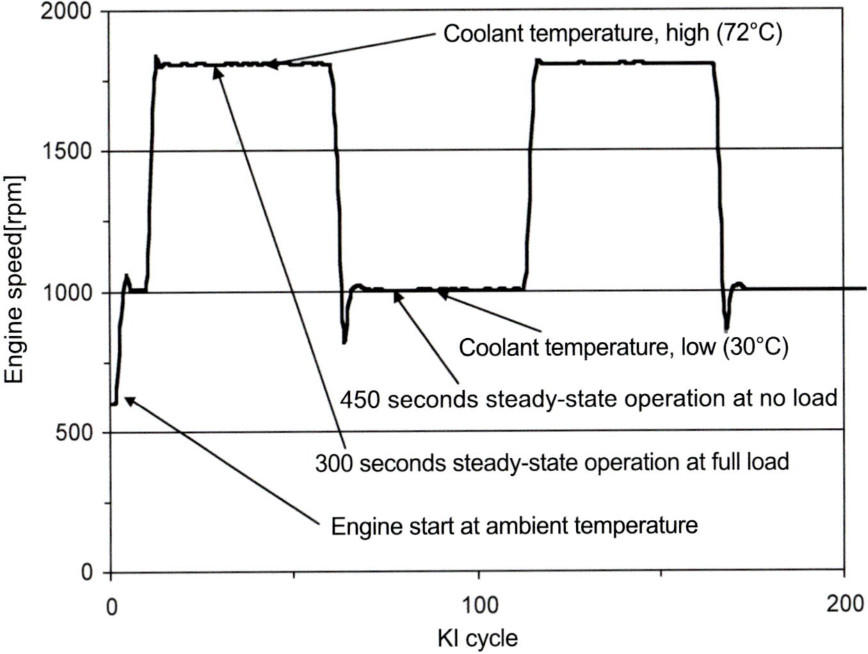

Figure 7.148: Transient test run program for cavitation measurements

in which the various factors affecting cavitation intensity overlap. **Figure 7.148** shows an example of such a program.

7.8.10 Relationship of cavitation intensity to the arrangement of the cylinder and the position on the cylinder

Depending on the design of the cooling gallery in the engine and the installed engine auxiliaries, **Figure 7.147**, which are partially integrated in the cooling circuit, the coolant's flow conditions can be very different from cylinder to cylinder. The cavitation intensities, and thus the cavitation damage, may therefore vary. Occasionally only one or two cylinder liners in an engine will show cavitation damage, whereas the others have only slight or no damage. The position of the cavitation damage can also vary from cylinder to cylinder. **Figure 7.149** shows the cavitation intensities measured on various cylinders at various positions. It also shows the results from a steady-state point measurement. As can be seen, the cavitation intensities determined for the lower thrust side measurement positions are significantly greater than for the upper thrust side positions or those on the antithrust side.

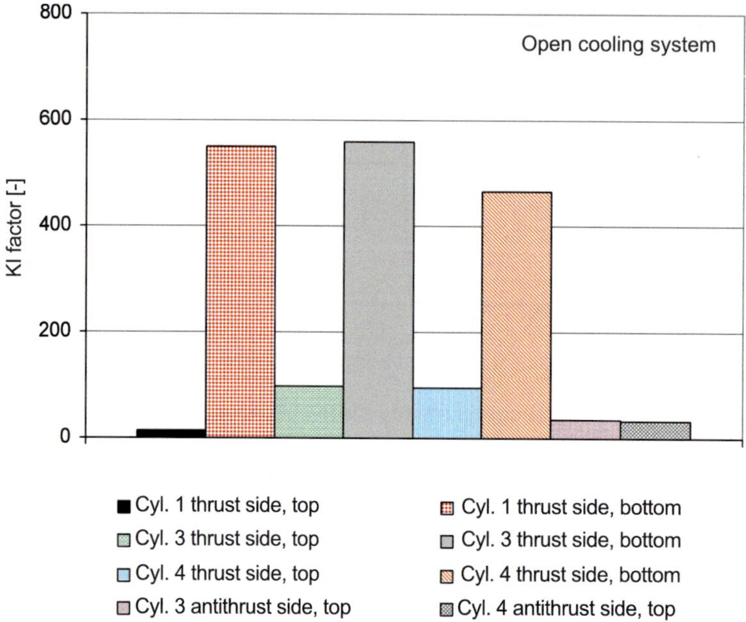

Figure 7.149: Relationship of cavitation intensity to the arrangement of the cylinders and the measurement position on the cylinder

7.8.11 Influencing parameters

Cavitation in wet cylinder liners can be affected by various parameters. The parameters can be fundamentally divided into two main groups:

■ The main group of engine operating parameters includes all the parameters that can be affected by the engine operating conditions. Engine operation parameters include, in particular

   – engine speed;

   – load;

   – maximum peak cylinder pressure and characteristic cylinder pressure curve;

   – coolant pressure;

   – coolant temperature;

   – coolant flow velocity;

   – chemical composition of the coolant.

■ The main group of design parameters includes all the design layout options for the piston, the cylinder liner, and the engine block, which minimize the excitation of the cylinder liners and optimize the cooling gallery design with regard to cavitation. These parameters include, in particular

   – various piston types;

   – design of the piston skirt profile and the ring belt;

- piston pin offset;
- installation clearance between the piston and the cylinder liner, and between the cylinder liner and the engine block;
- cylinder liner guidance (top stop, mid stop, bottom stop);
- shape of the outside surface of the cylinder liners;
- shape of the cooling gallery contour in the engine block.

7.8.11.1 Influence of engine operating parameters on cavitation

7.8.11.1.1 Influence of engine speed

Figure 7.150 shows the influence of engine speed on cavitation propensity. Owing to the fact that the experimental engine is set up on the test bench under near-series production conditions, not all of the influencing factors related to the engine speed could be completely eliminated.

The influence of the coolant pump, for example, and therefore the coolant flow behavior, also depends on the engine speed.

Another factor is that, depending on the dataset of the ECU (electronic control unit), various characteristic combustion chamber pressure curves and peak pressure values may be set at different engine speeds. The excitation of the cylinder liner, caused by the secondary piston motion and affecting the cavitation, can therefore be different at different speeds.

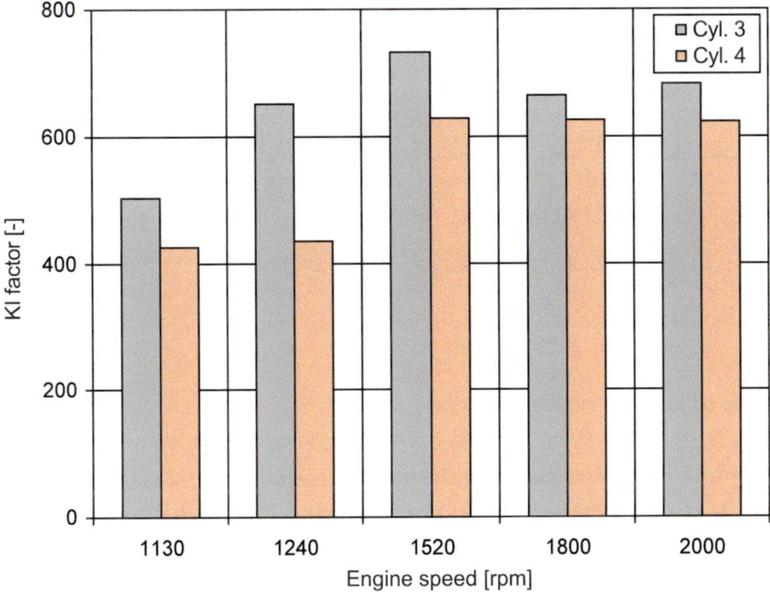

Figure 7.150: Influence of engine speed on the KI factor

In the experimental engine under test, the changes in the influence variables presented were small in the higher speed range, and therefore played a secondary role. As **Figure 7.150** shows, the engine speed therefore has almost no direct influence on the cavitation intensity. Only at low and medium engine speeds can a significant reduction in cavitation intensity be observed. This, however, is mainly the result of the significant changes in the indirect influence variables, and thus does not represent a direct influence of the engine speed.

7.8.11.1.2 Influence of engine load

Figure 7.151 shows the influence of engine load on cavitation intensity at constant engine speed. A clear correlation between load and cavitation intensity can be seen here. With increasing load, and otherwise identical operating conditions, the cavitation intensity clearly increases as well.

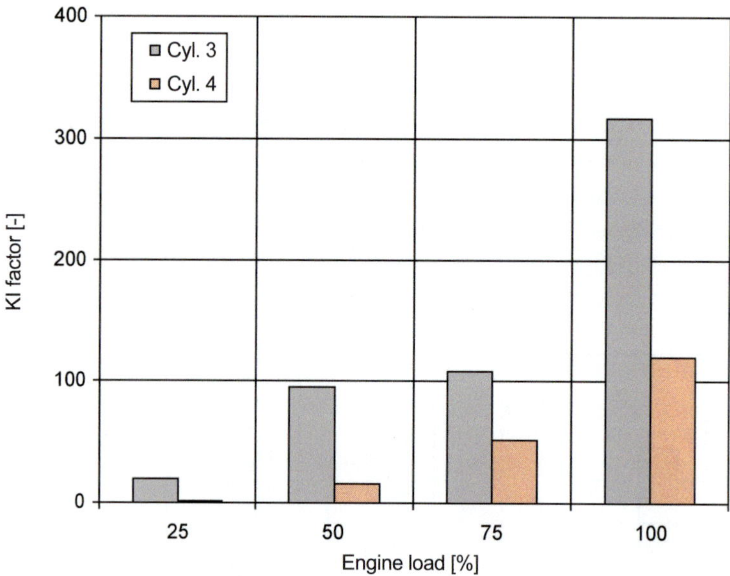

Figure 7.151: Influence of engine load on the KI factor at constant engine speed

7.8.11.1.3 Influence of cooling system pressure

On the basis of the Bernoulli equation and the cavitation number sigma derived from it, it is clear that an increase in static cooling system pressure would theoretically reduce the cavitation intensity. This fact is also confirmed in practice; **Figure 7.152**. With increasing static cooling system pressure, a significant decrease in cavitation intensity can be observed. Depending on the severity of the cavitation that does occur, the static pressure level necessary to achieve a significant reduction may vary greatly from engine to engine. A generally

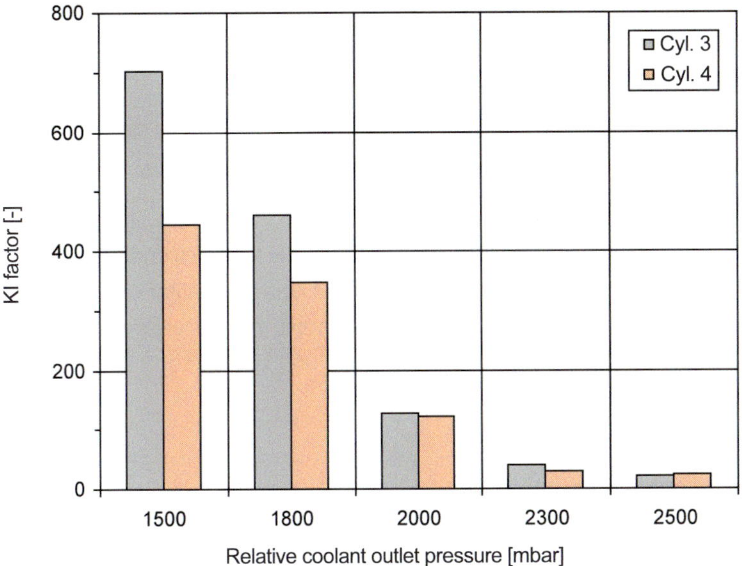

Figure 7.152: Influence of relative coolant pressure on the KI factor

applicable value from which cavitation propensity is significantly reduced can therefore not be provided. It must be determined for each engine individually. In practice, a static cooling system pressure that is higher than the pressure resulting from thermal expansion of the coolant can be achieved only with great technical effort, in the form of an additional pressurizing unit. In order to ensure operating safety of the gaskets connected to the cooling system and of the radiator, the maximum possible pressure is also limited. **Figure 7.152** shows the cavitation intensities for steady-state point measurements for various relative coolant outlet pressures at a full-load operating point.

7.8.11.1.4 Influence of coolant volume flow rate

The influence of the coolant volume flow rate, and that of the cooling system pressure, are directly related to each other. A greater coolant volume flow leads to an increase in dynamic pressure, and thus to a reduction in static pressure. This, in turn, is associated with an increase in cavitation propensity.

7.8.11.1.5 Influence of coolant temperature

If it were possible to consider coolant temperature in full isolation, then, theoretically, there would be no direct influence on cavitation. In practice, however, the coolant's temperature cannot be considered in full isolation. Indirect influence variables bring about changes in the temperature.

If the coolant temperature is increased to close to the boiling point in an open system, then vapor bubbles can form. These generally have a damping effect on the occurrence of imploding cavitation bubbles, which decreases the cavitation intensity.

If the coolant temperature is increased in a closed cooling system, then the static pressure rises in the system, because of the thermal expansion of the coolant, thus reducing the cavitation propensity.

Also to be considered is the fact that higher coolant temperatures lead to higher component temperatures, and therefore to lower operating clearances. Lower operating clearances, in turn, have a positive effect in preventing cavitation. In this context, the critical operating clearances are those between the piston and the cylinder liner, and between the engine block and the cylinder liner.

7.8.11.1.6 Influence of coolant composition

It has been empirically demonstrated that, under otherwise identical boundary conditions, visible cavitation damage is often very different depending on the coolant used. As measurements have shown, there are no significant differences in measured cavitation intensities in experiments with different coolants. It is, however, possible that different coolants with the same measured cavitation intensity cause very different levels of cavitation damage. The reason is that a protective coating can form on the cylinder liner depending on the coolant and coolant additives used. This protective coating can have a damping effect on the imploding cavitation bubbles. Depending on the composition of the coolant additives, the damping effect of the protective coating can be more or less effective.

7.8.11.1.7 Influence of combustion chamber pressure

There is a direct correlation between the peak combustion chamber pressure and the cavitation intensity; **Figure 7.153**. The cavitation intensity increases with increasing pressure. The peak combustion chamber pressure from which a significant increase in cavitation intensity occurs is different for every engine. Therefore, a general threshold value cannot be provided.

7.8.11.2 Influence of design parameters on cavitation

The main reason for the occurrence of cavitation is the vibrations of the cylinder liner induced by the piston and its secondary piston motion.

The group of design parameters that can be used to directly or indirectly affect cavitation includes all those engine component measures that can reduce vibrations of the cylinder liner and allow optimization of the coolant circuit with regard to cavitation. These engine components include, in particular, the pistons, the cylinder liners, the coolant pump, and the engine block itself. Measures involving the guiding concept of the cylinder liners, the cooling

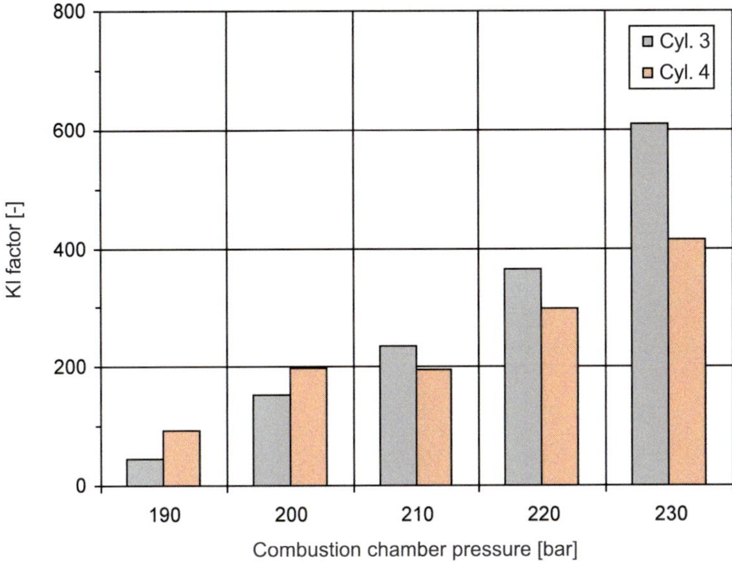

Figure 7.153: Influence of combustion chamber pressure on the KI factor

gallery shape in the engine block, and the installation clearances between the piston and cylinder liner, and between the cylinder liner and engine block are also part of this group.

For comparisons of the influence of various design parameters on cavitation, the transient engine test program has proven to be effective.

In the transient engine test program, negative effects promoting cavitation are captured in an overlapping manner. This program allows the cavitation behavior to be analyzed very quickly at various engine speeds, loads, and coolant temperatures. Potential dynamic effects are also included in the analysis. The highest cavitation intensities can be captured using the transient engine test program.

7.8.11.2.1 Influence of piston and cylinder liner installation clearance

There is a clear correlation between installation clearance and cavitation intensity; **Figure 7.154**. Accordingly, a reduction in the installation clearance leads to a lower cavitation intensity. In general, this also applies to the installation clearance between the cylinder liner and the engine block. The reduction in installation clearance is limited, however, by seizure resistance.

Initial stages of development serve to determine the minimum possible installation clearance and to reduce cavitation intensity with additional design measures on the component. Potential measures include changes to the piston profile, piston skirt ovality, ovality of the piston ring land, piston pin offset, piston crown offset, and ring land clearance. Changes in material can also contribute to a reduction in cavitation intensity.

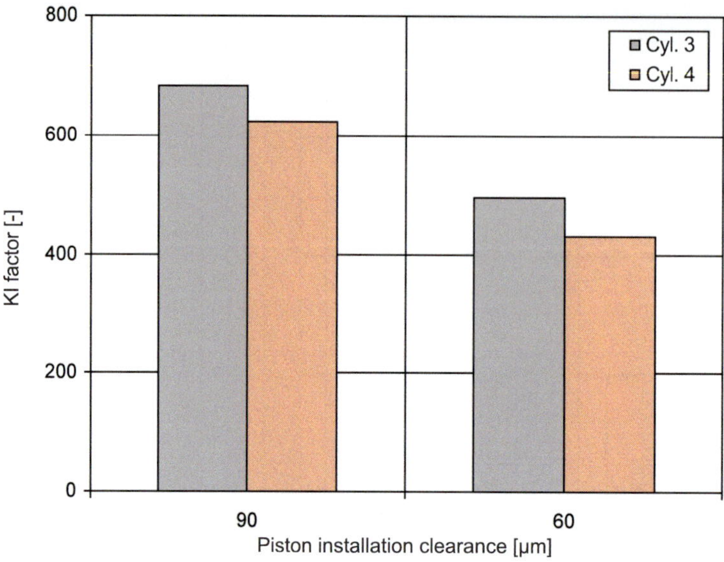

Figure 7.154: Influence of piston installation clearance

7.8.11.2.2 Influence of piston type and piston profile

A typical cavitation behavior can often be determined on the basis of the piston type. **Figure 7.155** shows measurement results for an aluminum piston and an articulated piston (formerly widespread type, with steel crown and aluminum skirt). Compared with other concepts, both piston types have a relatively large installation clearance in a cold engine on account of the thermal expansion behavior of their aluminum skirts. These two piston types in particular, therefore, have very high cavitation intensities in the first stage change in the transient engine test program, immediately after starting the engine. This unfavorable set of conditions is reduced as the engine load and engine coolant temperature increase, because the operating clearances become significantly smaller. The conditions are no longer present in a warm engine. According to **Figure 7.155**, there is a fundamental difference in level between the aluminum piston and the articulated piston. It is evident in all heating stages and is not determined by the piston type here, but rather by a different level of optimization.

As **Figure 7.156** indicates, there are no significant increases in cavitation intensity for steel pistons during the first stage change, immediately after starting the engine. Among the steel pistons tested, the MONOTHERM® piston with a closed skirt had the lowest cavitation intensities. The optimized MONOTHERM® piston with the open skirt, however, also had measured cavitation intensities that were not in the critical range.

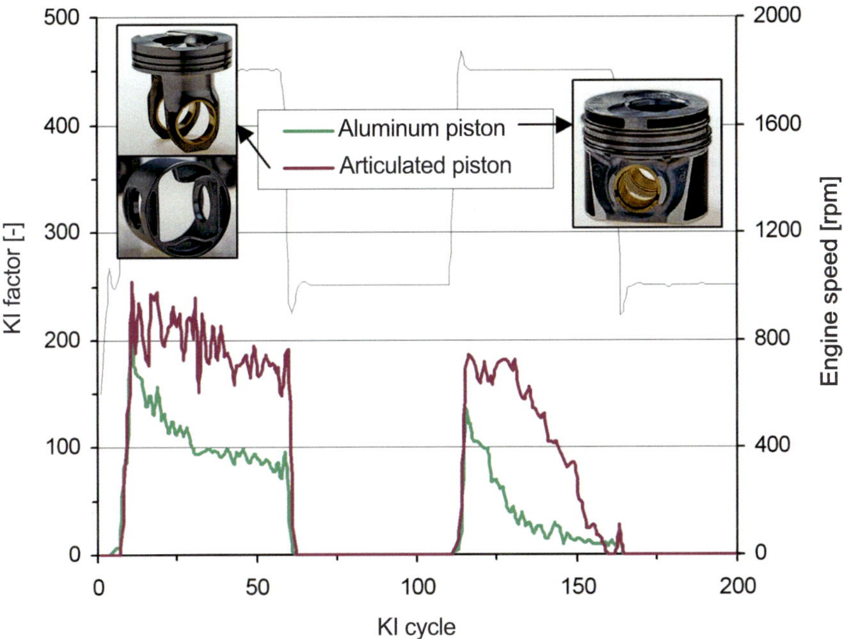

Figure 7.155: Influence of piston type on cavitation behavior

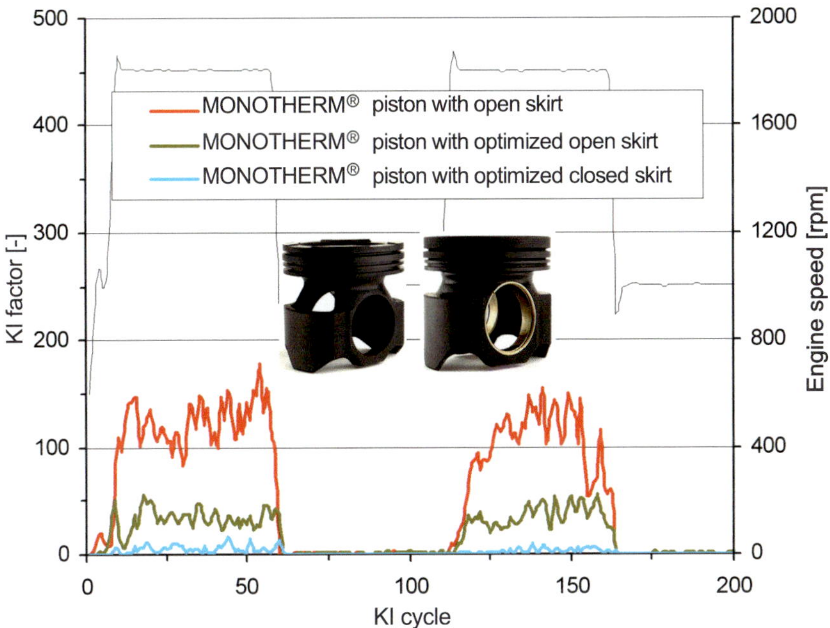

Figure 7.156: Influence of piston type and piston profile on cavitation behavior

7.8.11.2.3 Influence of other piston design characteristics

The design of the piston allows a whole series of options for positively affecting cavitation intensity. They include

- reducing installation clearance (while ensuring seizure resistance);
- altering the piston skirt (area, gating, profile, ovality);
- piston pin offset;
- piston crown offset;
- ring land ovality;
- ring land clearance;
- material selection;
- casting process.

Such changes, however, must not have any negative effects on the stiffness and deformation characteristics.

Generally applicable standard values for a cavitation-free design cannot be provided. Each change to a design parameter is a specific individual solution for the engine type being optimized. Therefore, successful optimization measures can be applied directly to a different test engine only to a limited degree.

Promising results for reducing the cavitation intensity can be obtained by adapting the ovality (piston skirt and crown), introducing piston pin offset, **Figure 7.157**, and altering the ring land.

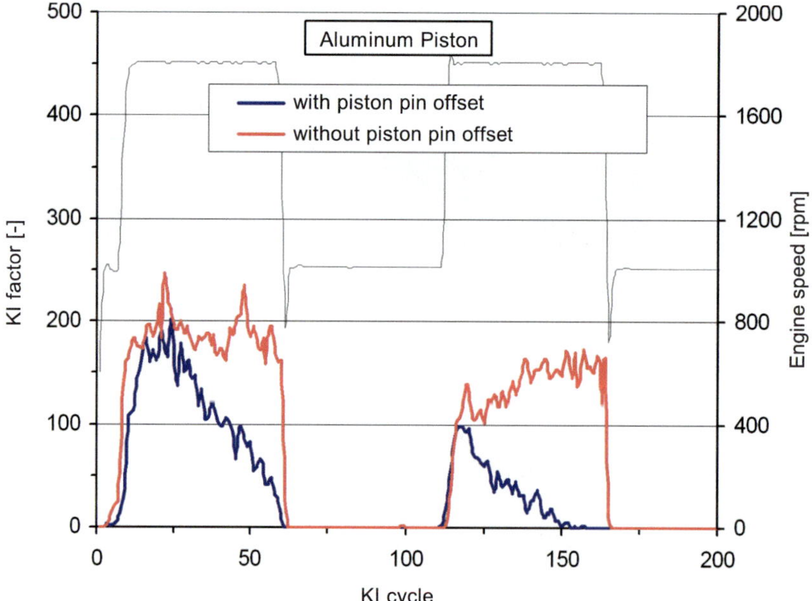

Figure 7.157: Influence of piston pin offset on cavitation behavior

7.8.11.2.4 Influence of design characteristics of the cylinder liner and cooling gallery shape

In addition to design changes to the piston, the cavitation intensity can also be influenced by changes to the cylinder liner and cooling gallery. Design parameters that have the potential to reduce cavitation intensity include

- optimizing the cylinder liner guidance;
- optimizing the cylinder liner installation clearance;
- optimizing the cylinder liner geometry with regard to reducing vibration;
- optimizing the cylinder liner geometry, in combination with changes to the engine block's cooling gallery, under the aspect of optimizing coolant flow conditions;
- optimizing the coolant pump, under the aspect of optimizing coolant volume flow rates and achieving the highest possible static pressure component.

7.9 Lube oil consumption and blow-by in the combustion engine

The important goals of mechanical development of a combustion engine include low blow-by and minimal lube oil consumption, and thus minimizing raw exhaust gas emissions.

The oil components entering the combustion chamber increase the particle concentration when they are carried out in the exhaust gas, which makes exhaust gas aftertreatment more difficult [38].

Further optimization of the current low levels of lube oil consumption in combustion engines absolutely requires precise understanding of the lube oil consumption mechanisms inside the engine. Such knowledge also helps to expand simulation approaches and their validation.

This article is intended to demonstrate, with examples, the mechanisms leading to lube oil consumption and effective actions that can be taken in the development stage.

7.9.1 Lube oil consumption mechanisms in the combustion engine

The term "lube oil consumption" generally refers to the amount of engine oil by which the amount of oil in the engine is reduced over a specified period of time. Oil dilution with fuel and combustion products and residues into the engine oil can give the appearance of a positive lubricating oil balance, despite an actual consumption of oil. Particularly when using biodiesel, an increased fuel dilution of the engine oil should be expected. Less volatile components in biodiesel do not sufficiently vaporize in the combustion chamber, and can therefore enter the engine oil.

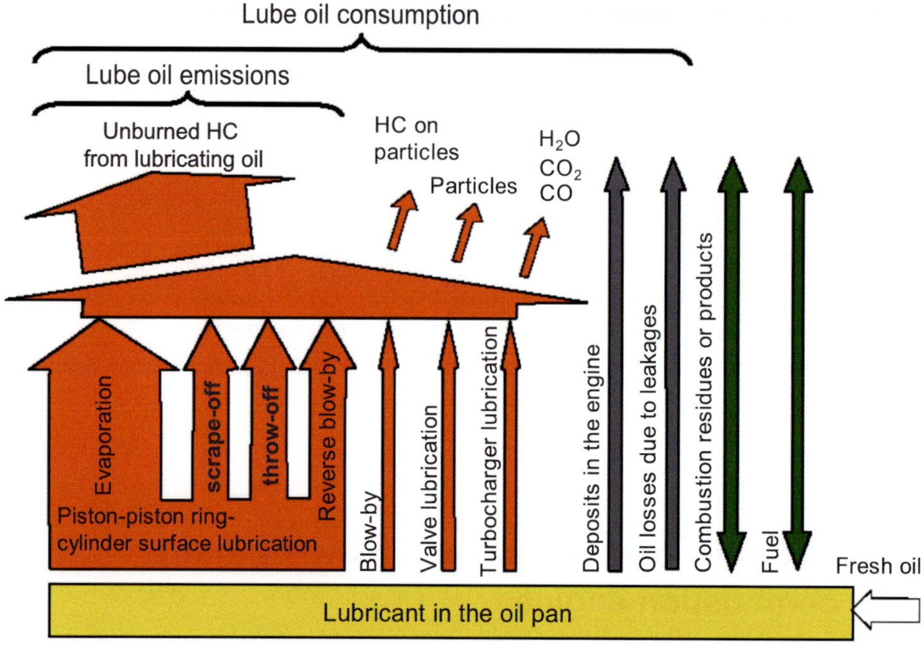

Figure 7.158: Lubricant balance in the combustion engine [39, 40]

Figure 7.158 shows schematically the various mechanisms as part of a lubricant balance. The exact levels of individual components cannot be provided across the board, because they vary from engine to engine, giving rise to a wide range of boundary conditions. The greatest, most definitive component of consumption, at 80 to 90%, results from the lubrication of the group of friction partners consisting of the piston, piston ring, and cylinder wall. It necessarily ends up in the exhaust gas in the form of unburned hydrocarbons (HC). Only a small portion of the lubricating oil is burned during combustion and contributes to particle formation with other combustion residues [38, 39].

The proportion of hydrocarbons (HC) in the exhaust gas is referred to as "lube oil emissions," because they originate from the lubricating oil. The oil enters the combustion chamber through various mechanisms, which are described in more detail in this chapter.

Figure 7.159 shows paths through which engine oil enters the combustion chamber. The largest portion comes from the cylinder wall (1), which is sealed by the piston and piston rings. Additional oil enters the combustion chamber through the crankcase ventilation system (blow-by)(2), and through the compressor side seal between the center housing and the shaft of the turbocharger (2). Another potential path for oil into the combustion chamber is through the valve stem seal (3).

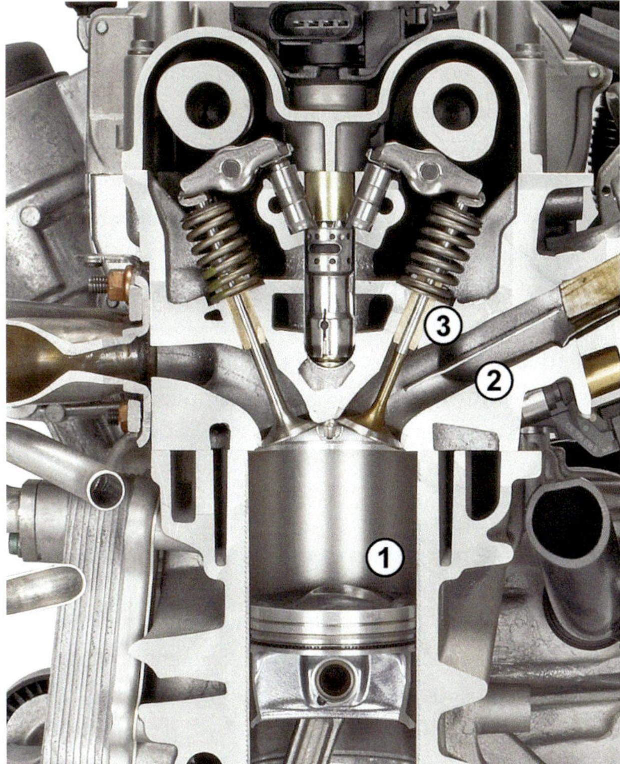

Figure 7.159:
Paths through which engine
oil enters the combustion
chamber [42]

① Cylinder wall-piston rings
② Crankcase ventilation
② Turbocharger, compressor
   side
③ Valve stem seal

The causes and mechanisms of lube oil consumption and blow-by can be very complex
and interconnected. Individual interactions between lube oil consumption and blow-by in
the area of the cylinder wall and piston rings, **Figure 7.159**, (1), are examined in the following:

- Severe cylinder distortion can lead to the piston rings sealing insufficiently during engine
  operation if they can no longer track the contour of the cylinder because of insufficient ring
  conformability. The results can be high lube oil consumption and/or blow-by.
- The surface roughness and honing pattern of the cylinder surface permanently affect lube
  oil consumption via the functional behavior of the piston rings.
- Piston ring geometries that are optimized for low lube oil consumption, such as sharp
  scraping first piston rings, or piston rings with very barrel-shaped running surfaces, can
  lead to increased blow-by–particularly under full load–if the gas force acting on the running surface forces the ring away from the cylinder wall.
- Piston ring geometries that are optimally tuned for low blow-by values, such as a low axial
  width of the first piston ring, have a positive effect on blow-by behavior at zero load, on
  account of their low ring mass, but can cause increased lube oil consumption under full
  load, because of their lower stiffness.

- The use of compression rings with a deliberate cross-sectional perturbance, in the form of an internal bevel or an interior angle, can strongly influence the twisting behavior of the piston rings. An additional inclined position in the ring grooves (disk-shaped or roof-shaped) and optimization of the axial ring installation clearance can have a positive effect on blow-by behavior, in particular, as well as on lube oil consumption.

- In order to optimize lube oil consumption and blow-by, particularly during dynamic engine operation, the targeted application of chamfers to the top land and ring lands can affect the pressure buildup behind the first piston ring in a controlled manner.

- Brief periods of excessively high pressure between the first and second piston rings can cause the first piston ring to be forced away from the lower flank of the groove, and thereby increase the lube oil consumption due to "reverse blow-by," Chapter 7.9.1.1. A change in volume between these two rings, in the form of grooves, chamfers, or undercuts, and optimization of the gap clearances can lead to faster and more uniform pressure equalization in the intermediate ring space.

7.9.1.1 Lube oil consumption in the frictional system of the piston, piston rings, and cylinder wall

As previously described, the tribological system of the piston, piston rings, and cylinder wall is mainly responsible for overall lube oil consumption in the engine, and therefore for a significant portion of lube oil emissions. This portion is determined by the three lube oil consumption mechanisms depicted in **Figure 7.160**.

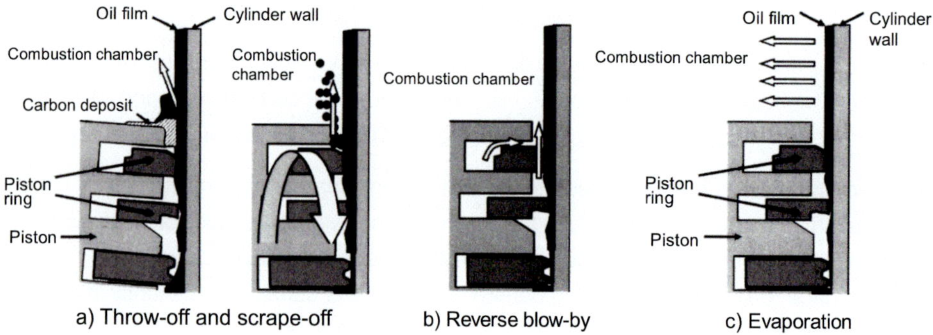

Figure 7.160: Schematic diagram of lube oil consumption mechanisms in the tribological system of the piston, piston rings, and cylinder wall [44]

At operating points with low load, the oil can collect on the top land and above the first piston ring. The high acceleration and deceleration forces "throw-off" these oil drops directly into the combustion chamber or increase the oil film thickness on the cylinder wall. Carbon buildup on the top land can also lead to scrape-off of the oil from the cylinder wall during the upward stroke; **Figure 7.160 a)**.

The second way for lubricating oil to enter the combustion chamber is known as "reverse blow-by"; **Figure 7.160 b)**. The oil enters the combustion chamber in the form of a liquid or a mist. The pressure in the area between the first and second piston ring can sometimes be higher than the pressure in the top land gap or the combustion chamber. This pressure differential causes the gas to flow back through the ring gap of the first piston ring into the combustion chamber. This portion of oil can also increase the oil film thickness on the cylinder wall.

Another potential way for lubricating oil to enter the combustion chamber is "evaporation" of the lubricant from the hot cylinder wall; **Figure 7.160 c)**. During combustion, expansion, and exhaust strokes, the high wall temperatures in the combustion chamber briefly cause the oil film to heat up. This causes partial evaporation of the oil. This behavior is highly dependent on the lubricating film thickness and the condition of the oil on the cylinder wall. A higher fuel content in the oil film causes oil to be evaporated to a greater degree. This effect contributes significantly to lube oil emissions [41].

7.9.1.2 Lube oil consumption through valve stem seals

With the development of effective and wear-resistant valve stem guides and seals, the problem of lube oil consumption in this area is of secondary significance in modern engines.

7.9.1.3 Lube oil consumption through crankcase ventilation (blow-by)

The gas entering the crankcase from the combustion chamber through the piston rings (blow-by) is partially charged with unburned fuel and oil from the cylinder wall. After the liquid components have largely been separated, the gas, with the remaining oil components, is fed back to the combustion cycle. This means that the engine's lube oil consumption is directly affected by the blow-by quantity and the quality of the oil separation.

Typical lube oil consumption proportions through crankcase ventilation are currently in the range of 1 to 2 g/h. They are definitely interesting because the total lube oil consumption in a modern engine can be well below 10 g/h at individual operating points. Further optimization of the oil separator in the blow-by recirculation systems thus remains an unchanged goal.

7.9.1.4 Lube oil consumption and blow-by in the turbocharger

Blow-by gases enter the oil return of the turbocharger because of the pressure gradient between the compressor or turbine and the crankcase, and thus collect with those from the combustion chamber; **Figure 7.161**. Lube oil consumption through the shaft seal takes place on the compressor side, where the oil enters the combustion chamber with the charge air, and on the turbine side, where it enters the exhaust gas. This sealing behavior is dependent on the pressure ratios at the shaft seal and the wear condition of the turbocharger.

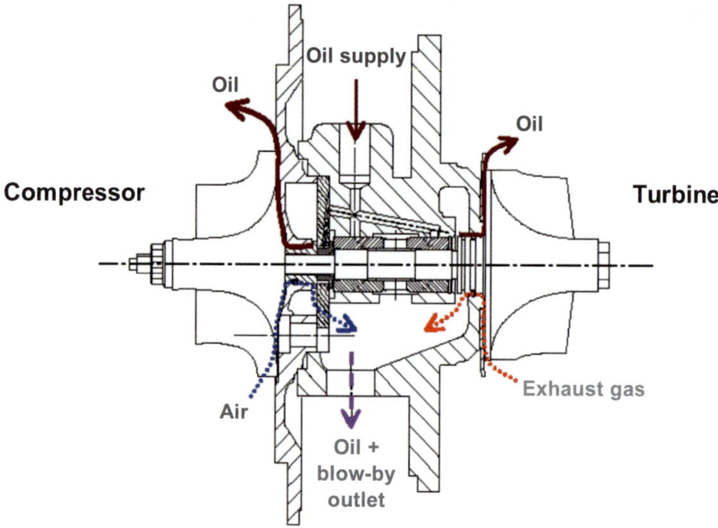

Figure 7.161: Leakage paths for oil and blow-by of the turbocharger

Figure 7.162 shows the results of an endurance test on a test bench, where the engine was operated with an external lubrication system for the turbocharger at the end of the endurance run. This allowed the lube oil consumption and blow-by in the turbocharger to be measured.

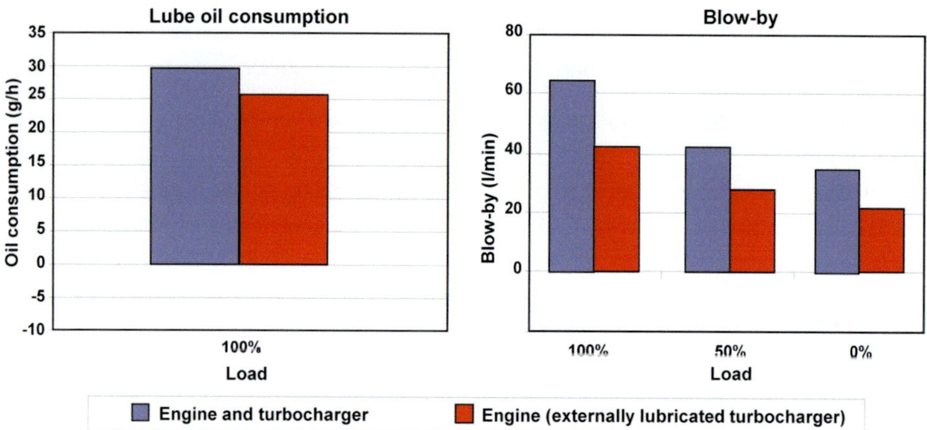

Figure 7.162: Lube oil consumption and blow-by in engine and turbocharger (2.0-liter, 4-cylinder diesel engine after a 500 h endurance run)

7.9.2 Lube oil consumption measurement methods

A basic distinction is made between conventional (gravimetric and volumetric) and analytical measurement methods.

Figure 7.163 shows an overview of potential lube oil consumption measurement methods.

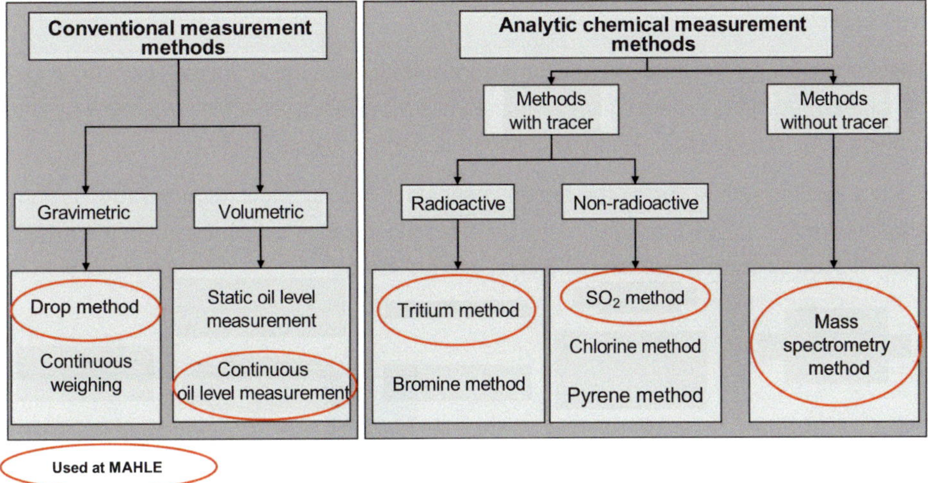

Figure 7.163: Overview and classification of lube oil consumption measurement methods

The conventional measurement methods include the following:
- Gravimetric:
  - Drain-and-weigh method over a specific time unit
  - Continuous weighing
- Volumetric:
  - Static oil level measurement
  - Continuous oil level measurement

The analytical measurement methods are divided into those with tracers (indicators) and those without tracers. The tracer methods can be divided into radioactive and nonradioactive:
- Radioactive:
  - Tritium method
  - Bromine method
  - …

■ Nonradioactive:
  – SO_2 method
  – Chlorine method
  – Pyrene method
  – ...

In conventional measurement methods, the consumed oil is measured. Particularly at low levels of lube oil consumption, therefore, very long running times are needed, which means high costs.

The analytical tracer methods, in which the concentration of a tracer (marker) from the engine oil is measured in the exhaust gas, can sometimes provide high accuracy in a relatively short running time. The creation of lube oil consumption maps is thus ideal here.

Such a system, with sulfur as the tracer, is shown schematically in **Figure 7.164**, using an engine oil with a defined sulfur content. By detecting the SO_2 concentration in the exhaust gas, the lube oil consumption in the engine can be derived.

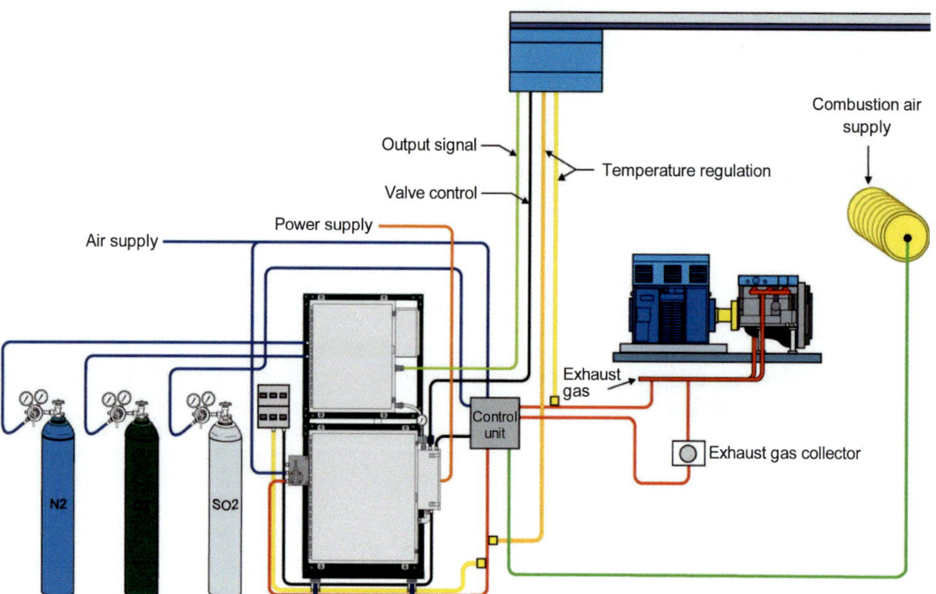

Figure 7.164: System for lube oil consumption measurement with sulfur as a tracer

Direct, fast online measurement, however, is nearly or completely impossible for transient engine operation when using tracer methods.

Experiments on modern engines show that highly dynamic operation of an engine is exactly what can lead to increased lube oil consumption, and that the most significant potential for improvement is hidden right here. Lube oil consumption maps based on steady-state operations do not reveal this potential. Only knowledge of the influence on lube oil consumption of rapid changes in load and speed can enable correct design measures to be derived and properly implemented for the piston or piston rings, for example.

With tracerless analytical measurement methods, the lube oil emissions in the exhaust gas can be determined directly. The underlying measurement principle is based on the condition that the hydrocarbons present in the exhaust gas, which can be associated with the fuel and the engine oil, have different molecular chain lengths. In order to measure lube oil emissions, a mass spectrometer optimized for this application is used, and the long-chain, low-volatility HC components associated with the engine oil are separated out [39, 40]. The HC molecules from the fuel and the engine oil are then measured separately using adjustable mass filters.

Figure 7.165 schematically shows a typical mass spectrum for exhaust gas with the usable measurement ranges for gasoline and diesel fuel [39, 40]. Because of the chemical similarities between HC molecules from engine oil and diesel fuel, measurements on diesel engines are particularly challenging. Implementing suitable calibration of the measurement system using the oil present in the engine and a known exhaust gas mass flow rate, the oil concentrations measured in the exhaust gas can be converted to a mass of lube oil emitted per unit of time.

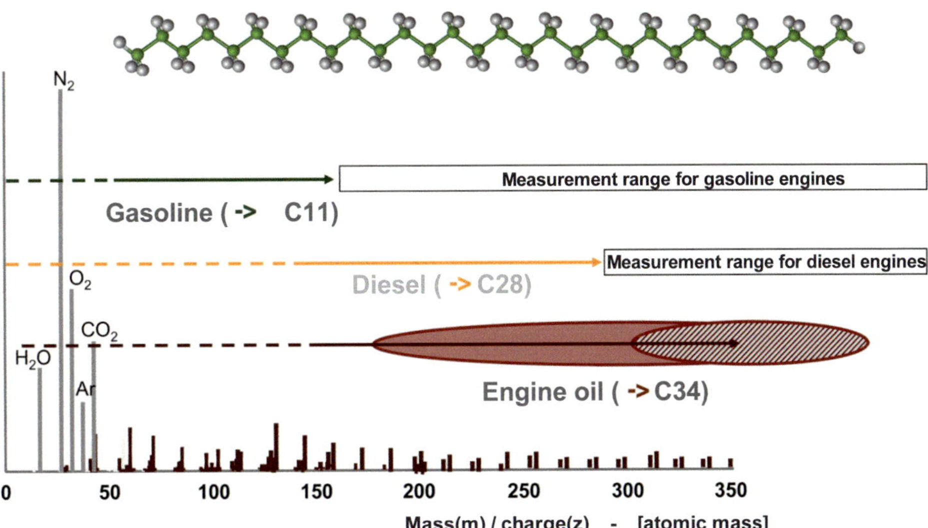

Figure 7.165: Mass spectrum of exhaust gas and usable measurement ranges for gasoline and diesel fuel [40]

The derived lube oil emissions indicate the amount of oil that is carried off unburned in the exhaust gas. Fundamentally, however, the measured lube oil emission values should always be viewed in relation to the total lube oil consumption. This also includes the burned oil

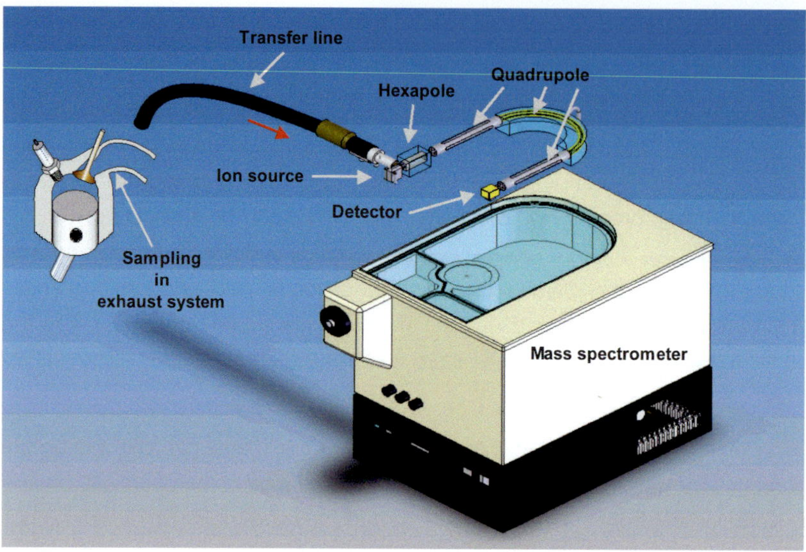

Figure 7.166: Schematic diagram of the mass spectrometric measurement system for analyzing lube oil emissions in exhaust gas [43]

components that are not captured as well as any leakages. In the case of a lubricating oil quantity balance, the fuel and water components that enter the oil and the residues arising from combustion must also be taken into consideration.

Figure 7.166 sketches out the Lubrisense 1200 measurement system with heated direct inlet system [43], which is used at MAHLE.

Since the exhaust gas can be collected immediately downstream of the exhaust valve, it is possible to determine lube oil emissions selectively for each cylinder, and thus to detect cylinder-specific lube oil consumption problems. This rapid, online measurement of lube oil emissions can capture the oil mass that is carried off in near real time, even for transient test cycles with rapid load and speed changes.

MAHLE has developed special transient test cycles for such online lube oil consumption measurements, which are used in engine development to reliably identify the operating ranges that are critical for lube oil consumption.

7.9.3 Lube oil consumption map and dynamic lube oil consumption

Figure 7.167 shows an example of the lube oil consumption map of a 4-cylinder gasoline engine. Lube oil consumption was determined using the described rapid lube oil emission measurement system based at 42 quasi-steady-state measurement points (2 to 4 minutes per operating point).

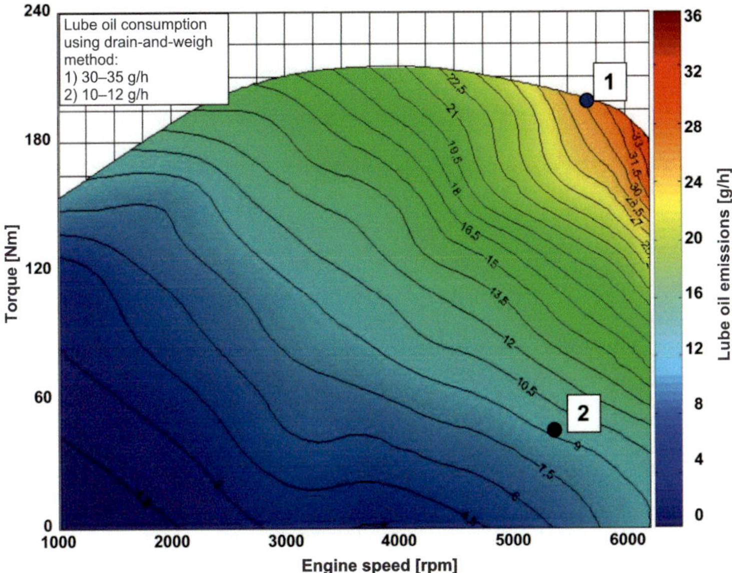

Figure 7.167: Lube oil consumption map for a 2.0-liter, 4-cylinder gasoline engine, using rapid lube oil consumption measurement; measurement points 1 and 2 measured conventionally

The long-term lube oil consumption was also determined during an endurance test using the drain-and-weigh and weighing methods at two selected operating points. In order to make definitive statements about the actual quantity of lube oil consumed, the engine was subjected to a steady-state endurance test for 3 times 5 hours at each operating point. This established a good correlation between the lube oil consumption values and the lube oil consumption map.

Precise matching of these values cannot be achieved in principle, because the drain-and-weigh method collects not only the oil emitted with the exhaust gas but also all the components from the entire lubricating oil balance, including all the fuel components that enter the engine oil.

This relationship between lube oil emissions and actual lube oil consumption can be influenced by a number of external boundary conditions. This includes, for example, fuel quality. Admixing biocomponents to fuels affects the evaporation behavior and therefore the mechanisms that are most critical for lube oil consumption. Combustion methods with associated mixture formation can also play a decisive role in evaporation behavior because of a different fuel amount in the cylinder wall oil film. A correlation between the air-fuel mixture and the measured lube oil emissions has already been demonstrated [41].

Figure 7.168 shows the lube oil emissions of individual cylinders over time during a transient test cycle. Here again, the lube oil consumption as measured by the drain-and-weigh method matches the cylinder-specific lube oil emissions very well, and no concrete lube oil consumption problem can be detected.

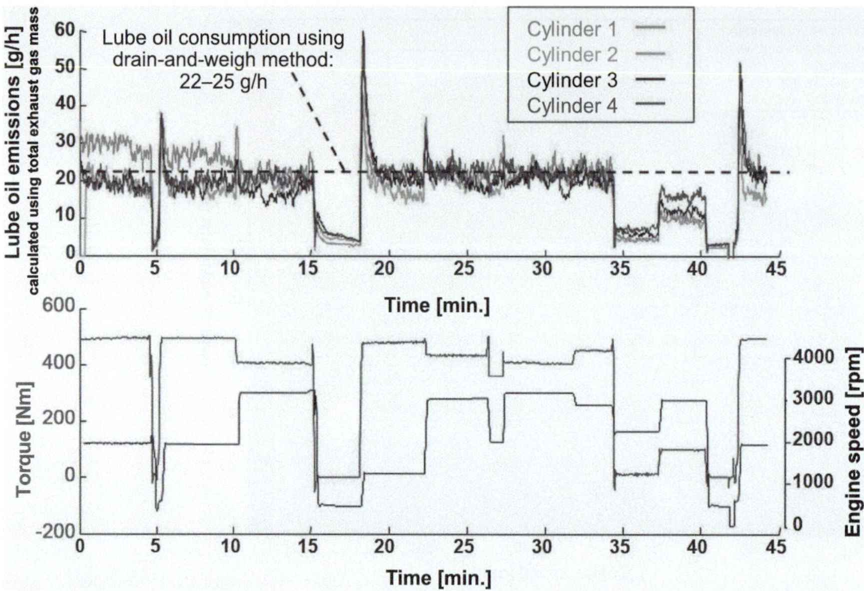

Figure 7.168: Lube oil emissions of individual cylinders over time during a transient test cycle

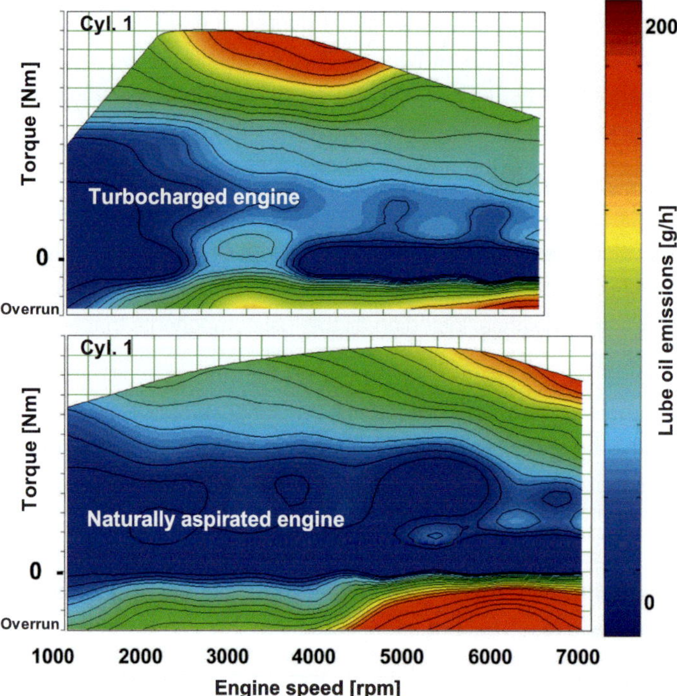

Figure 7.169: Cylinder-specific lube oil emission maps for a turbocharged (top) and naturally aspirated (bottom) gasoline engine (lube oil emission levels relative to the exhaust gas mass of the complete engine)

Figure 7.169 shows additional typical lube oil emission maps, as they can be measured for gasoline engines with and without turbocharging. Cylinder-selective measurements were performed on engines of similar types and with similar displacements.

Turbocharged engines typically reach their maximum lube oil emissions at medium speeds in the range of maximum torque. Naturally aspirated engines, on the other hand, generally reach their maximum lube oil emissions at full load in the range of maximum engine speed. These phenomena can be observed with measurements of various turbocharged and naturally aspirated engines with different displacements, and are presumably best explained by the location of the maximum combustion chamber pressures in the operating map.

Another maximum point for lube oil emissions is often found during overrun at higher speeds and, as a rule, is significantly more distinct for naturally aspirated engines than for turbocharged engines because of the different pressure ratios in the intake section.

For engines with concrete lube oil consumption problems, experience often shows that it is difficult to identify an operating range that is most responsible for the average lube oil consumption. Indications from actual vehicle operation can typically be helpful here.

Critical operating ranges can be clearly identified in a transient lube oil consumption measurement run using the online lube oil emission measurement system presented here.

Figure 7.170 shows the results of measurements from such a transient lube oil consumption measurement run on a naturally aspirated gasoline engine, which was performed with different piston ring packs. The measurement in the initial state (piston ring pack 1) shows extremely high lube oil emissions in overrun mode and at low speeds; **Figure 7.170** (see operating range B). In this case, this knowledge allowed targeted changes to be made to the ring pack. A direct comparison of the results of both test variants, running the same measurement program, shows that the optimized piston ring pack 2 no longer shows any significant lube oil emissions in operating range B.

The increased lube oil emissions that occurred in overrun mode and at high speeds, **Figure 7.170** (operating range A), is present at the same level in the measurements for both piston ring packs. The changes made to the piston rings, therefore, did not lead to any significant reduction in lube oil emissions for the indicated operating range A, and these must therefore be addressed with other measures.

The average total lube oil consumption calculated from the lube oil emission curves does, however, show a significant drop in lube oil consumption for piston ring pack 2. This result was confirmed in actual vehicle operation.

Beyond such quasi-steady-state phenomena, there is a further potential for improvement by analyzing dynamic lube oil emission processes. Spikes in lube oil emissions can often be found under highly dynamic changes in load and speed, and their amplitudes and damping times also affect the average lube oil emissions.

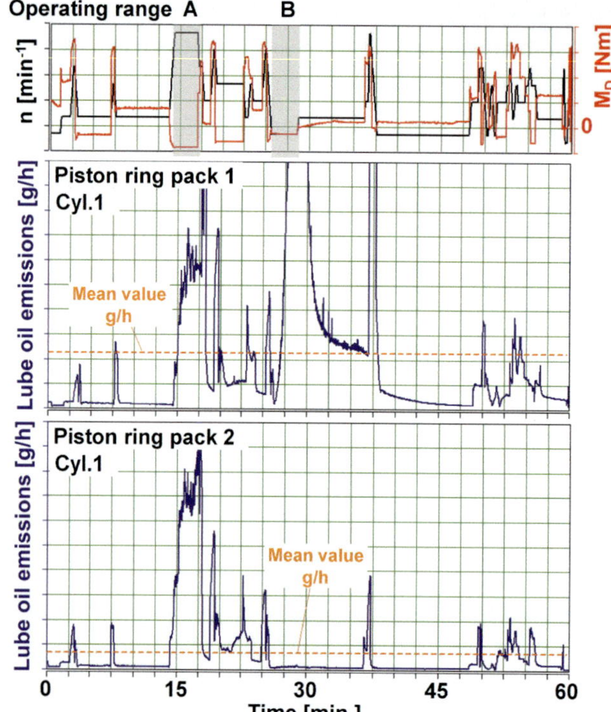

Figure 7.170:
Lube oil emission curve in cylinder 1 of a naturally aspirated gasoline engine in a transient program using different piston ring packs

The online lube oil emission measurement system presented here allows extensive parameter studies to be performed systematically. Suitable measures can be derived from these results, allowing the lube oil consumption to be minimized through targeted changes to the piston and piston rings, optimizing them as a system.

7.9.4 Influence of intake manifold vacuum on lube oil consumption in the gasoline engine

In the past, the influence of intake manifold vacuum on lube oil consumption was typically marked by visible blue smoke in vehicles without exhaust gas aftertreatment systems. In overrun phases at high speed, the intake manifold vacuum extends into the combustion chamber during the valve overlap phase, which increases "reverse blow-by" and thus the quantity of oil drawn into the combustion chamber.

Figure 7.171 shows the result of a series of tests in which the lube oil consumption behavior of a naturally aspirated V8 engine was measured as a function of the intake manifold vacuum occurring in the overrun phase of a dynamic running program with engine braking. The intake manifold vacuum is varied by the throttle valve position in the overrun phase.

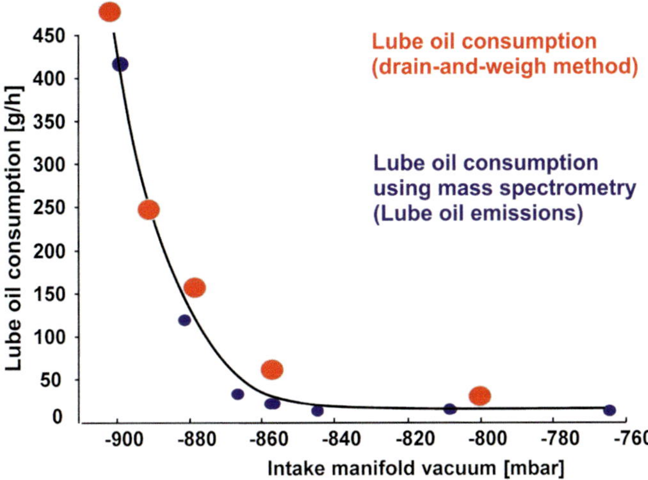

Figure 7.171: Influence of intake manifold vacuum on lube oil consumption and lube oil emissions for a dynamic running program with engine braking on a naturally aspirated V8 gasoline engine

The lube oil consumptions shown here were determined using two different measurement methods. In order to measure the lube oil consumption using the conventional drain-and-weigh method, the engine ran the dynamic running program with engine braking for a long period of time. In parallel, the lube oil emissions in the exhaust gas were captured online using mass spectrometry.

In this case, a strong increase in lube oil consumption occurred at very low absolute intake manifold pressure levels. This relationship was confirmed by further measurements on other naturally aspirated engines.

The advantage of the online lube oil emission measurement system is the online support for the application engineer responsible for the operating map, who can thus make appropriate changes with regard to lube oil emission levels, such as the throttle valve control in this example. The necessary boundary conditions can thereby be established for further reductions in lube oil consumption in order to meet future requirements.

7.9.5 Trade-off between friction power loss and lube oil consumption using the example of tangential force reduction on the oil control ring of a passenger car diesel engine

The combined application of the new measurement methods described here for measuring lube oil consumption and friction power loss is a valuable development tool for optimally resolving conflicting objectives when selecting design parameters. This is demonstrated below using the example of tangential force F_t on the oil control ring of a passenger car diesel engine [45].

The influence of tangential force changes in the piston rings on friction power loss is derived from the results presented in Chapter 7.3. The relatively high tangential force on the oil control ring provides the greatest potential for reducing friction here.

The results show an almost linear influence of changes to the tangential force on friction mean effective pressure, which makes it possible to calculate additional support points (see Chapter 7.3.3.5).

7.9.5.1 Influence of tangential force on the oil control ring on lube oil emissions

In order to investigate the influence of a change in tangential force on the oil control ring on lube oil consumption, cylinder-specific lube oil emission measurements were performed on a passenger car diesel engine with mixed equipment. The test engine is identical to the engine used in the friction power loss measurements described in Chapter 7.3.

Starting with a base tangential force on the oil control ring of 28.0 N, variants of the two-piece coil spring loaded beveled ring with identical geometry and reduced tangential forces of 17.7 N, 13.5 N, and 9.1 N were installed. All other components of the piston group, such as the piston, piston rings in the first and second groove, and the cylinder bore remain unchanged in comparison with the base experiments.

Figure 7.172 shows the results of the measurements as difference maps for the oil control ring variants with reduced tangential forces, wherein the values shown represent the change in lube oil emissions for the complete engine in comparison with the base ring variant.

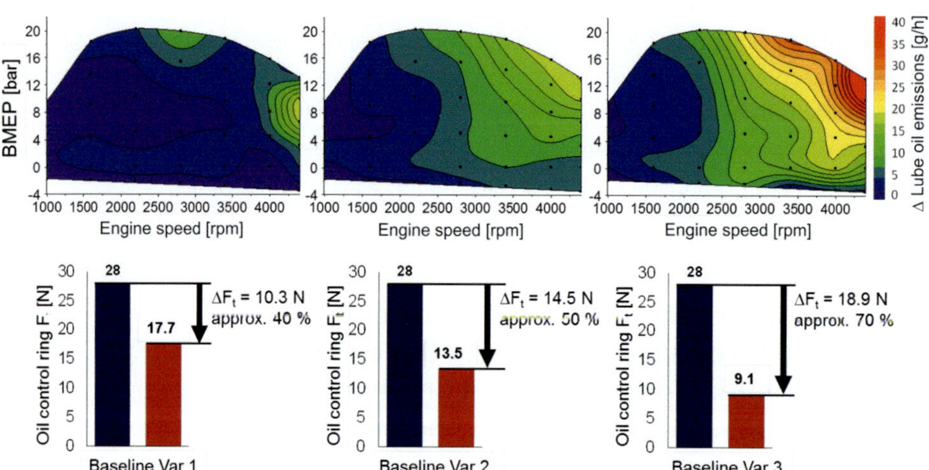

Figure 7.172: Difference maps for lube oil emissions for the tested oil control ring variants with reduced tangential force in comparison with the base variant (lube oil emissions values are representative of the complete engine)

A reduction in tangential force to 17.7 N leads to a maximum increase in lube oil emissions of 18 g/h at maximum speed and 75% relative load. This represents nearly double the oil mass emitted in this operating range, while the lube oil emission behavior in large areas of the operating map exhibits much smaller or negligible changes. With a further reduction of the tangential force to 13.5 N, an increase in lube oil emissions can be seen over nearly the entire operating map. In the low speed range, however, the increase is very slight. The maximum increase in lube oil emissions in comparison with the base occurs under full load in the high speed range. The lube oil emission difference map for the oil control ring variant with the smallest tested tangential force of 9.1 N shows the greatest increase in lube oil emissions in comparison with the base variant. Particularly in the area of high loads and speeds, above around 3,000 rpm, there is a disproportionately large increase in lube oil emissions relative to the other variants tested. This variant also shows the highest lube oil emissions values in the other areas of the operating map. The degradation is much more moderate here, however.

7.9.5.2 Comparison of the influence of tangential force on the oil control ring on lube oil emissions and frictional behavior

A reduction in tangential force of the oil control ring has varying effects on lube oil consumption and friction, depending on the operating range.

Figure 7.173 shows the changes in friction mean effective pressure and lube oil emissions in comparison with the base variant as a function of the reduction in tangential force on the

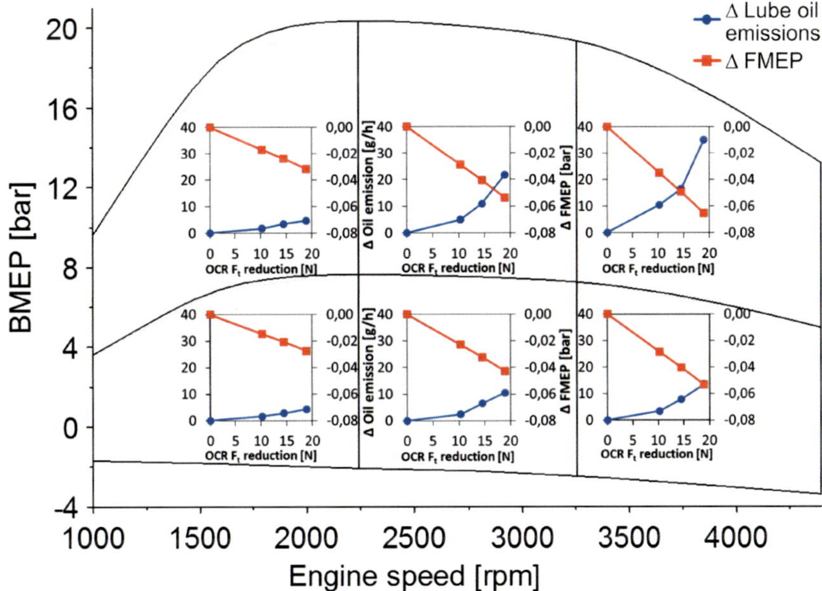

Figure 7.173: Changes in friction mean effective pressure and lube oil emissions as a function of the reduction in tangential force on the oil control ring (values averaged within the six operating map areas)

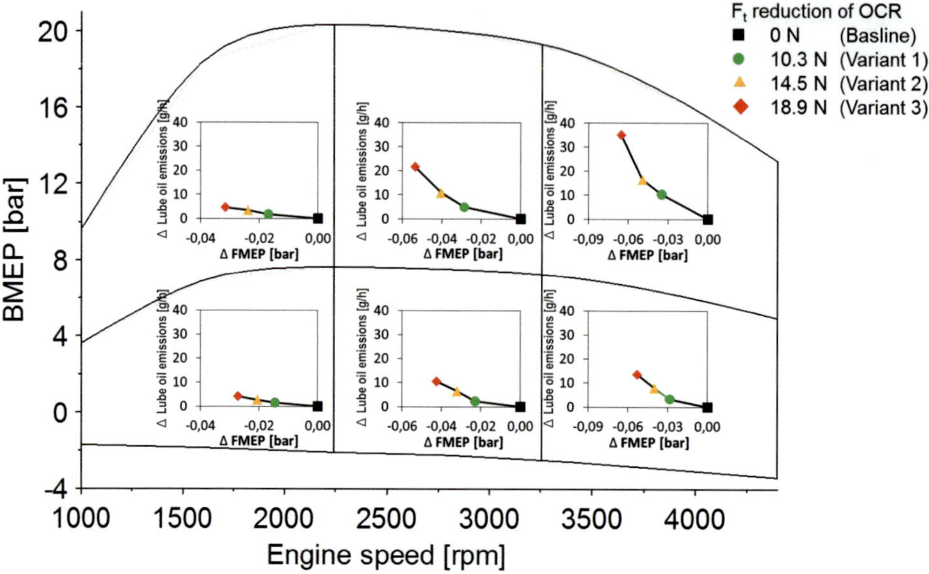

Figure 7.174: Correlation of changes in lube oil emissions and friction mean effective pressure for reductions in tangential force on the oil control ring (values averaged within the six operating map areas)

oil control ring. The operating map is divided into six adjacent load and speed ranges. The curves shown are the average of changes within each area.

A comparison of the curves for changes in friction mean effective pressure in the defined areas shows a significant influence due to speed and a slight influence due to load. A reduction in tangential force on the oil control ring leads to a reduction in friction mean effective pressure across the entire operating map. The curves for changes in lube oil emissions, in contrast, show the opposite behavior. Lowering the tangential force leads to increased lube oil emissions over the entire operating map. The curves for changes in lube oil emissions in the defined areas demonstrate a strong influence due to load and speed on the observed amount of change in lube oil emissions. The increased lube oil emissions at lower speeds, up to about 2,000 rpm, rise almost linearly as a function of the tangential force, and are relatively moderate at 2–4 g/h. In the area of medium and high speeds, the change in lube oil emissions shows exponential behavior as a function of the tangential force, which is further amplified as the load and speed increase.

The curves shown in **Figure 7.174** describe the degradation of lube oil emissions for each change in friction mean effective pressure that results for each of the tested variants. This makes the conflict of objectives between the two target parameters evident in the various load and speed ranges.

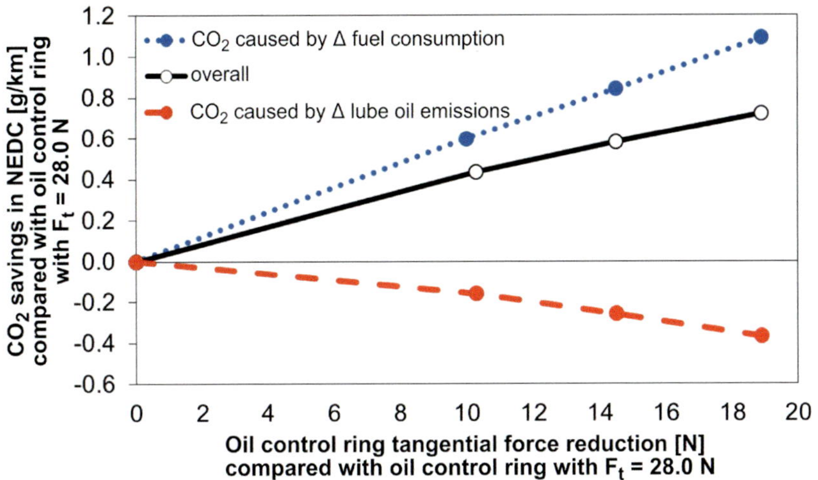

Figure 7.175: Influence of reduction of tangential force on the oil control ring on the specific contributions of fuel consumption and lube oil emissions to CO_2 savings in the NEDC

7.9.5.3 Influence of tangential force of the oil control ring on fuel consumption, lube oil emissions, and the CO_2 balance in the NEDC

A crucial question is the influence of the results shown on fuel consumption and CO_2 emissions in the relevant statutory test cycles. In addition to fuel, lube oil emissions also contribute to the CO_2 balance because of the oxidation of unburned hydrocarbon compounds in the exhaust gas system, particularly in the exhaust gas aftertreatment system. Using the results of friction power loss and lube oil emission measurements, the simulation processes described in Chapter 7.3.4 and [12] were used to calculate the effects of changes in tangential force on fuel consumption, lube oil emissions, and the CO_2 balance in the NEDC. Under the theoretical assumption of complete conversion of the emitted oil mass into CO_2, the increases in lube oil emissions in the NEDC and the associated changes in CO_2 emissions were computed.

Figure 7.175 shows the specific contributions of both fuel consumption and lube oil emissions to the reduction in CO_2, as well as the total CO_2 savings as the sum of the two. A change in the contribution to CO_2 emissions caused by burned lubricating oil components is not considered here. The fuel consumption savings for the maximum tested reduction in tangential force of 18.9 N (Σ engine Δ F$_t$ = 75.6 N) result in CO_2 savings of about 1.1 g/km in the NEDC. This is accompanied by increased CO_2 emissions of about 0.4 g/km due to the simultaneous increase in lube oil emissions. In the present case, a significant proportion, about 30%, of the CO_2 savings resulting from the reduction in fuel consumption achieved are canceled out.

As the present case demonstrates for the example of the tangential force on the oil control ring, design measures for reducing friction losses in the piston group can make a crucial contribution to reducing CO_2 emissions in combustion engines. For all measures taken to reduce friction in the piston group, however, the effects on lube oil consumption behavior must always be considered as well. The disadvantageous lube oil consumption resulting from the reduced tangential force on the oil control ring can be minimized–for example, by adjusting the ring running surfaces to improve scraping performance, or by optimizing the ring cross section to increase ring conformability. Such measures are always considered for a systematic optimization of the piston ring pack during series production development.

References

[1] Aluminium-Taschenbuch, 14th edition, Düsseldorf 1983

[2] Böhm, H.: Aushärtung. In: Aluminium Heft (1963), no. 12

[3] Wellinger, K.; Stähli G.: Verhalten von Leichtmetallkolben bei betriebsähnlicher Beanspruchung. In: VDI Z, (1978) no. 6

[4] Koch, E.: Charakteristik von Kolbenmaterialien unter Berücksichtigung des Verschleißwertes, TH Aachen 1931

[5] Müller-Schwelling, D.; Röhrle, M.: Verstärkung von Aluminiumkolben durch neuartige Verbundwerkstoffe. In: MTZ – Motortechnische Zeitschrift (1988) no. 2

[6] Zustandsschaubild Eisen-Kohlenstoff und die Grundlagen der Wärmebehandlung des Stahles. Revised by D. Horstmann, 4th edition, Düsseldorf 1961

[7] Nickel, O.: Austenitische Gusseisenwerkstoffe. In: Gießerei (1969), no. 18

[8] Kaluza, E.: Die Wärmebehandlung von Baustählen. In: Industrieanzeiger (1964), no. 67/72

[9] Frodl, D.; Gulden, H.: Neue mikrolegierte Stähle für den Fahrzeugbau. In: Der Konstrukteur (1989, May)

[10] Brochure of the Musashi Institute of Technology, Japan (2001)

[11] Deuß, T.: Reibverhalten der Kolbengruppe eines Pkw-Dieselmotors. Dissertation, University of Magdeburg 2013

[12] Deuß, T.; Ehnis, H.; Bassett, M.; Bisordi. A.; Reibleistungsmessungen am befeuertem Dieselmotor – Bestimmung der zyklusrelevanten CO2-Ersparnis. In: MTZ – Motortechnische Zeitschrift 12/2011

[13] Deuß, T.; Ehnis, H.; Freier, R.; Künzel, R.: Reibleistungsmessungen am befeuerten Dieselmotor – Potenzial der Kolbengruppe. In: MTZ – Motortechnische Zeitschrift 05/2010

[14] Deuß, T.; Ehnis, H.; Rose, R.; Künzel, R.: Reibleistungsmessungen am befeuerten Dieselmotor – Einfluss von Kolbenschaftbeschichtungen. In: MTZ – Motortechnische Zeitschrift 04/2011

[15] MAHLE talk at HDT Conference: Klopfregelung für Ottomotoren. Berlin 2003

[16] MAHLE brochure: KI-Meter. (1985)

[17] FVV proposal no. 816: Extremklopfer. Closing report volume 836, 2007

[18] FVV proposal no. 931: Vorentflammung bei Ottomotoren. 2009

[19] Künzel, R.; Essers, U.: Neues Verfahren zur Ermittlung der Kolbenbewegung in Motor-
      quer- und Motorlängsrichtung. In: MTZ – Motortechnische Zeitschrift 11/1994
      pp. 636–643

[20] Künzel, R.; Tunsch, M.; Werkmann, M.: Piston Related Noise with Spark Ignition Engines
      – Characterization, Quantification and Mechanisms of Excitation. In: 9th Congresso
      SAE Brasil, SAE Paper 2000-01-3311

[21] Gabele, H.; Beitrag zur Klärung der Entstehungsursachen von Kolbengeräuschen bei
      Pkw-Ottomotoren. Dissertation, University of Stuttgart, 1994

[22] Künzel, R.; Werkmann, M.; Tunsch M.: Piston Related Noise with Diesel Engines –
      Parameters of Influence and Optimization. In: ATT Congress Barcelona 2001, SAE
      Paper 2001-01-3335

[23] Künzel, R.: Die Kolbenquerbewegung in Motorquer- und Motorlängsrichtung, Teil 2:
      Einfluss der Kolbenbolzendesachsierung und der Kolbenform. In: MTZ – Motortechni-
      sche Zeitschrift 09/1995 pp. 534–541

[24] Haller, H.; Spessert, B.; Joerres, M.: Möglichkeiten der Geräuschquellenanalyse bei
      direkteinspritzenden Dieselmotoren. In: VDI-Berichte no. 904, 1991

[25] Helfer, M.: Zur Anregung und Ausbreitung des vom Kolben erregten Geräusches. Dis-
      sertation, University of Stuttgart, 1994

[26] Werkmann, M.; Tunsch, M.; Künzel, R.: Piston Pin Related Noise – Quantification Pro-
      cedure and Influence of Pin Bore Geometry. Congresso SAE Brasil, São Paulo, 2005,
      SAE Paper 2005-01-3967

[27] o. V.: Von der Kavitation zur Sonotechnologie, Department Future Technology of the VDI
      Technology Center 2000

[28] Hosny, D. M. et al.: A System Approach for the Assessment of Cavitation Corrosion
      Damage of Cylinder Liners in Internal Combustion Engines. In: SAE Paper 930581 1993

[29] Steck, B.: Cavitation on Wet Cylinder Liners of Heavy Duty Diesel Engines. SAE Paper
      2006-01-3477 2006

[30] Steck, B.: Kavitation an nassen Zylinderlaufbuchsen in Nutzfahrzeug-Dieselmotoren –
      mögliche Einflussparameter zur Vermeidung. In: 9th Stuttgart International Symposium
      on Automotive and Engine Technology 2009

[31] Brennen, C. E.: Cavitation and Bubble Dynamics. In: Oxford University Press
      ISBN 0-19-509409-3 2006, Chapter 1, Pages 1–33, Chapter 3, Pages 80–106

[32] Sauer, J.: Instationär kavitierende Strömungen. Ein neues Modell, basierend auf Front
      Capturing (VoF) und Blasendynamik. Dissertation, Engineering University Karlsruhe,
      2000, pp. 4–5

[33] Katragadda, S. and Bata, R.: Cavitation Problem in Heavy Duty Diesel Engines: A Literature Review, Heavy Vehicle Systems, International Journal of Vehicle Design, Vol. 1, No. 3, 1994, pp. 324–346

[34] Benjamin, T. B. and Ellis A. T.: The Collapse of Cavitation Bubbles and the Pressures thereby Produced against Solid Boundaries. In: Phil. Trans. Roy. Soc. London, Ser. A, 260, 1996, pp. 221–240

[35] Frost, D. and Stuetevant, B.: Effects of Ambient Pressure on the Instability of a Liquid Boiling Explosively at the Superheat Limit. In: ASME Journal of Heat Transfer, 108, 1986, pp. 418–424

[36] Knapp, R. T., Daily, J. W., and Hammitt, F. G.: Cavitation, McGraw-Hill, New York 1970

[37] Tandara, V.: Beitrag zur Kavitation an Zylinderlaufbüchsen von Dieselmotoren. Dissertation, Engineering University Berlin 1968, pp. 13–17

[38] Tritthart, P.: Dieselpartikelemissionen: Analysetechniken und Ergebnisse. In: Mineralöltechnik, volume 8, July 1994;
Beratungsgesellschaft für Mineralöl-Anwendungstechnik mbH, Hamburg

[39] Gohl, M.; Ihme, H.: Massenspektrometrische Bestimmung des Ölverbrauchs von Verbrennungsmotoren und dessen Einfluss auf die HC-Emission. FVV closing report, proposal no.707, volume 691, 2000

[40] Gohl, M.: Massenspektrometrische Bestimmung der Ölemission im Abgas von Otto- und Dieselmotoren. FVV closing report, proposal no.758, volume 764, 2003

[41] Krause, S.; Stein, C.; Brandt, S.: Beeinflussung der Schmierölemission durch die Gemischbildung im Brennraum von Verbrennungsmotoren. FVV closing report, proposal no. 933, volume 901, 2010

[42] Püffel, P.: Eine neue Methode zur schnellen Ölverbrauchsmessung. In: MTZ – Motortechnische Zeitschrift 11/1999

[43] Airsense Automotive: Information brochure on the Lubrisense 1200 system, Schwerin 2005

[44] Krause, S.: Massenspektrometrisches Verfahren zur Charakterisierung der Ölverdampfung im Brennraum von Ottomotoren. Dissertation, Hamburg University of Technology, 2009

[45] Frommer, A.; Freier, R.; Ehnis, H.; Künzel. R.: Trade-off between Friction Reduction and Lube Oil Consumption – Tangential Force Reduction on Oil Control Rings. In: 13th Stuttgart International Symposium on Automotive and Engine Technology 2013

Dictionary/Glossary

Piston designations and dimensions

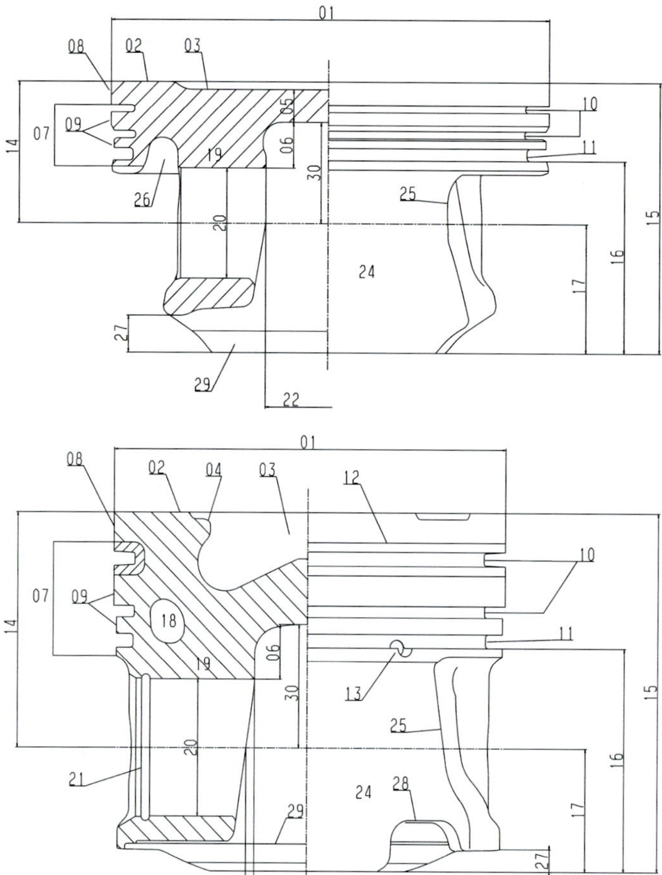

01 Piston diameter	11 Oil ring groove	22 Pin boss spacing
02 Piston crown	12 Ring carrier	23 Upper pin boss spacing (upper distance between bosses)
03 Combustion bowl	13 Oil return notch	24 Piston skirt
04 Bowl rim	14 Compression height	25 Window
05 Crown thickness	15 Total height	26 Cast undercut
06 Elongation length	16 Skirt length	27 Skirt bottom recess
07 Ring belt	17 Lower height	28 Fuel injector recess
08 Top land	18 Cooling gallery	29 Register
09 Ring land	19 Piston pin boss	30 Inner height
10 Compression ring groove	20 Pin bore	
	21 Piston pin circlip groove	

Shapes of piston crowns

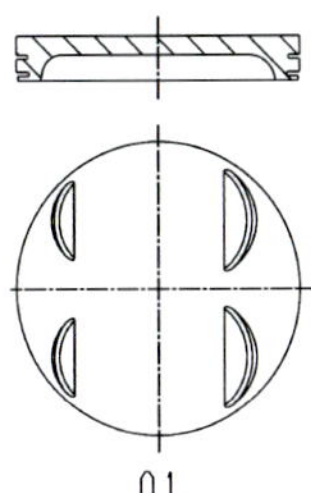

01

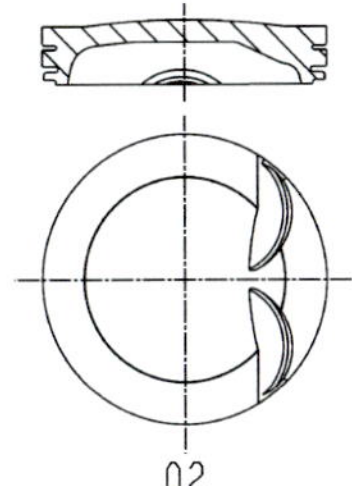

02

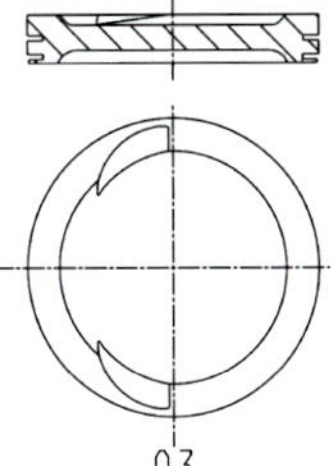

03

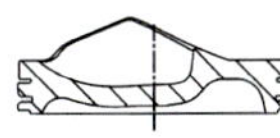

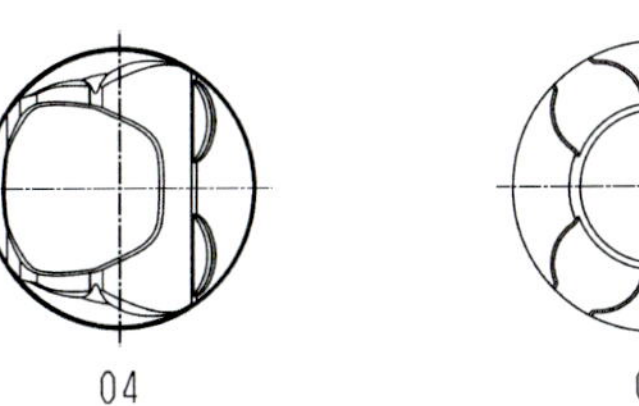

04

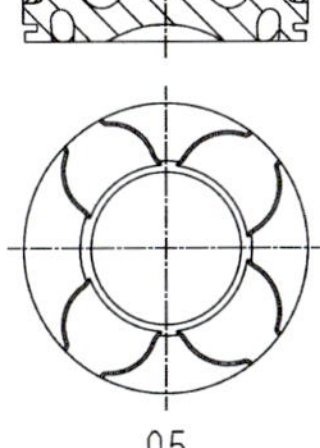

05

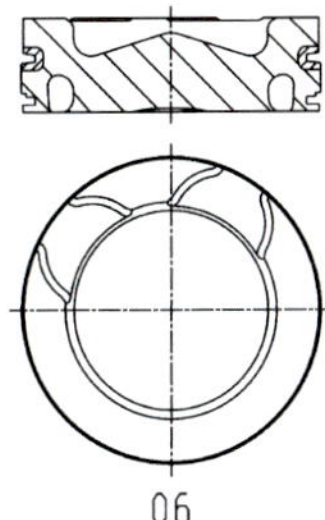

06

01 Flat crown with valve
    pockets

02 Cap

03 Flat bowl

04 Bowl for direct gasoline
    injection piston with strati-
    fied charge

05 Direct diesel injection
    piston with deep, undercut
    bowl

06 Direct diesel injection
    piston with flat bowl

Top land, ring belt

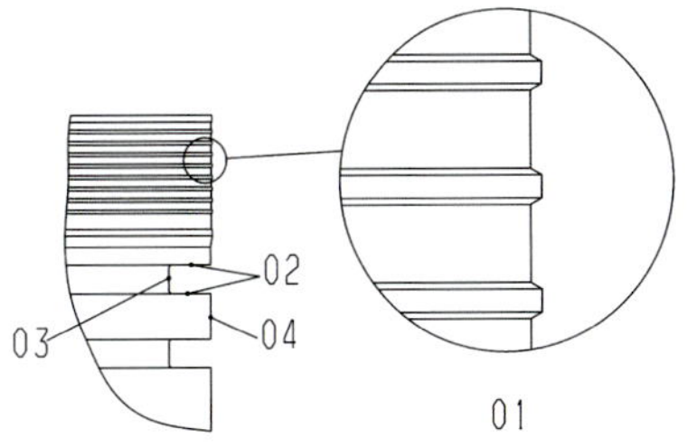

01

05

06

01 Scuff band of top land

02 Groove flank

03 Groove root

04 Ring land

05 Single ring insert

06 Double ring insert

Pin boss support

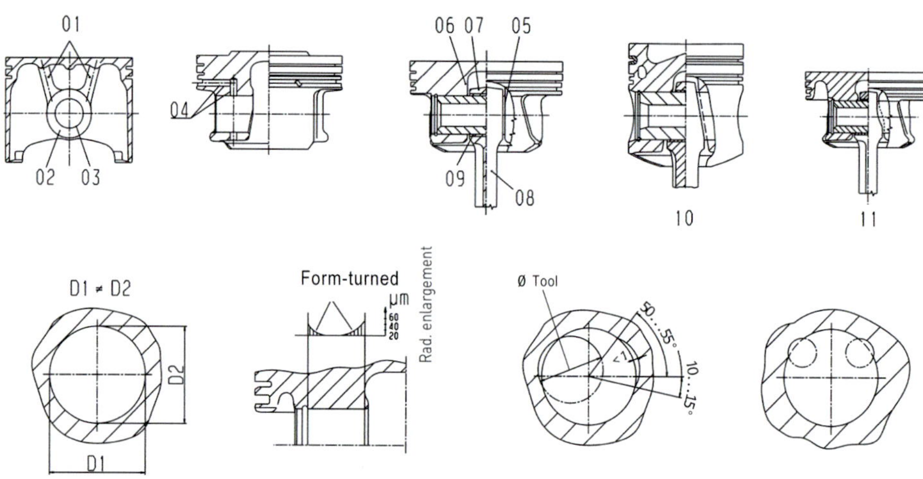

01 Supporting ribs	08 Connecting rod shank	14 Pin bore side reliefs
02 Piston pin boss	09 Connecting rod bushing	15 Oil pockets/slots (generally deeper than side relief, made with a smaller tool and larger tolerances), only for improved lubrication
03 Pin bore	10 Tapered support	
04 Pin lubricating holes	11 Stepped support	
05 Parallel bin boss support	12 Oval pin bore	
06 Solid bin boss support	13 Shaped pin bore, barrel-shaped expansion at one or both sides	
07 Connecting rod small end (small end bore)		

Piston pins, piston pin circlips

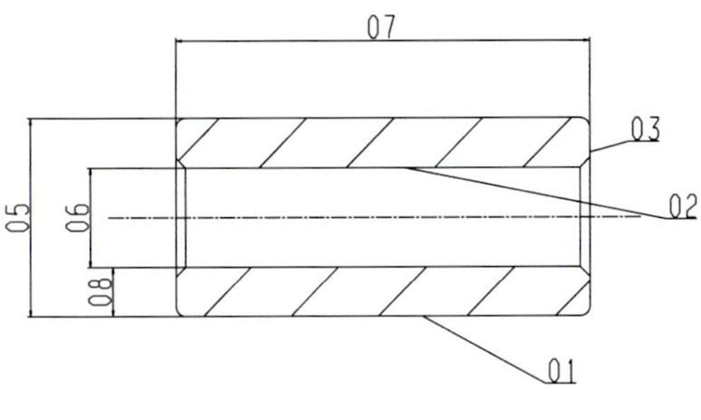
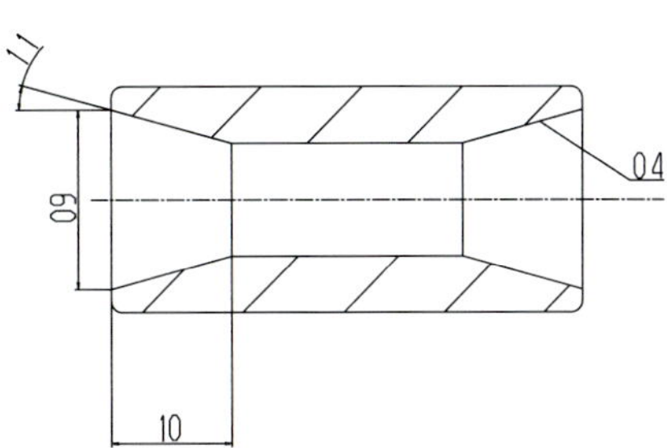

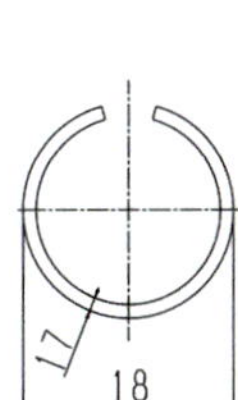
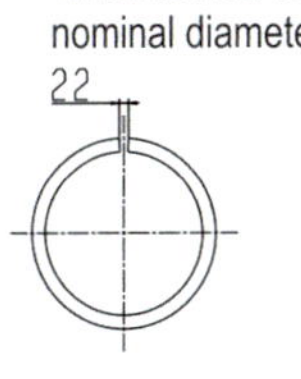
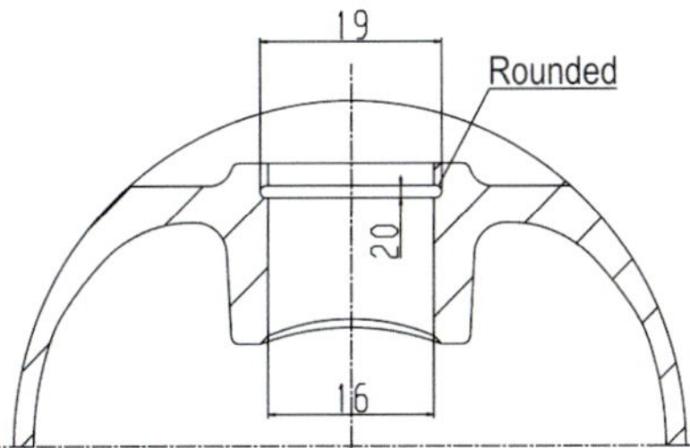

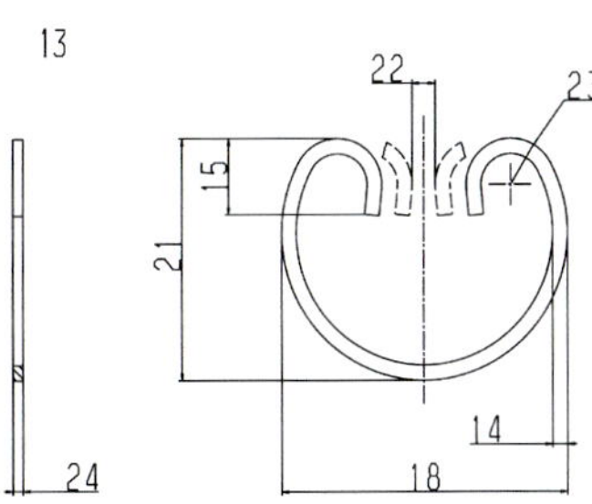
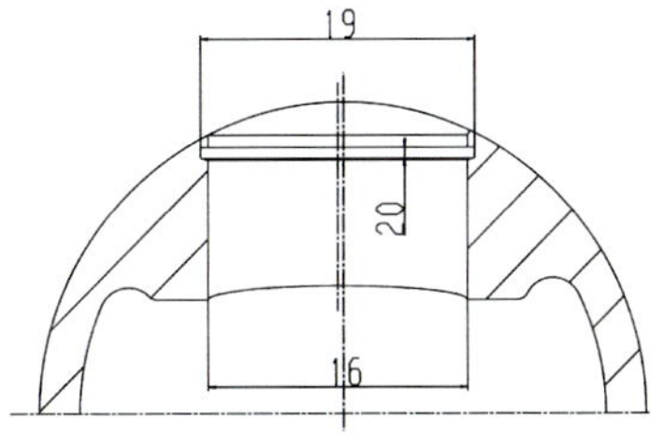

01 Outer surface	11 Taper angle	17 Wire diameter
02 Inner surface	12 Circlip made of round wire, no hook, wire snap ring with no hook	18 Circlip diameter, unassembled
03 End surface		19 Circlip groove diameter
04 Taper	13 Circlip made of flat wire, with hook, square wire snap ring with hook	20 Circlip groove width
05 Outer diameter		21 Circlip height
06 Inner diameter	14 Radial wall thickness	22 Gap width when assembled to nominal diameter (16)
07 Length	15 Hook length	
08 Wall thickness	16 Pin bore diameter, nominal diameter	23 Hook radius
09 Taper diameter		24 Axial wall thickness
10 Taper length		

Piston types

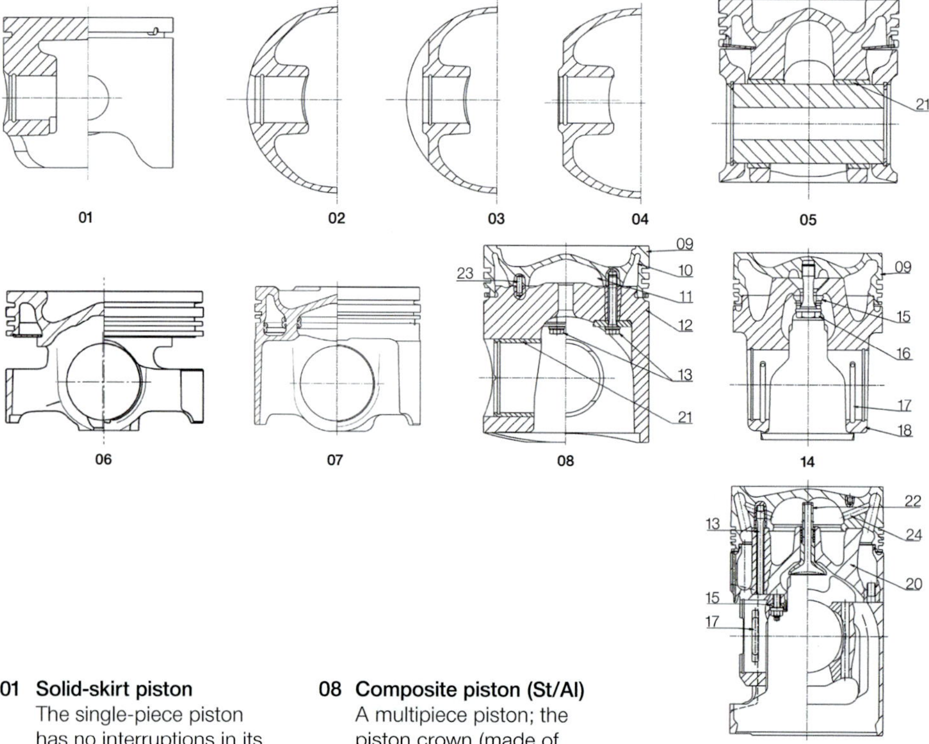

01 Solid-skirt piston
The single-piece piston has no interruptions in its supporting cross sections between the piston crown, ring belt, and skirt, except for any oil return notches.

02 Full-skirt piston
The skirt has a continuous rotational surface.

03 Window-type piston
The skirt has a recess near the pin bores.

04 Box-type piston
Skirt recess outside of the pin bores, all the way to the end of the skirt.

05 FERROTHERM® piston
Articulated piston, multi-piece piston; piston crown/ring belt and skirt are connected by the piston pin.

06 MONOTHERM® piston
Single-piece steel piston

07 MonoWeld® piston
Friction-welded steel piston

08 Composite piston (St/Al)
A multipiece piston; the piston crown (made of forged steel) is connected to the piston skirt (made of aluminum alloy) using high-strength bolts.

09 Steel crown

10 Outer cooling cavity

11 Inner cooling cavity

12 Aluminum skirt

13 Antifatigue bolt
Multiple screw joint

14 Composite piston (St/St)
A multipiece piston; the piston crown (made of forged steel) is connected to the piston skirt (here also made of forged steel) using high-strength bolts.

15 Thrust piece

16 Antifatigue bolt
Central screw joint

17 Oil groove
Provides the oil supply

18 Steel skirt

19 Composite piston (St/NCI)
A multipiece piston; the bore-cooled piston crown (made of forged steel) is connected to the piston skirt (here made of nodular cast iron) using high-strength bolts.

20 Nodular cast iron skirt

21 Pin bore bushing

22 Slide shoe
Provides the cooling oil supply

23 Locator pin
Locates crown with respect to the skirt

24 Crossflow holes
Oil passage between outer and inner cooling cavities

Piston cooling

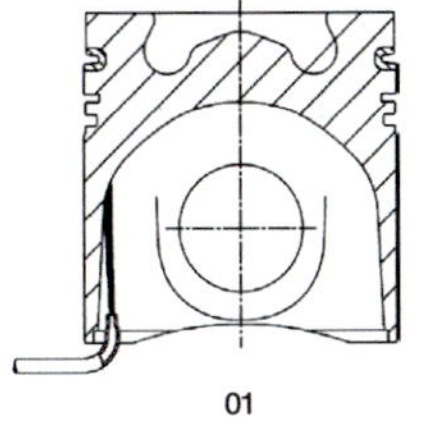

01

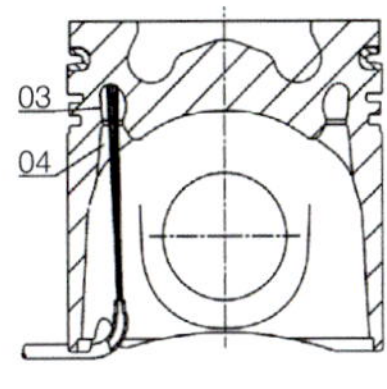

02

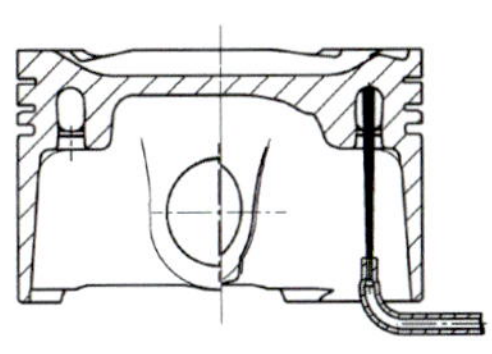

05

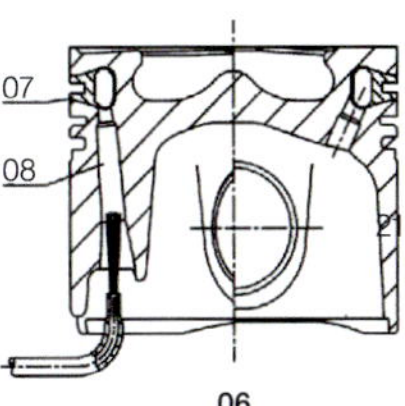

06

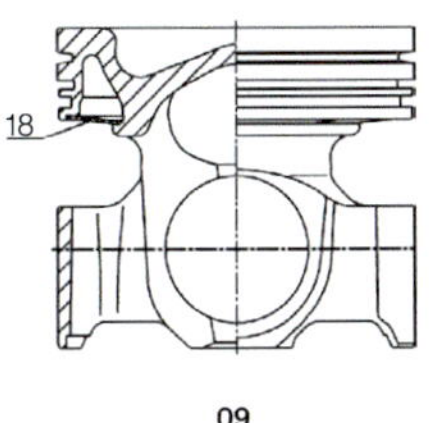

09

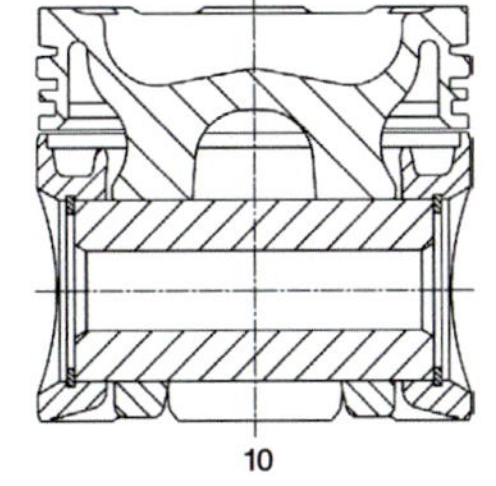

10

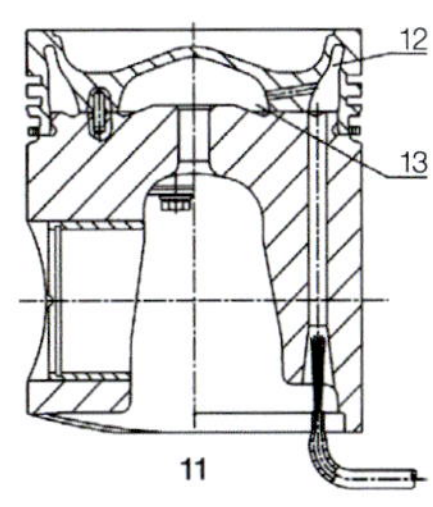

11

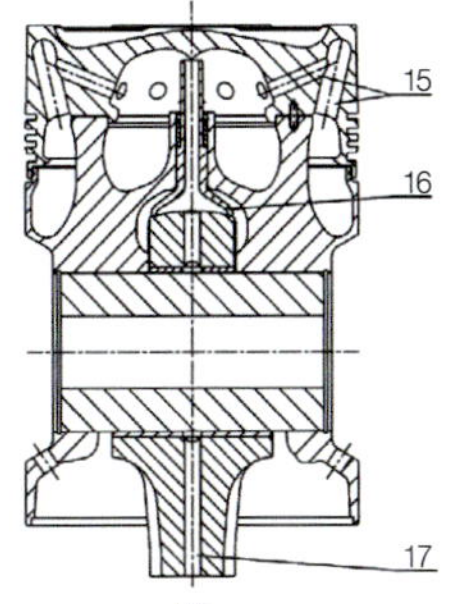

14

01 Spray jet cooling
Oil is fed via nozzle from the crankcase, or via conrod and bore in the small end

02 Diesel piston with salt core cooling gallery
Infeed through cast funnel; oil is fed via nozzle from the crankcase

03 Salt core cooling gallery

04 Funnel (cast)

05 Gasoline piston with salt core cooling gallery

06 Piston with cooled ring carrier
Drilled infeed bore; oil is fed via nozzle from the crankcase

07 Cooled ring carrier

08 Drilled oil infeed

09 MONOTHERM® piston
With closed cooling gallery

10 FERROTHERM® piston
With shaker pockets in the skirt

11 Composite piston
Oil is fed via nozzle from the crankcase

12 Outer cooling cavity

13 Inner cooling cavity

14 Bore-cooled cooling cavity piston (composite piston)
Cooling oil is fed via conrod and slide shoe to the inner piston cooling cavity

15 Cooling oil bore

16 Slide shoe

17 Pressure oil bore

18 Cover plate

Piston pin offset (schematic)

Offset of the pin axis from the longitudinal piston axis

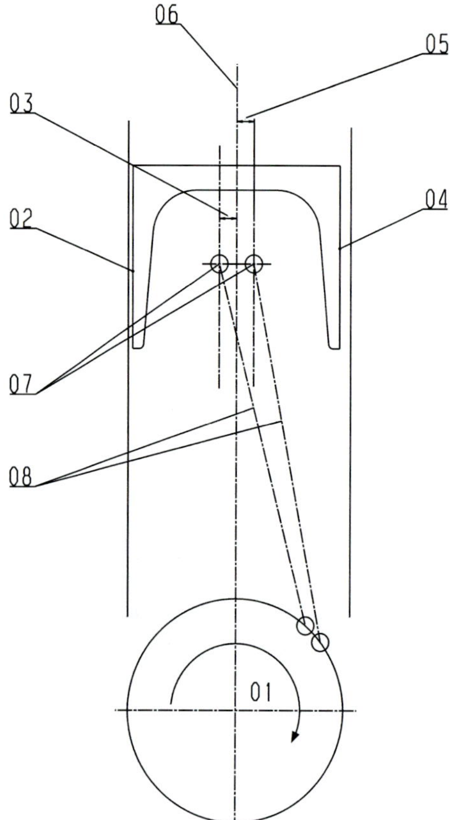

01 Crankshaft rotation, as seen from the control side

02 Thrust side

03 Piston pin offset toward the thrust side; typically noise offset and/or to prevent wet liner cavitation

04 Antithrust side

05 Offset toward the antithrust side; typically thermal offset

06 Longitudinal piston axis (cylinder axis)

07 Offset piston pin axis

08 Connecting rod

Technical terminology

Blow-by	see *Blow-by gas*
Blow-by gas	Gas that passes from the combustion chamber, past the piston rings, into the crankcase
Burning mark	Piston ring face marks indicative of local overheating between the piston ring and the cylinder running surface due to lack of oil
Catching efficiency	For evaluating piston cooling, indicates the percentage of the oil delivered by the cooling oil nozzle that flows back into the crankcase from the drain hole(s) on the piston
Cavitation	Local material loss on the water side of a wet cylinder liner due to the interaction of several physical processes

Cold-start scuffing	Local wear between the piston and the cylinder liner, due to lack of clearance or lubricating oil when the engine is started
Compression height	Distance between the centerline of the piston pin and the edge of the piston crown (top edge of the top land)
Compression ring	see *1st piston ring* and *2nd piston ring*
Cooling gallery	Cooling oil cavity(ies) in the piston crown / ring belt area for thermally relieving the piston
Crankcase ventilation	Gas passage for the combustion gases that enter the crankcase
Cylinder liner	Liner inserted in the engine block
Cylinder surface	Inner surface of the cylinder bore
Gap clearance	Spacing of the piston ring ends in the installed condition
Honing	Machining process to produce a defined topographical structure of the cylinder surface
Installation clearance	Difference between the inner diameter of the cylinder liner and the largest piston diameter at room temperature
Keystone ring	Piston ring with tapered side faces on one or both sides
KI factor	Knock intensity factor
KI meter	Device for quantitatively evaluating extreme variations in pressure, such as knock strength or knocking intensity in the combustion chamber or during a cavitation attack
Knock control	Dynamic control of the ignition point to prevent knocking combustion
Knocking	Irregular combustion of the fuel-air mixture, with some high pressure peaks
Knocking damage	Damage, particularly to the piston top land, due to knocking combustion
Lateral force	Part of the combustion force exerted on the piston that acts perpendicularly on the cylinder via the piston skirt
Normal force	see *Lateral force*
Oil control ring	see *3rd piston ring*
Ovality	Shape of the relaxed piston ring, such that the contact pressure on the cylinder surface is uniform in the installed state
Pin bore	Bore for supporting the piston pin in the piston
Piston	Movable part of the combustion chamber
Piston coating	Protective coating on the piston skirt, tribologically matched to the cylinder bore
Piston cooling	Heat dissipation and temperature reduction of the piston using engine oil
Piston crown	The part of the piston that bounds the combustion chamber

Piston pin	Connecting member between the piston and the connecting rod
Piston pin offset	Offset of the piston pin axis relative to the longitudinal piston axis
Piston ring	Slotted, self-tensioning ring
1st piston ring	First piston ring toward the combustion chamber side, also called top ring. Compression of combustion air or gas mixture, and support of gas pressure in the operating cycle, conduction of generated heat to the cylinder surface, and, to a slight degree, scraping of the residual oil from the cylinder surface
2nd piston ring	Middle ring, support of the remaining gas pressure due to blow-by past the first piston ring, control of pressure ratios in the ring belt, scraping of oil from and conduction of generated heat to the cylinder surface
3rd piston ring	Homogeneous distribution of the oil for lubrication of the piston group/cylinder bore tribological system and scraping of excess oil
Rectangular ring	Basic shape of a piston ring with a rectangular cross section
Ring carrier	Cast-in part made of Ni-resist in the area of the top ring groove in diesel pistons in order to improve the ring groove wear resistance
Ring conformability	Ability of a piston ring to adapt to the noncircularity of the cylinder surface
Ring gap	Open ends of the piston ring
Running surface	see *Cylinder surface*
Scraper ring	see *Oil control ring*
Secondary piston motion	Motion of the piston transverse to its longitudinal axis during an expansion stroke see also *Skirt ovality*
Shaped pin bore	Pin bore with defined convex end(s)
Shrink-fit connecting rod	Also called fixed-pin connecting rod; connecting rod with a piston pin that is firmly seated in the small end of the connecting rod thanks to a controlled interference fit
Side faces	Axial surfaces of the piston ring or piston ring grooves
Skirt collapse	Permanent deformation of the piston skirt
Skirt ovality	Noncircular shape of the piston skirt, so that a defined contact behavior is achieved in the cylinder
Top dead center wear	Wear on the cylinder surface at the top dead center (combustion chamber side) of the first piston ring
Top land	Area between the edge of the piston crown and the upper flank of the top ring groove
Twist	Torsional deformation of a piston ring cross section due to a groove or chamfer on one ring side of the inner diameter

Keyword index

MIX
Papier aus verantwortungsvollen Quellen
Paper from responsible sources
FSC® C105338
FSC
www.fsc.org